METHODS IN MOLECULAR BIOLOGY

T0335414

Series Editor
John M. Walker
School of Life and Medical Sciences
University of Hertfordshire
Hatfield, Hertfordshire, AL10 9AB, UK

For further volumes:
http://www.springer.com/series/7651

Nonribosomal Peptide and Polyketide Biosynthesis

Methods and Protocols

Editor

Bradley S. Evans

Donald Danforth Plant Science Center, Saint Louis, Missouri, USA

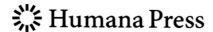 Humana Press

Editor
Bradley S. Evans
Donald Danforth Plant Science Center
Saint Louis, Missouri
USA

ISSN 1064-3745   ISSN 1940-6029 (electronic)
Methods in Molecular Biology
ISBN 978-1-4939-3373-0   ISBN 978-1-4939-3375-4 (eBook)
DOI 10.1007/978-1-4939-3375-4

Library of Congress Control Number: 2015957395

Springer New York, NY Heidelberg New York Dordrecht London

Printed on acid-free paper

Humana Press is a brand of Springer
Springer Science+Business Media LLC New York is part of Springer Science+Business Media (www.springer.com)

Preface

Nonribosomal peptide synthetases (NRPSs) and polyketide synthases (PKSs) are fascinating families of enzymes from many standpoints. These biosynthetic enzymes share a common logic with fatty acid synthases (FASs), thiotemplate-guided assembly of metabolites. The NRPSs, PKSs, and FASs can be combined in pathways (and singular enzymes) to biosynthesize hybrid molecules. They are often very large, multienzyme polypeptides that form some of the largest protein structures in the cell. These molecular assembly lines utilize complicated substrate shuttling routines to form their products from an array of over 500 documented monomer subunits. NRPS and PKS biosynthetic pathways produce a wide range of bioactive compounds from the well-known medicines penicillin, tetracycline, erythromycin, lovastatin, cyclosporine, and vancomycin, to the widely used herbicide phosphinothricin tripeptide (Bialaphos, Basta, Glufosinate) to deadly toxins such as microcystin, aflatoxin, and gliotoxin. The variety and importance of the products of thiotemplate pathways has spurred investigation into NRPS and PKS biosynthesis. The modular nature of these pathways and the complex, high value and sometimes deadly nature of their pathways make them tantalizing targets for bioengineering. Research in NRPS and PKS biosynthesis has accelerated in recent years with the rapid rise in available genome sequence data. It is within this context that this volume is presented. We have assembled an overview of NRPS and PKS structure and function (Part I) followed by methods for analysis of these pathways including conventional enzymological assays, contemporary mass spectrometric analysis techniques, specialized molecular biological approaches applicable to NRPSs and PKSs, and small molecule analysis tools tailored to this very special class of natural products (Part II), and finally bioinformatics tools for the analysis of these enzymes, pathways, and molecules (Part III). We anticipate reaching researchers in the field whose backgrounds are in the disciplines of enzymology, microbiology, structural biology, genomics, proteomics, bioinformatics, and natural products chemistry. This compilation should serve as a reference for those experienced in studying NRPS and PKS enzymes, pathways, and natural products as well as a gateway for those just entering the field.

Saint Louis, MO, USA   *Bradley S. Evans*

Contents

Contributors

GUILLERMIN AGÜERO-CHAPIN • *CIMAR/CIIMAR, Centro Interdisciplinar de Investigação Marinha e Ambiental, Universidade do Porto, Porto, Portugal; Centro de Bioactivos Químicos, Universidad Central "Marta Abreu" de Las Villas (UCLV), Santa Clara, Cuba*

COURTNEY C. ALDRICH • *Department of Medicinal Chemistry, University of Minnesota, Minneapolis, MN, USA*

AGOSTINHO ANTUNES • *CIMAR/CIIMAR, Centro Interdisciplinar de Investigação Marinha e Ambiental, Universidade do Porto, Porto, Portugal; Departamento de Biologia, Faculdade de Ciências, Universidade do Porto, Porto, Portugal*

CHRISTOPHER N. BODDY • *Department of Chemistry and the Center for Advanced Research on Environmental Genomics, University of Ottawa, Ottawa, ON, Canada; Department of Biology and the Center for Advanced Research on Environmental Genomics, University of Ottawa, Ottawa, ON, Canada*

CLARA BRIEKE • *Department of Biomolecular Mechanisms, Max Planck Institute for Medical Research, Heidelberg, Germany*

RALPH A. CACHO • *Department of Chemical and Biomolecular Engineering, University of California, Los Angeles, CA, USA*

HÉLÈNE CAWOY • *Gembloux Agro-Bio Tech, University of Liège, Gembloux, Belgium*

YUNQIU CHEN • *Department of Chemistry, Northwestern University, Evanston, IL, USA*

KYLE R. CONWAY • *Department of Chemistry and the Center for Advanced Research on Environmental Genomics, University of Ottawa, Ottawa, ON, Canada*

JASON M. CRAWFORD • *Department of Chemistry, Yale University, New Haven, CT, USA; Chemical Biology Institute, Yale University, New Haven, CT, USA; Department of Microbial Pathogenesis, Yale School of Medicine, New Haven, CT, USA; Department of Chemistry, Yale University, New Haven, CT, USA*

MAX J. CRYLE • *Department of Biomolecular Mechanisms, Max Planck Institute for Medical Research, Heidelberg, Germany*

DELPHINE DEBOIS • *Mass Spectrometry Laboratory (LSM-GIGA-R), Chemistry Department, University of Liege - Allee du 6 aout, Liege-1 (Sart Tilman), Belgium*

BENJAMIN P. DUCKWORTH • *Department of Medicinal Chemistry, University of Minnesota, Minneapolis, MN, USA*

LEI FANG • *Department of Chemical and Biological Engineering, The State University of New York at Buffalo, Buffalo, NY, USA*

ANDREW M. GULICK • *Hauptman-Woodward Medical Research Institute, Buffalo, NY, USA; Department of Structural Biology, University at Buffalo, Buffalo, NY, USA*

YOSHIMITSU HAMANO • *Department of Bioscience, Fukui Prefectural University, Fukui, Japan*

PATRICK HILL • *Department of Biology and the Center for Advanced Research on Environmental Genomics, University of Ottawa, Ottawa, ON, Canada*

FUMIHIRO ISHIKAWA • *Department of System Chemotherapy and Molecular Sciences, Division of Bioinformatics and Chemical Genomics, Graduate School of Pharmaceutical Science, Kyoto University, Kyoto, Japan*

PHILIPPE JACQUES • *ProBioGEM, Institut Charles Viollette, Polytech'Lille, University of Lille 1, Villeneuve d'Ascq, France*

HIDEAKI KAKEYA • *Department of System Chemotherapy and Molecular Sciences, Division of Bioinformatics and Chemical Genomics, Graduate School of Pharmaceutical Science, Kyoto University, Kyoto, Japan*

HAJIME KATANO • *Department of Bioscience, Fukui Prefectural University, Fukui, Japan*

NEIL L. KELLEHER • *Department of Chemistry, Northwestern University, Evanston, IL, USA*

NANCY P. KELLER • *Department of Bacteriology, University of Wisconsin-Madison, Madison, WI, USA; Department of Medical Microbiology and Immunology, University of Wisconsin-Madison, Madison, WI, USA*

VERONIKA KRATZIG • *Department of Biomolecular Mechanisms, Max Planck Institute for Medical Research, Heidelberg, Germany*

VALÉRIE LECLÈRE • *ProBioGEM, Institut Charles Viollette, Polytech'Lille, University of Lille 1, Villeneuve d'Ascq, France; CRIStAL, UMR 9189, Univ Lille, Villeneuve d'Ascq, France; Inria Lille Nord-Europe, Villeneuve d'Ascq, France*

CHITOSE MARUYAMA • *Department of Bioscience, Fukui Prefectural University, Fukui, Japan*

RYAN A. MCCLURE • *Department of Chemistry, Northwestern University, Evanston, IL, USA*

BRADLEY R. MILLER • *Hauptman-Woodward Medical Research Institute, Buffalo, NY, USA; Department of Structural Biology, University at Buffalo, Buffalo, NY, USA*

TIMO HORST JOHANNES NIEDERMEYER • *Interfaculty Institute for Microbiology and Infection Medicine, Eberhard Karls University, Tübingen, Germany; German Centre for Infection Research (DZIF), Tübingen, Germany*

HARUKA NIIKURA • *Department of Bioscience, Fukui Prefectural University, Fukui, Japan*

MARC ONGENA • *Gembloux Agro-Bio Tech, University of Liège, Gembloux, Belgium*

MARCIA S. OSBURNE • *EarthGenes Pharmaceuticals, Lexington, MA, USA; Department of Molecular Biology and Microbiology, Tufts University School of Medicine, Boston, MA, USA*

EDWIN DE PAUW • *Mass Spectrometry Laboratory, University of Liège, Liège, Belgium*

GISSELLE PÉREZ-MACHADO • *Centro de Bioactivos Químicos, Universidad Central "Marta Abreu" de Las Villas (UCLV), Santa Clara, Cuba*

MADELEINE PESCHKE • *Department of Biomolecular Mechanisms, Max Planck Institute for Medical Research, Heidelberg, Germany*

BLAINE A. PFEIFER • *Department of Chemical and Biological Engineering, The State University of New York at Buffalo, Buffalo, NY, USA*

CARLOS PRIETO • *Bioinformatics Service, Nucleus, University of Salamanca (USAL), Edificio I+D+i, C/ Espejo 2, Salamanca, 37007, Spain*

MAUDE PUPIN • *CRIStAL, UMR 9189, Univ Lille, Villeneuve d'Ascq, France; Inria Lille Nord-Europe, Villeneuve d'Ascq, France*

PAUL R. RACE • *School of Biochemistry, University of Bristol, Bristol, UK*

BRISSYNBIO SYNTHETIC BIOLOGY RESEARCH CENTREUNIVERSITY OF BRISTOL, BRISTOL, UK

AMINAEL SÁNCHEZ-RODRÍGUEZ • *Departamento de Ciencias Naturales, Universidad Técnica Particular de Loja, Loja, Ecuador*

MIGUEL MACHADO SANTOS • *CIMAR/CIIMAR, Centro Interdisciplinar de Investigação Marinha e Ambiental, Universidade do Porto, Porto, Portugal; Departamento de Biologia, Faculdade de Ciências, Universidade do Porto, Porto, Portugal*

ALEXANDRA A. SOUKUP • *Department of Genetics, University of Wisconsin-Madison, Madison, WI, USA*

MASAHIRO TAKAKUWA • *Department of Bioscience, Fukui Prefectural University, Fukui, Japan*

YI TANG • *Department of Chemical and Biomolecular Engineering, University of California, Los Angeles, CA, USA; Department of Chemistry and Biochemistry, University of California, Los Angeles, CA, USA; Department of Bioengineering, University of California, Los Angeles, CA, USA*

MARISA TILL • *School of Biochemistry, University of Bristol, Bristol, UK; BrissynBio Synthetic Biology Research Centre, University of Bristol, Bristol, UK*

NICOLAS TREMBLAY • *Department of Chemistry and the Center for Advanced Research on Environmental Genomics, University of Ottawa, Ottawa, ON, Canada*

MARIA I. VIZCAINO • *Department of Chemistry, Yale University, New Haven, CT, USA; Chemical Biology Institute, Yale University, New Haven, CT, USA*

TILMANN WEBER • *The Novo Nordisk Foundation Center for Biosustainability, Technical University of Denmark, Hørsholm, Denmark*

PHILIPP WIEMANN • *Department of Medical Microbiology and Immunology, University of Wisconsin-Madison, Madison, WI, USA*

DANIEL J. WILSON • *Center for Drug Design, University of Minnesota, Minneapolis, MN, USA*

HAORAN ZHANG • *Department of Chemical and Biological Engineering, The State University of New York at Buffalo, Buffalo, NY, USA*

Part I

Background and Overview

Chapter 1

Structural Biology of Nonribosomal Peptide Synthetases

Bradley R. Miller and Andrew M. Gulick

Abstract

The nonribosomal peptide synthetases are modular enzymes that catalyze synthesis of important peptide products from a variety of standard and non-proteinogenic amino acid substrates. Within a single module are multiple catalytic domains that are responsible for incorporation of a single residue. After the amino acid is activated and covalently attached to an integrated carrier protein domain, the substrates and intermediates are delivered to neighboring catalytic domains for peptide bond formation or, in some modules, chemical modification. In the final module, the peptide is delivered to a terminal thioesterase domain that catalyzes release of the peptide product. This multi-domain modular architecture raises questions about the structural features that enable this assembly line synthesis in an efficient manner. The structures of the core component domains have been determined and demonstrate insights into the catalytic activity. More recently, multi-domain structures have been determined and are providing clues to the features of these enzyme systems that govern the functional interaction between multiple domains. This chapter describes the structures of NRPS proteins and the strategies that are being used to assist structural studies of these dynamic proteins, including careful consideration of domain boundaries for generation of truncated proteins and the use of mechanism-based inhibitors that trap interactions between the catalytic and carrier protein domains.

Key words Structural biology, Nonribosomal peptide synthetase, Enzymology, Modular enzymes, Peptides, Siderophores, Metabolic pathways

1 Introduction

Nonribosomal peptide synthetases deliver amino acid and peptide intermediates, covalently bound to the pantetheine cofactor of a peptidyl carrier protein, to different catalytic domains where the nascent peptide chain is elongated, modified, and ultimately released [1]. The primary catalytic domains are the adenylation domain that activates and loads the pantetheine thiol group and a condensation domain that catalyzes peptide bond formation. During the latter reaction, the condensation domain catalyzes the transfer of the aminoacyl or peptidyl group from an upstream carrier protein domain to the primary amine of an amino acid that has been previously loaded onto the downstream carrier domain. The

Bradley S. Evans (ed.), *Nonribosomal Peptide and Polyketide Biosynthesis: Methods and Protocols*, Methods in Molecular Biology, vol. 1401, DOI 10.1007/978-1-4939-3375-4_1, © Springer Science+Business Media New York 2016

final core catalytic domain, present only in the terminal module of an NRPS, is a thioesterase domain that catalyzes either hydrolysis or, more commonly, cyclization of the peptide to catalyze release from the final carrier protein domain.

This modular architecture poses important questions for the mechanisms that allow the synthesis to occur in an efficient manner [2]. Namely, for proper peptide synthesis, it is necessary that each carrier protein domain visits the respective catalytic domains in an organized manner. One could easily see how the delivery of a carrier protein domain to the incorrect adenylation domain or the delivery of an amino acid to a downstream condensation domain rather than the upstream would lead to incorrect peptide products. The NRPSs have therefore attracted the attention of many structural biologists who have determined structures of individual catalytic domains and multi-domain components [2, 3]. The cumulative structural understanding is beginning to provide clues to the strategies that NRPS enzymes use to coordinate peptide synthesis. This chapter first presents the structures of NRPS domains and multi-domain complexes (Table 1). The fundamental structural mechanism of NRPS enzymes, requiring that the carrier proteins migrate between neighboring catalytic sites, mandates a degree of conformational flexibility in the NRPS systems. We therefore describe the strategies that have been used to obtain meaningful multi-domain structures that provide insights into the function of these assembly line enzymes.

2   PCP Domain

Similar to the carrier domains of fatty acid synthesis, the NRPS enzymes use a peptidyl carrier protein (PCP) domain that is used to shuttle the substrates and peptide intermediates between different catalytic domains [4]. The PCP domains are the smallest NRPS domains, usually only 70–90 amino acids in length. The PCP domains contain a conserved serine residue that serves as the site for covalent modification with a phosphopantetheine cofactor that is derived from coenzyme A (Fig. 1). This posttranslational modification converts the *apo*-carrier protein to a *holo*-state and is catalyzed by a specific phosphopantetheinyl transferase (PPTase) that is often co-expressed with the NRPS cluster [5]. The thiol of the phosphopantetheine group binds covalently to the amino acid and peptide substrates through a thioester linkage with the carboxyl group of the amino acid.

Like other acyl carrier proteins, the NRPS PCP domains are composed of four α-helices (Fig. 2a). Helices 1, 2, and 4 are longer, and mostly parallel, while the third helix is shorter and runs approximately perpendicular to the axes of the other three. The serine residue that is the site of addition of the phosphopantethe-

Table 1
List of NRPS proteins that have been structurally characterized

PDB	Protein	Ligands/comments	Ref.
Carrier protein domains			
1DNY	TycC PCP from third module	Also structures 2GDW, 2GDX, and 2GDY.	[6, 9]
4I4D	BlmI	Free-standing (Type II)	[8]
2FQ1	EntB	Fused to isochorismatase domain	[62]
Adenylation domains			
1AMU	Truncated phenylalanine activating domain of GrsA	AMP and phenylalanine	[19]
1MDF	DhbE, free-standing	Also structures 1MD9 (DHB and AMP) and 1MDB (DHB-adenylate)	[20]
3O82	BasE, free-standing	Also structures 3O83, 3O84, 3U16, 3U17, bound to a variety of inhibitors	[30, 31]
3ITE	SidN	Eukaryotic NRPS	[63]
Condensation domains			
1L5A	VibH		[43]
4JN3	CDA synthetase	Also 4JN5	[46]
Thioesterase domains			
1JMK	SrfA-C thioesterase		[48]
2CB9	Fengycin Biosynthesis		[49]
Reductase domains			
4DQV	R_{NRP} from *M. tuberculosis*		[58]
4F6L	AusA Reductase		[59]
Multi-domain NRPS structures			
2VSQ	SrfA-C (Cond-Aden-PCP-TE)	Leucine in adenylation domain	[45]
3TEJ	EntF (PCP-TE)	Also 2ROQ (NMR)	[60, 61]
2JGP	TycC (PCP-cond domain)		[44]
4IZ6	EntE-B (Aden-PCP)	Mechanism-based inhibitor	[28, 29]
4DG9	PA1221 (Aden-PCP)	Mechanism-based inhibitor	[55]
4GR5	SlgN1 (MLP-Aden)	AMPCPP	[38]

ine group is located at the start of helix α2. This helix is preceded by a long loop that is diverse in sequence and structure between the different NRPS PCP domains.

The NMR structure of the PCP domain from the third module of the TycC NRPS protein of tyrocidin synthesis [6] demonstrated that the NRPS PCP domains share the prototypic fold of

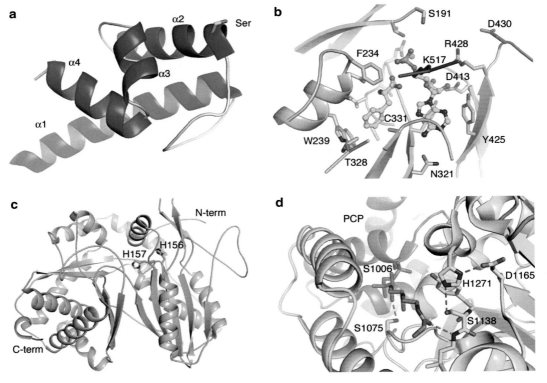

Fig. 1 Chemical structure of the phosphopantetheine cofactor attached to a conserved serine residue of the peptidyl carrier protein

Fig. 2 Structures of core NRPS domains. (**a**) The structure of the Type II PCP domain BlmI (PDB **4I4D**). The four helices are shown along with Ser44, the site of phosphopantetheinylation. (**b**) The active site of PheA (PDB **1AMU**), the adenylation domain from the gramicidin synthetase NRPS. The ligand molecules AMP and phenylalanine are shown in *ball*-and-*stick* representation. Protein side chains are labeled including the residues that form the phenylalanine binding pocket and residues that interact with the nucleotide. (**c**). The condensation domain of the CDA synthetase (PDB **4JN3**) is shown in *ribbon* representation. The two subdomains are shown, along with active site residues His156, His157, and Asp161, which is partially obscured by His157. (**d**) The active site of the thioesterase domain from EntF is shown (PDB **3TEJ**). The pantetheine, covalently bound to Ser1006, is directed from the PCP domain to the active site, which is composed of the catalytic triad Asp1165, His1271, and Ser1138. The different NRPS domains are shown in specific colors, which are maintained throughout the chapter. PCP domains are shown in *blue*. Adenylation domains are shown in *pink* for the N-terminal sub-domain and *maroon* for the C-terminal sub-domain. Condensation domain is shown in *light green* and the thioesterase domain is shown in *yellow*

the carrier proteins from related modular polyketide and fatty acid synthesis [7]. A recent crystal structure of a free-standing PCP domain [8] confirms the overall fold in an independent carrier domain. Additional studies have demonstrated that the core helical structure in the *apo* and *holo* states [9].

As described below, the structures of PCP domains in complexes with catalytic domains demonstrate the regions of the carrier domains that interact with partner proteins. Not surprisingly, given the presence of the phosphopantetheine cofactor at the start of helix α2, this helix and the loop that joins helix α1 to α2 appear to be the primary determinants for interactions with the catalytic domains. Shotgun mutagenesis of the carrier protein of the EntB protein from enterobactin biosynthesis in *E. coli* followed by screening to test function in vivo identified regions of the PCP that are involved in interactions with catalytic domains. In addition to the loop and helix α2 mentioned above, these studies also identified residues from the short orthogonal helix α3 that also formed part of the hydrophobic patch that governed interactions with the downstream condensation domain [10, 11].

3 Adenylation Domain

NRPS adenylation domains play a key role in peptide natural product biosynthesis. In the assembly line-like choreography, the adenylation domain is the first domain the substrate encounters before it is added to the nascent peptide natural product. The adenylation domains catalyze a two-step reaction that activates the amino acyl substrate as an adenylate, followed by transfer of the amino acid to the thiol of the pantetheine cofactor of the carrier protein domain (Fig. 3).

Adenylation domains belong to a larger adenylate-forming enzyme superfamily containing Acyl-CoA synthetases, NRPS adenylation domains, and beetle luciferase [12]. These enzymes are structural homologs and utilize a similar reaction mechanism that comprises two half reactions. Structural and kinetic results obtained from acyl-CoA synthetases [13–15] and luciferase enzymes [16, 17] have aided in the understanding of the adenylate-forming enzyme family. We focus here specifically on the adenylation domains of NRPS.

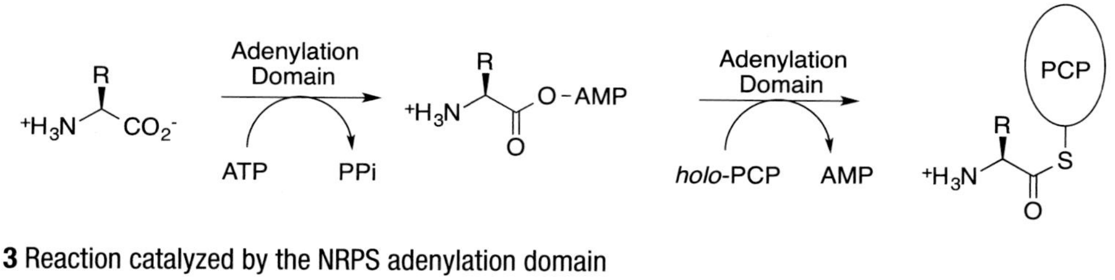

Fig. 3 Reaction catalyzed by the NRPS adenylation domain

NRPS adenylation domains consist of approximately 500 residues. The bulk of the enzyme, residues 1–400, makes up the N-terminal subdomain while the final 100 residues form the C-terminal subdomain that sits atop the N-terminal subdomain. Several consensus sequences were identified in adenylation domains and designated A1 through A10 [1, 18]. These regions impart both structural and substrate stabilizing roles. The two-step reaction (Fig. 3) is carried out in a Bi Uni Uni Bi ping-pong mechanism. First Mg-ATP and the carboxylic acid bind to form an acyl-adenylate. After PP_i from the ATP leaves the active site, a reorganization of the active site occurs where the C-terminal subdomain rotates changing the active site for the second half reaction. This domain alternation strategy transitions the adenylation domain between the two half reaction conformations, adenylate-forming and thioester-forming [12].

The first two structures of NRPS adenylation domains were PheA (Fig. 2b), a phenylalanine activating adenylation domain dissected from the multi-domain gramicidin synthetase 1, and the free-standing 2,3-dihydroxybenzoic acid (DHB) specific DhbE [19, 20]. Both of these structures are in the adenylate-forming conformation with Phe and AMP in the active site of PheA and no substrate, a DHB-adenylate, and DHB and AMP in the active site of the three DhbE structures. While the bulk of the active site is located in the N-terminal subdomain, a Lys found on the A10 loop of the C-terminal subdomain is required for acyl-adenylate formation [21, 22]. In both PheA and DhbE the Lys is poised in the active site to interact with both the carboxylic acid and the phosphate of the AMP (Fig. 2b). Important N-terminal regions to note are: the phosphate-loop (A3) that orients the β and γ phosphates of ATP and is often unresolved when ATP is not in the active site demonstrating its flexibility, the aromatic residue of the A4 motif (Phe234 in PheA and His207 is DhbE) which interacts with the carboxylic acid, and the aspartic acid of A7 motif that interacts with the ATP ribose hydroxyls. Once the high-energy acyl-adenylate is formed and PP_i leaves the active site, domain alternation occurs to prepare the active site for the thioester-forming reaction.

The structures of a related acyl-CoA synthetase [13] provided the first view of a distinct catalytic conformation of a member of this adenylate-forming family. Compared with the previous structures of PheA and DhbE, the C-terminal subdomain of bacterial acetyl-CoA synthetase (Acs) was rotated by ~140° to a new position that created a binding pocket for the CoA nucleotide and a tunnel through which the pantetheine thiol approaches the adenylate of the active site. The hypothesis that all members of this family adopt both catalytic conformations, an adenylate-forming conformation as seen in PheA and a thioester-forming conformation seen in Acs, has now been thoroughly tested. In particular, extensive structural and biochemical analyses with Acs [13, 22] and the related protein 4-chlorobenzoyl-CoA ligase [14, 15, 23, 24] have demonstrated the specific involvement of residues on opposite faces of the C-terminal domain in catalyzing the respective partial reactions.

The 140° domain alternation occurs around a conserved hinge residue, an Asp or a Lys, located in the A8 motif. The importance of the hinge residue and its ability to change rotamers was demonstrated when the hinge in 4-chlorobenzoyl-CoA ligase was mutated to a Pro which essentially trapped it in the adenylate forming conformation [15, 22]. Domain alternation changes the active site without moving the substrate. Notably, the A10 catalytic Lys is removed from the active site by ~25 Å. Also the A8 loop interacts with the N-terminal subdomain where PP$_i$ exited and hydrogen bonds with the aromatic residue of the A4 motif rotating it away from the adenylate. This makes room for the pantetheine thiol to attack the carboxylic carbon thus displacing AMP and loading the pantetheine arm of the PCP with the amino acid substrate.

Another model protein that is closely related to NRPS adenylation domains offers more evidence for the role of the rotation of the C-terminal sub-domain. The DltA protein from *B. subtilis* is involved in cell wall biosynthesis, where it activates a molecule of L-Ala and loads into onto the partner carrier protein DltC. Thus, while not strictly an NRPS adenylation domain, the protein is highly homologous and serves as a useful model for understanding NRPS adenylation domains. The structures of the DltA have been solved with AMP [25] and Mg-ATP [26] in the active site and illustrate the distinct adenylate- and thioester-forming conformations.

In addition to DhbE, many additional bacterial siderophores derived from NRPS systems contain a salicylate or 2,3-dihydroxybenzoate moiety that is involved in iron binding [27]. Often, the aryl acid is activated by an independent adenylation domain. The structures of the *E. coli* homolog EntE [28, 29], as part of an adenylation-PCP complex, and *A. baumannii* BasE [30, 31] have also been solved. While DhbE adopts the adenylate-forming conformation, EntE adopts the thioester-forming conformation (Fig. 4). BasE, like several other adenylation domains, shows no electron density for the C-terminal sub-domain, suggesting it is adopting multiple conformations in the crystal lattice.

Many NRPS clusters contain a small ~70 residue protein that plays a role in activation of the adenylation activity [32]. The first characterized protein was encoded by the *mbtH* gene of *M. tuberculosis* and these proteins are therefore known as MbtH-Like Proteins (MLPs). Biochemical evidence has demonstrated that some adenylation domains require the MLPs for acyl-adenylate formation [33–35]. To date three MLP structures are available: the founding member MbtH, PA2412, and SlgN1 [36–38]. MLPs are thin arrowhead-shaped proteins with three central antiparallel

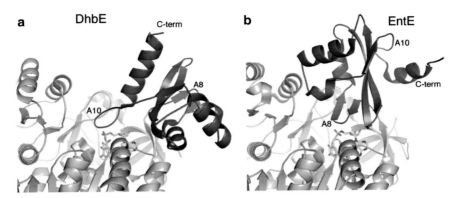

Fig. 4 Domain alternation of NRPS adenylation domains. The structures of two free-standing adenylation domains are shown from two bacterial siderophore synthesis, (**a**) DhbE from the bacillibactin NRPS of *B. sub-tilis* and (**b**) EntE from the enterobactin NRPS of *E. coli*. The DhbE structure (PDB **1MDB**) is in the adenylate-forming conformation, with the A10 motif of the C-terminal subdomain directed towards the active site. The EntE structure (PDB **4IZ6**) adopts the thioester-forming conformation with the A8 motif near the active site. The carrier protein and the pantetheine cofactor of structure 4IZ6 are not shown for clarity

β-sheets followed by two α-helices [37]. Defined in the MLP consensus sequence [32] and presented on one side of the MLP are two Trp residues that stack against each other. These Trp residues were shown to be required for activation of the adenylation domain by MLP [39]. A series of conserved proline residues have also been tested and appear to not be essential for activation of the adenylation domain [40]. A clear understanding of the mechanisms by which MLPs activate the adenylation domains is currently unknown.

4 Condensation Domain

Condensation domains, usually located at the N-terminus of a module, catalyze amide bond formation between two substrates. The condensation domains transfer the amino acid or peptide from an upstream carrier protein domain to the amino moiety of the substrate that has been previously loaded onto a downstream carrier protein domain (Fig. 5).

These 450 residue domains belong to the chloramphenicol acetyltransferase (CAT) superfamily. Similar to CAT, condensation domains contain a conserved HHxxxDG motif [1]. In CAT, the second His of this motif acts a general base that extracts a proton from chloramphenicol promoting nucleophilic attack and thus acyl transfer [41]. This His is also essential for condensation domain activity [42]; however, its exact role may depend on the substrates [43, 44]. Currently there are four crystal structures of condensation domains. They are: the standalone condensation domain VibH [43], the final condensation domain and its donor PCP dis-

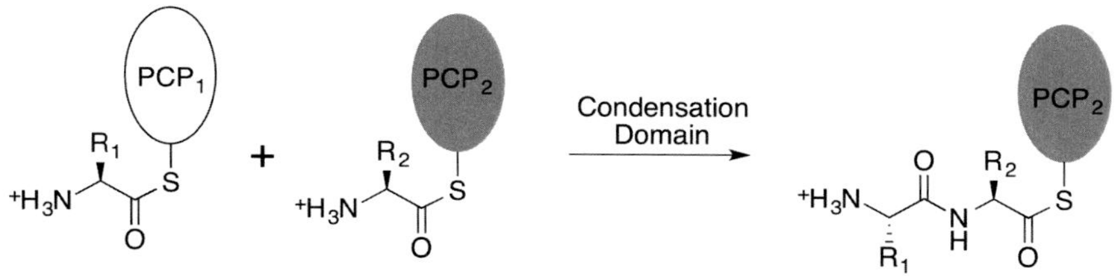

Fig. 5 Reaction catalyzed by the NRPS condensation domain

sected from the multi modular TycC [44], the condensation domain in the terminal module SrfA-C solved as a complete module [45], and finally the first condensation domain dissected from Calcium-Dependent Antibiotic synthetase (CDA-C1) [46].

Most commonly, a condensation domain will bind to an upstream donor and a downstream acceptor PCP. In some cases, especially with condensation starter domains, a substrate not bound to a PCP may be used. For example the standalone condensation domain VibH uses norspermidine as the acceptor which accepts DHB from the pantetheine arm of VibB [43]. Despite the current condensation domain structures lacking native ligands in the active site, much can be inferred from the current structures (Fig. 2c). While CAT forms a cyclic trimer [41], the monomeric condensation domains form a pseudo-dimer composed of two subdomains that contain CAT-like folds. The two subdomains adopt a V-shaped structure with a central cleft; at the base of this cleft, the two subdomains are linked by an α-helical linker. Also just above the linker between the two subdomains is the active site with the second His in the HHxxxDG motif located on a portion of the C-terminal subdomain that crosses over to the N-terminal subdomain. A second loop that spans the cleft between the two subdomains has been referred to as a lid or a latch. Despite the name, there is no evidence that this latch opens and closes [46]. Between these two crossovers, a tunnel is formed. As the PCPs bind to the condensation domain, their pantetheine arms must reach through this tunnel to the HHxxxDG active site in order for peptide bond formation to occur.

TycC and SrfA-C are both multi-domain structures each with a condensation domain that natively binds two PCPs. The TycC structure has an upstream donor PCP attached via a short linker. While the two domains are interacting, it does not appear that this is the catalytically active state. The Ser on the PCP which the pantetheine is loaded onto is 46.5 Å from the HHxxxDG motif, however the pantetheine is only ~16 Å long. Also several residues on the PCP that have been shown to be required for PCP–C domain interaction [47] are not involved in the interactions found in TycC. On the other hand, SrfA-C, as described below, does appear to form a valid domain interaction despite the pantetheine accepting Ser of the PCP being mutated to an Ala, thus making SrfA-C catalytically dead. This mutated residue is located ~16 Å from the HHxxxDG motif. Also the PCP residues required for PCP–C interactions are interacting with the condensation domain. Specifically, the PCP residues Met1007 and Phe1027 form hydrophobic interactions with Phe24 and Leu28 of the condensation domain [45].

The CDA-C1 structure [46] is the most recent condensation domain structure solved and provides a unique insight into the possible dynamics of NRPS condensation domains. CDA-C1 is in a distinctly more closed conformation than any of the other three condensation domains. In the CDA-C1 structure the N-terminal subdomain is 15°, 22°, and 25° more closed than VibH, TycC and SrfA-C respectively. Both SAXS and biochemical data suggest that this closed conformation seen in CDA-C1 is not due to crystal packing and is biochemically active [46]. Furthermore, normal mode analysis, morphing with energy minimization, and molecular dynamics all confirm that this opening and closing is possible. It is plausible however that condensation domains do not undergo this dynamic motion and are instead locked into a more opened or closed conformation based on the size of the substrate they need to accommodate and thus their location in the NRPS biosynthetic pathway. This, however, does not appear to be the case as SrfA-C is in a more opened conformation than TycC despite TycC being located on the ninth module of the system and SrfA-C being located on the seventh. More work is needed to fully understand the dynamics of condensation domains. Also since all current condensation domain structures lack ligands in the active site it is unclear how terminal condensation domain accommodates such large peptide substrates.

5   Thioesterase Domain

Within the final module of an NRPS pathway, the activity of the final condensation domain catalyzes the transfer of the upstream peptide to the amino acid substrate that is loaded onto the terminal PCP domain. To release the peptide and free the NRPS enzyme for another round of synthesis, the activity of a thioesterase domain is required (Fig. 6).

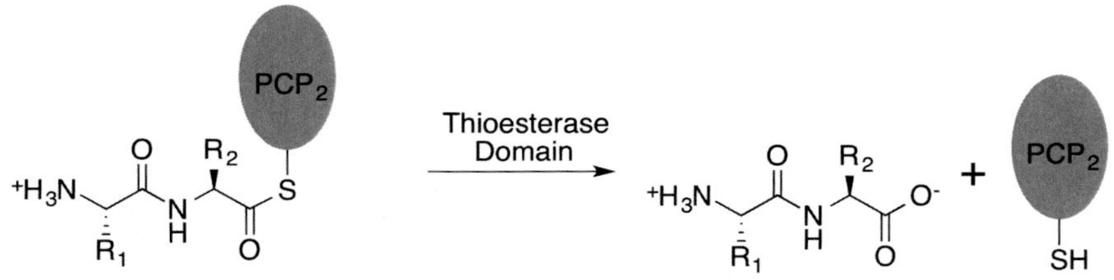

Fig. 6 Reaction catalyzed by the NRPS thioesterase domain

The thioesterase domains are approximately 30 kDa in size and, as a class, can function as either hydrolases (as shown in Fig. 6) or as cyclases, where they can catalyze either lactam or lactone formation with an upstream heteroatom from the peptide chain. The thioesterase domains form an acyl-enzyme intermediate with an active site serine residue that subsequently is released from the enzyme through either hydrolysis with a water nucleophile or cyclization. For enzymes that catalyze lactone or lactam formation, the active site pocket therefore must bind the peptide substrate in an orientation that favors cyclization over hydrolysis by positioning the nucleophilic group to resolve the acyl-enzyme intermediate.

Structures of the genetically truncated thioesterase domains from the SrfA-C subunit [48] of the surfactin NRPS cluster and the FenB protein [49] of the fengycin cluster have been determined. The structure of the thioesterase domain from SrfA-C (Srf-TE) showed the domain belongs to the family of α/β hydrolases composed of a central, mostly parallel β-sheet that is surrounded by α-helices. Srf-TE contains a catalytic serine residue that serves as a nucleophile, attacking the terminal carbonyl of the peptidyl thioester with the pantetheine on the terminal PCP (Fig. 2d).

Three helices form a lid that in Srf-TE adopts two different conformations in the two molecules of the asymmetric unit, referred to as open and closed. The closed state is suggested to be a ground state of the enzyme however the helices in this state also exhibit some degree of disorder. Surfactin, the product of the Srf NRPS system, is an acyl-heptapeptide that contains seven amino acids and an N-terminal β-hydroxy fatty acid. The molecule cyclizes via lactone formation between the C-terminal carboxylate and this β-hydroxyl moiety. The authors soaked an N-acylheptapeptide-*N-acetylcysteamine* analog lacking the β-hydroxyl into the crystal. Portions of this molecule could be observed in the active site pocket, although the density was not of sufficient quality to enable complete modeling. The density did demonstrate a bent configuration that suggested the contour of the active site directs the cyclization reaction [48].

The thioesterase domain of the fengycin FenB protein (Fen-TE) has also been structurally characterized [49]. Like Srf-TE, Fen-TE was also genetically truncated to enable crystallization of the thioesterase domain. Fengycin is an acyl decapeptide lactone that contains a tyrosine at the fourth position that cyclizes through the phenolic hydroxyl with the C-terminal carboxylate of the peptide. To further characterize the active site, the structure of the enzyme covalently acylated with phenylmethylsulfonyl fluoride, the common inhibitor of hydrolases and proteases containing a serine nucleophile, was determined. The phenyl group from the inhibitor bound in a pocket that likely is used by the last Leu residue of the fengycin peptide. An oxyanion hole that stabilizes the generation and cleavage of the acyl-enzyme intermediate is formed by the backbone amides and is conserved throughout the α/β-hydrolase family. A molecule of fengycin was modeled into the substrate binding pocket and examined with molecular dynamics that showed no dissociation and limited movement of the ligand after an initial equilibration period [49]. This supported the position derived from docking and identified residues that could form interactions with the peptide ligand. In particular, the Tyr hydroxyl from the fourth position of the fengycin peptide was directed through a hydrophobic ridge to a position that allowed attack on the acyl-enzyme intermediate for the cyclization portion of the reaction.

6   Additional Integrated Domains

Additional protein activities are used in NRPS pathways for the complete synthesis of the final product. These proteins often act upon the substrate precursors prior to incorporation into the NRPS assembly line or on the immature peptide in steps that result in the final product maturation [50]. Certain auxiliary proteins do catalyze reactions on the amino acyl or peptidyl intermediates that are bound to a carrier protein. Most of these domains are expressed from isolated genes and most function as independent single domain proteins. However, some proteins are integrated into the NRPS assembly line where they are co-expressed. The most common of these are epimerization and methylation domains, as well as the alternate termination domains that thioester cleavage via an NAD(P)H-dependent reduction.

6.1   Epimerization Domains

Two different types of epimerization domains have been identified in NRPS systems where they catalyze conversion of L-amino acids to D-amino acids. Canonical epimerization domains of ~450 residues are inserted between PCP and condensation domains. While several systems, notably the PchE protein of pyochelin biosynthesis [51] and the HMWP2 protein of yersiniabactin biosynthesis [52], have a shorter, ~350 residue, noncanonical epimerization domain.

Interestingly, the canonical epimerization domains show sequence and structural homology to condensation domains. Only a single structure exists of an NRPS epimerization domain of the epimerization domain from tyrocidin synthetase A [53]. The structure displays the same overall symmetrical fold as the condensation domains, with a large cavity in the center of two subdomains. The epimerization domain contains the same conserved HHxxxDG motif present in the condensation domains. Mutation of the second histidine (His753) and the aspartic acid (Asp757) of this motif, as well as several downstream residues including a glutamate at position 892 and a conserved asparagine and tyrosine pair 975 and 976, were defective for proton wash-out in the epimerization domain of the first module of the gramicidin synthetase, GrsA

[54]. In the structure of the TycA epimerization domain, these histidine and the glutamate residues, His743 and Glu882, are positioned on opposite sides of the active site cavity and are potential candidates for a two-base epimerization reaction. The glutamate is part of a conserved EGHGRE motif that is common to epimerization domains, but not the homologous condensation domains [1].

The noncanonical epimerization domains are present in the pyochelin producing NRPS, protein PchE of *P. aeruginosa* and *Burkholderia pseudomallei*. It shares the interesting property with N-methylation domains of being inserted within the C-terminal sub-domain of the NRPS adenylation domain. There is no structural information regarding these epimerization domains, nor is there any information how the adenylation domain accommodates these insertions.

6.2 N-Methylation Domains

N-methylation of the peptides of NRPS products is seen primarily in fungal NRPS systems. Presumably, this confers proteolytic stability to the methylated peptide bonds. Like the noncanonical epimerization domains, the N-methylation domains are inserted into the C-terminal subdomain of the NRPS adenylation domains. The methylation domain is ~420 residues in length and shows limited homology with other *S*-adenosylmethionine dependent transferase enzymes. The C-terminal subdomain of the NRPS adenylation domain contains a central three-stranded β-sheet that contains two long anti-parallel strands that each range in length from 7 to 11 residues. Following these two strands is a loop that leads to an amphipathic helix that packs against the central β-sheet. It is into this large loop that the N-methylation domains are inserted. In the structure of the complex between the adenylation and the PCP [28, 29, 55], this loop from the adenylation C-terminal subdomain is directed towards the PCP, suggesting that perhaps the amino acid substrate could be directed to the neighboring methyl transferase domain without a major structural rearrangement of the catalytic domains.

There are no reported structures of NRPS methyl transferase domains. A comparison with the Protein Data Bank finds only short stretches of homology (80–100 residues) to related SAM-dependent methyltransferases, for example the NodS protein of *Bradyrhizobium japonicum* [56]. Additionally, the structure of a SAM dependent N-methyltransferase from the glycopeptide NRPS cluster of *Amycolatopsis orientalis* has been determined [57] however this protein is an independent tailoring enzyme with limited homology to the integrated methyl transferase domains of NRPSs.

6.3 Reductase Domains

Certain NRPS clusters terminate not with a thioesterase domain but rather with a NAD(P)H-dependent reductase domain that cleaves the bound peptide to release a C-terminal alcohol or aldehyde rather than a carboxylate. The ~280 residue reductase domains show homology to nucleotide cofactor binding short-chain dehydrogenase/reductase (SDR) enzymes. The structure of a reductase domain from a functionally uncharacterized NRPS protein from *Mycobacterium tuberculosis* (designated R_{NRP}) was determined to confirm the expected Rossmann fold common to SDR enzymes [58]. A small helical C-terminal domain was identified as well as a short helix-turn-helix insertion in the standard Rossmann fold. Although the structure of a liganded reductase domain was not obtained, examination of cofactor binding by small-angle X-ray scattering identified a small change in the radius of gyration of the domain, suggesting a conformational transition from an open to a closed state. A second structure of the aldehyde-producing reductase domain of AusA, catalyzing therefore only a 2 e^- reduction rather than the sequential 4 e^- reduction of R_{NRP}, has also been structurally characterized showing a similar overall structure [59]. A di-domain construct, composed of the carrier protein and reductase domain was also crystallized; however, no electron density for the carrier protein domain was observed in this structure preventing insights into the nature of the interaction between the catalytic and carrier domains.

7   Multi-domain NRPS Crystal Structures

Due to the plasticity of NRPSs, crystal structures of multi-domain constructs are difficult to obtain. Similarly, the large size of these proteins presents challenges for NMR. Despite these difficulties there are currently five multi-domain NRPS structures that are in or appear to be in a catalytically relevant state. They are: the four-domain module of SrfA-C [45], the PCP-thioesterase domains of EntF [60, 61], the chimeric adenylation-Carrier Protein EntE-B [28, 29], the native adenylation-PCP protein PA1221 [55], and the MLP-adenylation domain SlgN1 [38]. As noted earlier, the structure of the excised PCP-condensation domain from TycC has been determined [44]; however, the two domains do not appear to interact in a functional manner. These crystal structures help to shed light on how the domains of NRPSs interact to carry out natural product biosynthesis.

7.1 SrfA-C:
A Complete NRPS
Module

SrfA-C is the largest NRPS crystal structure to date and offers a complete view of an NRPS module. Composed of a domain architecture of condensation-adenylation-PCP-thioesterase, the 144 kDa SrfA-C is the terminal module from the surfactin biosynthetic cluster of *Bacillussubtilis* [45]. In order to crystallize SrfA-C the Ser of the PCP that is pantetheinylated was mutated to an Ala to produce a homogenous protein sample. Nonetheless, the PCP is interacting with the condensation domain in a catalytically relevant way (Fig. 7a). The condensation and N-terminal subdomain of the

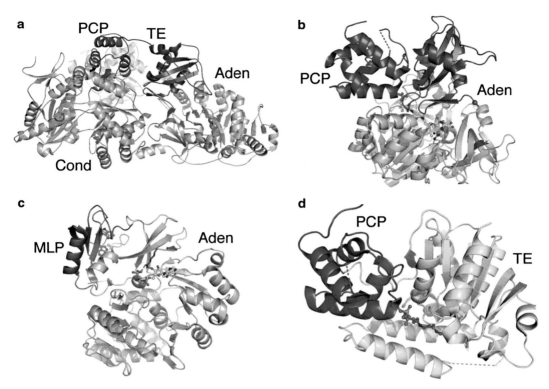

Fig. 7 Crystal Structures of multi-domain NRPS proteins. (**a**) The structure of the complete termination module of SrfA-C (PDB **2VSQ**) shows the PCP interactions with the downstream side of the condensation domain. (**b**) The adenylation-PCP structure of PA1221 (PDB **4DG9**) illustrates a functional interface between the PCP and the thioester-forming conformation of the adenylation domain. (**c**) The structure of the SlgN1 protein (PDB **4GN5**) shows the interaction between the MbtH-like domain (*forest green*) interacting with the N-terminal subdomain of the adenylation domain. The conserved tryptophan residues are highlighted in the MLP. (**d**) The PCP-thioesterase domain of EntF (PDB **3TEJ**) is shown illustrating the binding of the pantetheine into the active site of the thioesterase domain

adenylation domain share a large interdomain interface and are believed to form a stable *platform*, upon which the PCP is thought to migrate between different catalytic domains. The C-terminal subdomain of the adenylation domain is neither in the adenylate or thioester-forming conformations. Instead it has adopted an intermediate conformation that is closest to the adenylate forming conformation. Because of this intermediate conformation the catalytic Lys on the A10 motif is positioned slightly outside of the active site, which may prevent the adenylation domain from activating another substrate before the PCP is ready to accept it. The thioesterase domain is positioned near the condensation domain and, interestingly, the pantetheine channel on the thioesterase is facing directly into the core of the SrfA-C protein. A simple rotation of the PCP is therefore insufficient to allow a functional interaction between the PCP and thioesterase domains. This is evident from aligning SrfA-C with the EntF PCP-thioesterase structure [60, 61]. It therefore seems that rotation of the thioesterase domain relative to the core of the SrfA-C module is necessary to adopt a catalytic conformation.

7.2   EntE-B and PA1221: Adenylation-Carrier Protein Complexes

The structures of the adenylation-carrier protein complexes of the chimeric EntE-B and the native PA1221 (Fig. 7b) were solved in the thioester-forming conformation with the use of aryl/acyl-adenosine vinylsulfonamides [28, 29, 55]. In both structures, the carrier protein is interacting with the adenylation domain and the pantetheine is positioned in the adenylation domain tunnel forming a covalent bond with the adenosine inhibitors. The chimeric EntE-B, which was genetically combined using a linker of similar sequence length and composition to multi-domain NRPSs, formed a domain swapped dimer where the carrier protein from molecule 1 was interacting with the adenylation domain of molecule 2 and vice versa. A comparison to SrfA-C suggests that while the EntE-B linker was of similar sequence length as SrfA-C, the EntE-B linker formed an α-helix which rendered the linker too short for an intramolecular interaction.

Both the EntE-B and PA1221 structures reveal the various interactions between the adenylation and carrier protein domains which includes both hydrophobic interactions and hydrogen bonding [28]. Biochemical studies also help to confirm that while the EntE-B forms a domain swapped dimer due most likely to the linker, the interaction between the ArCP and the A domain seem to be the same interaction that would occur in the intra-domain interaction. In particular, the insights derived from the EntE-B crystal structure were used to guide mutations in the EntE homolog from *A. baumannii* to improve the ability to recognize the heterologous EntB carrier protein as a partner [29].

7.3   SlgN1: An MLP-Adenylation Domain Structure

As noted earlier, some adenylation domains require MLPs for acyl adenylate formation [32–35]. The crystal structure of SlgN1, containing a MLP fused to the N-terminus of an adenylation domain, showed for the first time were MLPs bind to adenylation domains [38]. The MLP binds the N-terminal subdomain of the adenylation domain distal from the active site (Fig. 7c). The closest active site motif appears to be A7 aspartic acid which binds the ribose hydroxyls of the ATP. A key interaction is an Ala residue presented on the surface of SlgN1 that is inserted between two stacked Trp on the MLP. For MLP dependent adenylation domain this Ala (Ala 433 in SlgN1) is highly conserved while for nondependent MLP adenylation domains this residue varies. Unfortunately the C-terminal subdomain of the adenylation domain of SlgN1 was removed to facilitate crystallization. Therefore it is unclear if the MLP is bound to the adenylation domain for just acyl adenylate formation or also for the subsequent thioester formation.

7.4 EntF: PCP-Thioesterase Domain Structures

Finally, the structures of the PCP-TE domains of EntF were solved by both X-ray diffraction and NMR [60, 61]. While the NMR structure was that of an apo PCP and shows a dynamic interaction, the crystal structure used an α-chloroacetyl amide coenzyme A analog which was loaded onto the PCP domain using the phosphopantetheinyl transferase Sfp. This made it possible to structurally analyze the thioesterase pantetheine channel along with the domain interactions (Fig. 7d). The pantetheine is inserted into the thioesterase channel which complements the pantetheine with a hydrophobic pocket around the di-methyls of the pantetheine, hydrogen bonding with the two amide carbonyls, and also van der Waals interactions [61]. At the end of the channel the loaded pantetheine encounters the thioesterase catalytic triad. The PCP and thioesterase form extended hydrophobic interactions which includes the thioesterase lid region which interacts with the PCP.

7.5 Modeling of Dynamics of the SrfA-C Module

While only a few structures of multi-domain NRPSs exist, the various conformations available assist in decoding the NRPS choreography for natural product biosynthesis. In order for substrate loading onto the pantetheinylated PCP the adenylation domain undergoes domain alternation after activating the substrate. Now in the thioester-forming conformation the PCP can interact with the adenylation domain and the pantetheine can enter the adenylation domain tunnel and attack the substrate (Fig. 8a). With the substrate loaded the PCP can now migrate over to the condensa-

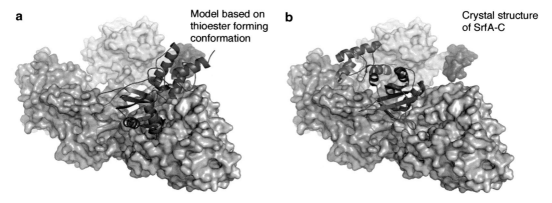

Fig. 8 Model for the delivery of the PCP to the adenylation domain active site. The SrfA-C protein is shown in (**a**). A model conformation that adopts the thioester-forming conformation where the PCP is bound to the adenylation domain and (**b**) the crystallographic structure where the PCP interacts with the condensation domain. The N-terminal subdomain of the adenylation domain, as well as the condensation and thioesterase domains, are all shown in surface representation, while the PCP and the C-terminal subdomain of the adenylation domain are shown as ribbons. The structure in panel (**a**) is derived by modeling the SrfA-C (PDB **2VSQ**) adenylation and PCP domains onto the thioester-forming conformation observed with EntE-B (PDB **4IZ6**) or PA1221 (PDB **4DG9**). The serine residue that is the site of phosphopantetheinylation is highlighted with the *yellow sphere*

tion domain where it awaits the delivery of the upstream peptide or amino acid substrate (Fig. 8b). Based on the SrfA-C structure it appears that A domain needs to adopt or come close to adopting the adenylate-forming conformation. It seems plausible that the movement of the C-terminal subdomain of the adenylation domain facilitates the movement of the PCP between the adenylation and condensation domains. The C-terminal subdomain and the PCP do not move as a rigid body; rather there appears to be two components to the movement to allow the PCP to adopt the proper position bound to the adenylation domain [55]. Finally the PCP must interact with the thioesterase domain. Since the thioesterase active site channel is directed toward the core of the condensation domain, the PCP and thioesterase domains must move relative to the condensation and adenylation domains, as also relative to each other, to allow a functional interaction to form.

8 Strategies to Crystallize NRPS Proteins

Structurally characterizing NRPS proteins through either NMR or X-ray crystallography is challenging given that a complete module is over 1000 residues in length and multi-domain proteins containing multiple modules are not uncommon. The large size limits the use of NMR and the size and flexibility make crystallization anything but routine. Despite these challenges, the previous sections have described numerous successful examples of structural characterization of the NRPS proteins. We present here a summary of the strategies that have been used to overcome the difficulties inherent to these large modular proteins.

8.1 Structural Studies of Individual Domains

Examination of single domains can provide insights into the action of multi-domain proteins. Initial attempts at structural characterization of NRPSs therefore focused on individual domains; only more recently have multi-domain structures been achieved. These early studies took advantage of both the rare type II nature of some NRPS clusters, where individual catalytic domains were expressed as a single domain protein, as well as the use of molecular tools to produce genetically truncated proteins.

Specific examples of the study of type II NRPS domains include the VibH condensation domain [43] and the DltA [25] and BasE [30] adenylation domains. Insights into the carrier domain architecture resulted from the type II PCP domain that has recently been determined [8], as well as the full length structure of EntB, a two-domain protein that contains a catalytic isochorismatase domain fused to a carrier protein domain [62]. Comparison of these structures with truncated and multi-domain proteins has failed to demonstrate any features that differ between the type I and type II protein domains.

Type IINRPS proteins are relatively rare, however, and most studies of single NRPS domains have used genetically truncated protein constructs. The production of isolated domains presents additional challenges as it is necessary to identify accurately the domain termini. A truncation that is too short can result in exposure of hydrophobic residues that normally reside within the protein core. Alternately, defining a domain boundary that is too large may result in the inclusion of linker sequences or portions of neighboring domains that are poorly ordered. Often, these extraneous regions can hinder protein folding or solubility and preclude functional or structural analysis. Likely for this reason, many of the genetically truncated NRPS domains that have been studied lie at the N- or the C-terminus of the native multi-domain protein. Indeed, of the many NRPS proteins that have been structurally characterized (Table 1), only three truncated proteins are truly "internal," where a choice needed to be made about the precise locations of both the N- and C-terminal truncations. Two structural studies on the TycC protein from tyrocidin synthesis, describing the PCP domain from the third module [6, 9] and the PCP-condensation di-domain crystal structure [44], were internal domain constructs. Additionally, an adenylation domain from SidN [63], a protein that is involved in the production of a fungal siderophore [64], was also excised from the middle of a larger NRPS protein. In this last example, the authors note that successful structure determination of SidN required optimization of the N- and C-termini and that changes of as few as 14 residues could impact protein solubility.

The choice of the boundaries between domains is therefore a critical decision to be made in expression of a genetically truncated NRPS domain. The availability of existing structures provides useful clues to guide the design of a protein construct. Identification of the boundary between the adenylation and PCP domains is aided by the presence of a conserved A10 catalytic motif near the C-terminus of the adenylation domain. This nearly identical motif is defined as PXXXXGK, where the X can represent any amino acid [1, 12]. Following this motif is a loop that leads to the start of helix 1 for the downstream PCP domain. In both the structures of SrfA-C and PA1221, the natural NRPS protein structures showing the adenylation-PCP linker, the 9 residues following the catalytic lysine pack against the C-terminal domain. In the structure of the PheA genetically truncated adenylation domain, the 11 residues following the A10 lysine interact with the body of the C-terminal subdomain. Therefore, at least ten residues following the A10 lysine should be considered to be part of the adenylation domain and should not be disrupted through a genetic truncation.

Not including the type II PCP domain [8], which has a longer N- and C-termini, the PCP domains are fairly consistent in size. The type I PCP domain structures, whether in isolation or as part of the multi-domain targets, all have 29–32 residues between the start of the helix α1 and the serine that is phosphopantetheinylated. This defines the N-terminal region of the PCPs that should not be disrupted. Similarly, the distance from the site of cofactor modification to the C-terminal end of helix α4 in the type I PCP domains is 37 or 38 residues in all but PA1221, which is slightly longer at 40 residues. This again defines the boundaries of the PCP fairly consistently.

The definition of the start of the condensation domains is more challenging, given that three of the four available structures are derived from protein domains that lie at the N-terminus of the protein. The shortest N-terminus comes from VibH, the self-standing condensation domain of vibriobactin biosynthesis [43]. An α-helix is the first secondary structural element that is shared by all four. While the VibH structure begins two residues before this helix, the remaining three structures share an additional six residues that should likely be maintained in a condensation domain.

To define the C-terminus of the condensation domain, the SrfA-C structure offers the best view of the linker between the condensation and adenylation domains. This linker, composed of 32 residues, interacts closely with residues from both domains. The first 13 residue interact with the condensation domain, while the remaining helix and loop interact with the adenylation domain. The tighter interaction between the linker and both catalytic domains relates to the fact that the condensation and N-terminal subdomain of the adenylation domain are expected to form a stable interface that likely does not change during the NRPS catalytic cycle.

The thioesterase domain, by definition, resides at the C-terminus of the NRPS protein so the C-terminal end will be defined. The thioesterase domains are nearly always preceded by a PCP domain and the boundary of the PCP is defined, as discussed above. This leaves a short loop, as is seen in the PCP-thioesterase structure that is a suitable site for truncation to produce an isolated thioesterase domain.

8.2 Mutations to the Phosphopantetheinylation Site

Crystallization requires a uniform protein sample that is suitable to form the crystal lattice. One source of heterogeneity in NRPS proteins derives from the PCP domain and their mixture of *apo-* and *holo-*proteins that bear the phosphopantetheine modification. The preparation of *holo*-PCP domains uses either co-expression of a phosphopantetheinyl transferase or biochemical incubation with a specific or general PPTase. While these reactions should proceed to completion, there are no good ways to separate the *apo-* from *holo-PCP* domains. Therefore, several structural studies, by both NMR and crystallography, have used mutated PCP domains in which the serine residue is replaced with an alanine. This mutation dictates that the *apo*-protein is present at 100 % and limits heterogeneity. Of course, there are certain limitations to this strategy as the *apo-*

protein is not the biologically active form of the PCP domain. The differences in overall structure of the TycC PCP domain that depend on the *apo-* and *holo-*state of the protein [9] should therefore be considered if this *apo-*PCP domain is used.

Nonetheless, the mutation of the serine to an alanine removes the potential for modification by the PPTase enzymes and prevents the presence of the flexible cofactor that may hinder crystallization. This strategy has been used in the crystallization of the SrfA-C module [45] and also in the investigation of the solution structure of the PCP-thioesterase domain [60].

8.3 Use of Selective Inhibitors to Minimize Conformational Flexibility

The use of specific ligands that bind in the active sites of NRPS domains to reduce the conformational dynamics has proven to be quite useful and additionally provides some of the most detailed views of ligands bound in the active sites of NRPS domains. The most effective of these inhibitors are mechanism-based inhibitors that mimic intermediates in the reaction pathway and therefore provide the added benefit of demonstrating how the catalytic domains recognize their substrates. Additionally, these compounds have been useful to trap the transient interactions between catalytic and carrier protein domains.

The first example of such a compound was the use of a chloro-acetyl thioester of the pantetheine cofactor that was used by Bruner and Liu to crystallize a PCP-thioesterase complex (Fig. 9a). An α-chloroacetyl amide derivative of amino-CoA, where the CoA terminal thiol is replaced with an amine was generated [65]. This modified CoA molecule was loaded onto the PCP domain via a PPTase and shown to inhibit the thioesterase reaction. The proper binding of the α-chloroacetyl moiety in the proximity of the nucleophilic serine of the thioesterase domain was proposed to facilitate the interaction between the PCP and thioesterase domains [65]. Subsequently, this inhibitor was indeed used as a tool to crystallize the di-domain PCP-thioesterase of EntF [61]. Although designed to form a covalent interaction between the pantetheine and the catalytic serine of the thioesterase domain by attack on the α-carbon [65], the inhibitor molecule in the active site was not covalently attached to the protein. The authors proposed a mechanism whereby the Ser1138 attacked the carbonyl carbon of the α-chloroacetyl amide pantetheine derivative to form an oxyanion that displaces the chloride ion and forms an epoxide intermediate. This was then hydrolyzed either directly or following reaction with another nucleophilic group on the enzyme.

A second strategy to use mechanism-based inhibitors has been used successfully to determine the structure of the interaction between adenylation and PCP domains. This strategy expanded upon the use of sulfonamide inhibitors of the adenylation domain in which the phosphate diester moiety of the acyl adenylate intermediate was replaced with sulfamate or sulfamide analogs [66–68].

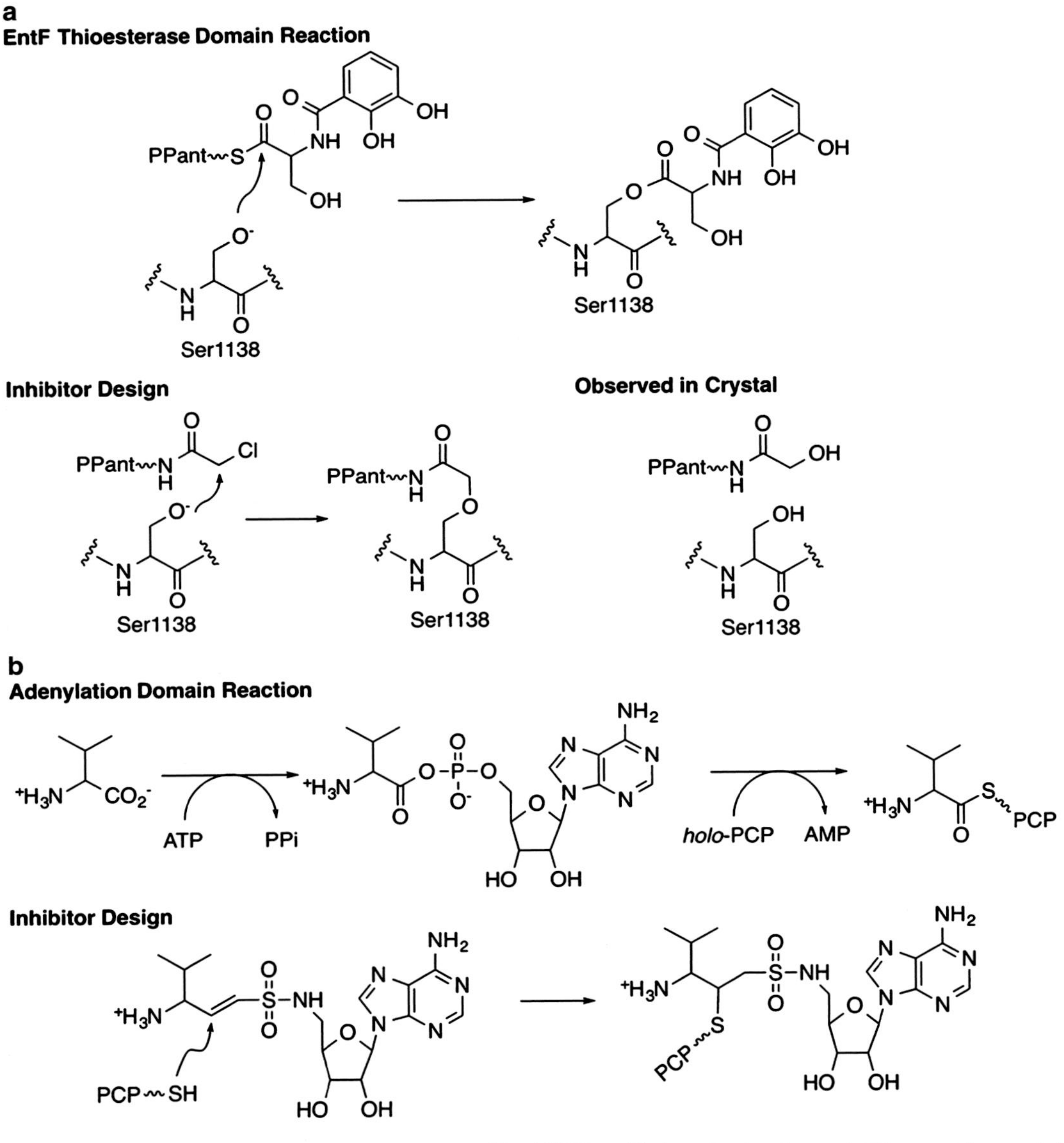

a
EntF Thioesterase Domain Reaction

Inhibitor Design

Observed in Crystal

b
Adenylation Domain Reaction

Inhibitor Design

Fig. 9 Mechanism-based inhibitors used to crystallize NRPS multi-domain structures. (**a**) The α-chloroacetyl-CoA derivative was designed to react with the catalytic serine of thioesterase domain. The catalytic reaction, the inhibitor rationale, and the observed structure are shown. (**b**) The vinylsulfonamide inhibitor is shown, along with the two-step reaction catalyzed by the adenylation domain. The covalent inhibitor complex has been observed in two crystal structures

To further optimize this inhibitor to react covalently with the pantetheine thiol of a partner carrier protein domain, Aldrich and colleagues introduced a linker containing a double bond between the sulfonamide moiety and the salicylic acid (Fig. 9b) on an inhibitor directed towards the aryl-activating type II adenylating enzyme MbtA [69]. The affinity of this vinylsulfonamide inhibitor is reduced by nearly five orders of magnitude with MbtA compared to the parental sulfonamide inhibitor [70], likely due to the loss of the carbonyl carbon and the nitrogen atom of the linker [69]. However, it still displayed apparent inhibition constants of 100–300 μM against the adenylation reaction and was deemed sufficient for biochemical and structural studies. Additionally, in the presence of the adenylating enzyme MbtA, the vinylsulfonamide inhibitor reacted covalently with the pantetheine moiety of the MbtB carrier protein and the adenylation and PCP proteins co-migrated on a native gel.

Suitable vinylsulfonamide analogs of adenylate intermediates have been used to trap crystallographically proteins from two different NRPS systems. It was first used to determine the structure of the adenylation-PCP interface between the EntE adenylation domain of enterobactin biosynthesis along with its partner EntB [28, 29]. Additionally, to facilitate crystallization of these proteins, a chimeric protein construct was designed that fused genetically the coding sequences for the EntE adenylation domain with the coding sequence for the carrier protein domain of the bifunctional EntB [29]. Subsequently, a vinylsulfonamide inhibitor was also used to determine the structure of an uncharacterized two domain adenylation-PCP protein from *P. aeruginosa* [55]. In the latter case, it was first necessary to identify the amino acid substrate preference to design the inhibitor with the appropriate amino acid side chain bound to the vinylsulfonamide linker.

Both the structures of PA1221 and the EntE-B chimeric protein demonstrate that the use of the mechanism-based inhibitor can serve to stabilize the domain interactions sufficiently to allow them to be observed crystallographically. In the case of PA1221, crystallization of the *apo*-protein in the absence of the inhibitor resulted in a structure in which no electron density could be observed for the carrier protein domain. Interestingly, the adenylation domain adopted the thioester-forming conformation that is identical to that observed to interact with the PCP in the structure of the *holo*-protein in the presence of the ligand yet still failed to bind the PCP. Thus the inhibitor was a critical reagent for sufficiently stabilizing the interaction to allow the full length protein to be observed in the crystal structure.

9   Conclusions

Despite the multitude of NRPS structures, including the structures of multi-domain proteins over the last 5 years, there remains much to be done to understand completely the structural basis for NRPS biosynthesis. In particular, more structures are needed of full modules of enzymes that will provide more insights into the nature of the interactions of the NRPS domains. The condensation domain lacks structures in the presence of ligands and more crystal or NMR structures will identify the roles of specific active site residues in binding and catalysis. Additionally, a superposition of the functional interaction between the PCP and thioesterase domains of EntF [60, 61] with the SrfA-C structure [45] shows that the PCP domain in the functional interaction is predicted to overlap with the adenylation-condensation core of SrfA-C. This suggests that the thioesterase domain of SrfA-C is not positioned in a "catalytic" orientation in the crystal lattice. Likely, the thioesterase position in SrfA-C is dynamic and adopts a different relative position when it interacts with the PCP in the course of the catalytic cycle. The use of inhibitors such as the α-chloroacetylamide pantetheine derivative [61, 65] with full-length EntF, for example, would allow the determination of the PCP-thioesterase interface in the context of a full NRPS module.

Another exciting target of NRPS structural biochemistry is the examination of the additional integrated domains, N-methyltransferases or epimerization domains, in the context of a full NRPS module. The carrier protein domains of SrfA-C, or the *E. coli* homolog EntF, must make their way to three consecutive catalytic domains (adenylation, condensation, and thioesterase domains). In modules that harbor additional integrated domains, the carrier protein must find its way to yet another catalytic domain and we can ask what structural features are required for these specialized modules.

Additional NRPS structures may also begin to address the question of what dynamic features, within or between catalytic domains, drive the proper delivery of the PCP along the catalytic assembly line. The large rotation of the two sub-domains within the NRPS adenylation domain offers a potential major rearrangement that could facilitate interactions with the PCP. Additional conformational changes may also compel the PCP to bind additional catalytic domains in a proper orientation.

Finally, now that the structure of the complete SrfA-C module [45] has toppled what was once the most sought-after structure in NRPS enzymology, a new high profile target is a structure of a multi-module protein. The SrfA-C structure suggests that the condensation domain and N-terminal subdomain of the adenylation domain form a stable interface and, relative to this platform, the PCP and possibly the other catalytic domains migrate to transit through the catalytic cycle. In a larger, multimodule NRPS, might this platform from one module form a larger complex with the condensation and adenylation domains of the next module as well. While atomic resolution crystal structures would provide answers to these questions, additional structural tools such as electron microscopy and small angle X-ray scattering may also provide insights into the larger macromolecular organization of the catalytic domains.

Acknowledgments

The research from our lab that is included in this review has been supported by the National Institutes of Health (GM-068440).

References

1. Marahiel MA, Stachelhaus T, Mootz HD (1997) Modular peptide synthetases involved in nonribosomal peptide synthesis. Chem Rev 97:2651–2674

2. Marahiel MA, Essen LO (2009) Chapter 13. Nonribosomal peptide synthetases mechanistic and structural aspects of essential domains. Methods Enzymol 458:337–351

3. Koglin A, Walsh CT (2009) Structural insights into nonribosomal peptide enzymatic assembly lines. Nat Prod Rep 26:987–1000

4. Mercer AC, Burkart MD (2007) The ubiquitous carrier protein--a window to metabolite biosynthesis. Nat Prod Rep 24:750–773

5. Beld J, Sonnenschein EC, Vickery CR et al (2014) The phosphopantetheinyl transferases: catalysis of a post-translational modification crucial for life. Nat Prod Rep 31:61–108

6. Weber T, Baumgartner R, Renner C et al (2000) Solution structure of PCP, a prototype for the peptidyl carrier domains of modular peptide synthetases. Struct Fold Des 8:407–418

7. Crosby J, Crump MP (2012) The structural role of the carrier protein--active controller or passive carrier. Nat Prod Rep 29:1111–1137

8. Lohman JR, Ma M, Cuff ME et al (2014) The crystal structure of BlmI as a model for nonribosomal peptide synthetase peptidyl carrier proteins. Proteins 82:1210

9. Koglin A, Mofid MR, Lohr F et al (2006) Conformational switches modulate protein interactions in peptide antibiotic synthetases. Science 312:273–276

10. Lai JR, Fischbach MA, Liu DR et al (2006) A protein interaction surface in nonribosomal peptide synthesis mapped by combinatorial mutagenesis and selection. Proc Natl Acad Sci U S A 103:5314–5319

11. Lai JR, Koglin A, Walsh CT (2006) Carrier protein structure and recognition in polyketide and nonribosomal peptide biosynthesis. Biochemistry 45:14869–14879

12. Gulick AM (2009) Conformational dynamics in the acyl-CoA synthetases, adenylation domains of non-ribosomal peptide synthetases, and firefly luciferase. ACS Chem Biol 4:811–827

13. Gulick AM, Starai VJ, Horswill AR et al (2003) The 1.75 A crystal structure of acetyl-CoA synthetase bound to adenosine-5′-propylphosphate and coenzyme A. Biochemistry 42:2866–2873

14. Reger AS, Wu R, Dunaway-Mariano D et al (2008) Structural characterization of a 140° domain movement in the two-step reaction catalyzed by 4-chlorobenzoate:CoA ligase. Biochemistry 47:8016–8025

15. Wu R, Cao J, Lu X et al (2008) Mechanism of 4-chlorobenzoate:coenzyme a ligase catalysis. Biochemistry 47:8026–8039

16. Branchini BR, Murtiashaw MH, Magyar RA et al (2000) The role of lysine 529, a conserved residue of the acyl-adenylate- forming enzyme superfamily, in firefly luciferase. Biochemistry 39:5433–5440

17. Branchini BR, Southworth TL, Murtiashaw MH et al (2005) Mutagenesis evidence that the partial reactions of firefly bioluminescence are catalyzed by different conformations of the luciferase C-terminal domain. Biochemistry 44:1385–1393

18. Mootz HD, Marahiel MA (1999) Design and application of multimodular peptide synthetases. Curr Opin Biotechnol 10:341–348

19. Conti E, Stachelhaus T, Marahiel MA et al (1997) Structural basis for the activation of phenylalanine in the non- ribosomal biosynthesis of gramicidin S. EMBO J 16:4174–4183

20. May JJ, Kessler N, Marahiel MA et al (2002) Crystal structure of DhbE, an archetype for aryl acid activating domains of modular nonribosomal peptide synthetases. Proc Natl Acad Sci U S A 99:12120–12125

21. Hamoen LW, Eshuis H, Jongbloed J et al (1995) A small gene, designated comS, located within the coding region of the fourth amino acid-activation domain of srfA, is required for competence development in Bacillus subtilis. Mol Microbiol 15:55–63

22. Reger AS, Carney JM, Gulick AM (2007) Biochemical and crystallographic analysis of substrate binding and conformational changes in acetyl-CoA synthetase. Biochemistry 46:6536–6546

23. Gulick AM, Lu X, Dunaway-Mariano D (2004) Crystal structure of 4-chlorobenzoate:CoA ligase/synthetase in the unliganded and aryl substrate-bound states. Biochemistry 43:8670–8679

24. Wu R, Reger AS, Lu X et al (2009) The mechanism of domain alternation in the acyl-adenylate forming ligase superfamily member 4-chlorobenzoate: coenzyme A ligase. Biochemistry 48:4115–4125

25. Yonus H, Neumann P, Zimmermann S et al (2008) Crystal structure of DltA. Implications for the reaction mechanism of non-ribosomal peptide synthetase adenylation domains. J Biol Chem 283:32484–32491

26. Du L, He Y, Luo Y (2008) Crystal structure and enantiomer selection by D-alanyl carrier protein ligase DltA from Bacillus cereus. Biochemistry 47:11473–11480

27. Quadri LE (2000) Assembly of aryl-capped siderophores by modular peptide synthetases and polyketide synthases. Mol Microbiol 37:1–12

28. Sundlov JA, Gulick AM (2013) Structure determination of the functional domain interaction of a chimeric nonribosomal peptide synthetase from a challenging crystal with noncrystallographic translational symmetry. Acta Crystallogr D Biol Crystallogr 69:1482–1492

29. Sundlov JA, Shi C, Wilson DJ et al (2012) Structural and functional investigation of the intermolecular interaction between NRPS adenylation and carrier protein domains. Chem Biol 19:188–198

30. Drake EJ, Duckworth BP, Neres J et al (2010) Biochemical and structural characterization of bisubstrate inhibitors of BasE, the self-standing nonribosomal peptide synthetase adenylate-forming enzyme of acinetobactin synthesis. Biochemistry 49:9292–9305

31. Neres J, Engelhart CA, Drake EJ et al (2013) Non-nucleoside inhibitors of BasE, an adenylating enzyme in the siderophore biosynthetic pathway of the opportunistic pathogen Acinetobacter baumannii. J Med Chem 56:2385–2405

32. Baltz RH (2011) Function of MbtH homologs in nonribosomal peptide biosynthesis and applications in secondary metabolite discovery. J Ind Microbiol Biotechnol 38:1747–1760

33. Wolpert M, Gust B, Kammerer B et al (2007) Effects of deletions of mbtH-like genes on clorobiocin biosynthesis in Streptomyces coelicolor. Microbiology 153:1413–1423

34. Boll B, Taubitz T, Heide L (2011) Role of MbtH-like proteins in the adenylation of tyrosine during aminocoumarin and vancomycin biosynthesis. J Biol Chem 286:36281–36290

35. Felnagle EA, Barkei JJ, Park H et al (2010) MbtH-like proteins as integral components of bacterial nonribosomal peptide synthetases. Biochemistry 49:8815–8817

36. Buchko GW, Kim CY, Terwilliger TC et al (2010) Solution structure of Rv2377c-founding member of the MbtH-like protein family. Tuberculosis (Edinb) 90:245–251

37. Drake EJ, Cao J, Qu J et al (2007) The 1.8 Å crystal structure of PA2412, an MbtH-like protein from the pyoverdine cluster of Pseudomonas aeruginosa. J Biol Chem 282:20425–20434

38. Herbst DA, Boll B, Zocher G et al (2013) Structural basis of the interaction of MbtH-like proteins, putative regulators of nonribosomal peptide biosynthesis, with adenylating enzymes. J Biol Chem 288:1991–2003

39. Zhang W, Heemstra JR Jr, Walsh CT et al (2010) Activation of the pacidamycin PacL adenylation domain by MbtH-like proteins. Biochemistry 49:9946–9947

40. Zolova OE, Garneau-Tsodikova S (2012) Importance of the MbtH-like protein TioT for production and activation of the thiocoraline adenylation domain of TioK. Med Chem Commun 3:950–955

41. Murray IA, Cann PA, Day PJ et al (1995) Steroid recognition by chloramphenicol acetyltransferase: engineering and structural analysis of a high affinity fusidic acid binding site. J Mol Biol 254:993–1005

42. Stachelhaus T, Mootz HD, Bergendahl V et al (1998) Peptide bond formation in nonribosomal peptide biosynthesis. Catalytic role of the condensation domain. J Biol Chem 273:22773–22781

43. Keating TA, Marshall CG, Walsh CT et al (2002) The structure of VibH represents nonribosomal peptide synthetase condensation, cyclization and epimerization domains. Nat Struct Biol 9:522–526

44. Samel SA, Schoenafinger G, Knappe TA et al (2007) Structural and functional insights into a peptide bond-forming bidomain from a nonribosomal peptide synthetase. Structure 15:781–792

45. Tanovic A, Samel SA, Essen LO et al (2008) Crystal structure of the termination module of a nonribosomal peptide synthetase. Science 321:659–663

46. Bloudoff K, Rodionov D, Schmeing TM (2013) Crystal structures of the first condensation domain of CDA synthetase suggest con-

formational changes during the synthetic cycle of nonribosomal peptide synthetases. J Mol Biol 425:3137–3150

47. Lai JR, Fischbach MA, Liu DR et al (2006) Localized protein interaction surfaces on the EntB carrier protein revealed by combinatorial mutagenesis and selection. J Am Chem Soc 128:11002–11003

48. Bruner SD, Weber T, Kohli RM et al (2002) Structural basis for the cyclization of the lipopeptide antibiotic surfactin by the thioesterase domain SrfTE. Structure 10:301–310

49. Samel SA, Wagner B, Marahiel MA et al (2006) The thioesterase domain of the fengycin biosynthesis cluster: a structural base for the macrocyclization of a non-ribosomal lipopeptide. J Mol Biol 359:876–889

50. Walsh CT, Chen H, Keating TA et al (2001) Tailoring enzymes that modify nonribosomal peptides during and after chain elongation on NRPS assembly lines. Curr Opin Chem Biol 5:525–534

51. Patel HM, Tao J, Walsh CT (2003) Epimerization of an L-cysteinyl to a D-cysteinyl residue during thiazoline ring formation in siderophore chain elongation by pyochelin synthetase from Pseudomonas aeruginosa. Biochemistry 42:10514–10527

52. Perry RD, Fetherston JD (2011) Yersiniabactin iron uptake: mechanisms and role in Yersinia pestis pathogenesis. Microbes Infect 13:808–817

53. Samel SA, Czodrowski P, Essen LO (2014) Structure of the epimerization domain of tyrocidine synthetase A. Acta Crystallogr D Biol Crystallogr 70:1442–1452

54. Stachelhaus T, Walsh CT (2000) Mutational analysis of the epimerization domain in the initiation module PheATE of gramicidin S synthetase. Biochemistry 39:5775–5787

55. Mitchell CA, Shi C, Aldrich CC et al (2012) Structure of PA1221, a nonribosomal peptide synthetase containing adenylation and peptidyl carrier protein domains. Biochemistry 51: 3252–3263

56. Cakici O, Sikorski M, Stepkowski T et al (2010) Crystal structures of NodS N-methyltransferase from Bradyrhizobium japonicum in ligand-free form and as SAH complex. J Mol Biol 404:874–889

57. Shi R, Lamb SS, Zakeri B et al (2009) Structure and function of the glycopeptide N-methyltransferase MtfA, a tool for the biosynthesis of modified glycopeptide antibiotics. Chem Biol 16:401–410

58. Chhabra A, Haque AS, Pal RK et al (2012) Nonprocessive [2 + 2]e- off-loading reductase domains from mycobacterial nonribosomal peptide synthetases. Proc Natl Acad Sci U S A 109:5681–5686

59. Wyatt MA, Mok MC, Junop M et al (2012) Heterologous expression and structural characterisation of a pyrazinone natural product assembly line. Chembiochem 13:2408–2415

60. Frueh DP, Arthanari H, Koglin A et al (2008) Dynamic thiolation-thioesterase structure of a non-ribosomal peptide synthetase. Nature 454:903–906

61. Liu Y, Zheng T, Bruner SD (2011) Structural basis for phosphopantetheinyl carrier domain interactions in the terminal module of nonribosomal peptide synthetases. Chem Biol 18:1482–1488

62. Drake EJ, Nicolai DA, Gulick AM (2006) Structure of the EntB multidomain nonribosomal peptide synthetase and functional analysis of its interaction with the EntE adenylation domain. Chem Biol 13:409–419

63. Lee TV, Johnson LJ, Johnson RD et al (2010) Structure of a eukaryotic nonribosomal peptide synthetase adenylation domain that activates a large hydroxamate amino acid in siderophore biosynthesis. J Biol Chem 285:2415–2427

64. Johnson LJ, Koulman A, Christensen M et al (2013) An extracellular siderophore is required to maintain the mutualistic interaction of Epichloe festucae with Lolium perenne. PLoS Pathog 9:e1003332

65. Liu Y, Bruner SD (2007) Rational manipulation of carrier-domain geometry in nonribosomal peptide synthetases. Chembiochem 8:617–621

66. Ferreras JA, Ryu JS, Di Lello F et al (2005) Small-molecule inhibition of siderophore biosynthesis in Mycobacterium tuberculosis and Yersinia pestis. Nat Chem Biol 1:29–32

67. Somu RV, Boshoff H, Qiao C et al (2006) Rationally designed nucleoside antibiotics that inhibit siderophore biosynthesis of Mycobacterium tuberculosis. J Med Chem 49:31–34

68. Miethke M, Bisseret P, Beckering CL et al (2006) Inhibition of aryl acid adenylation domains involved in bacterial siderophore synthesis. FEBS J 273:409–419

69. Qiao C, Wilson DJ, Bennett EM et al (2007) A mechanism-based aryl carrier protein/thiolation domain affinity probe. J Am Chem Soc 129:6350–6351

70. Vannada J, Bennett EM, Wilson DJ et al (2006) Design, synthesis, and biological evaluation of b-ketosulfonamide adenylation inhibitors as potential antitubercular agents. Org Lett 8:4707–4710

Chapter 2

The Assembly Line Enzymology of Polyketide Biosynthesis

Marisa Till and Paul R. Race

Abstract

Polyketides are a structurally and functionally diverse family of bioactive natural products that have found widespread application as pharmaceuticals, agrochemicals, and veterinary medicines. In bacteria complex polyketides are biosynthesized by giant multifunctional megaenzymes, termed modular polyketide synthases (PKSs), which construct their products in a highly coordinated assembly line-like fashion from a pool of simple precursor substrates. Not only is the multifaceted enzymology of PKSs a fascinating target for study, but it also presents considerable opportunities for the reengineering of these systems affording access to functionally optimized unnatural natural products. Here we provide an introductory primer to modular polyketide synthase structure and function, and highlight recent advances in the characterization and exploitation of these systems.

Key words Polyketide synthase, Biosynthesis, Structural enzymology, Natural products, Synthetic biology

1 Introduction

Polyketide natural products are a proven source of high-value bioactive small molecules [1]. These include compounds that have become mainstays for the treatment of human and animal diseases, and others that have found application as agrochemicals, flavors, fragrances, and nutraceuticals [2, 3]. Illustrative examples of clinically relevant polyketides include the statin family of cholesterol lowering agents, erythromycin A and related macrolide antibiotics, and the antiparasitic avermectins, which are used extensively in veterinary medicine (Fig. 1) [3–5]. The wealth of structural and functional diversity found within the polyketides is almost unsurpassed amongst natural products [6]. This, coupled to their exploitable bioactivities, has made both the compounds themselves, and the cellular machineries responsible for their biosynthesis, a source of intrigue and fascination for researchers worldwide.

Polyketides of bacterial origin represent a significant proportion of the natural product pool [1]. These compounds function to confer evolutionary fitness to the producing host, acting as defense

Bradley S. Evans (ed.), *Nonribosomal Peptide and Polyketide Biosynthesis: Methods and Protocols*, Methods in Molecular Biology, vol. 1401, DOI 10.1007/978-1-4939-3375-4_2, © Springer Science+Business Media New York 2016

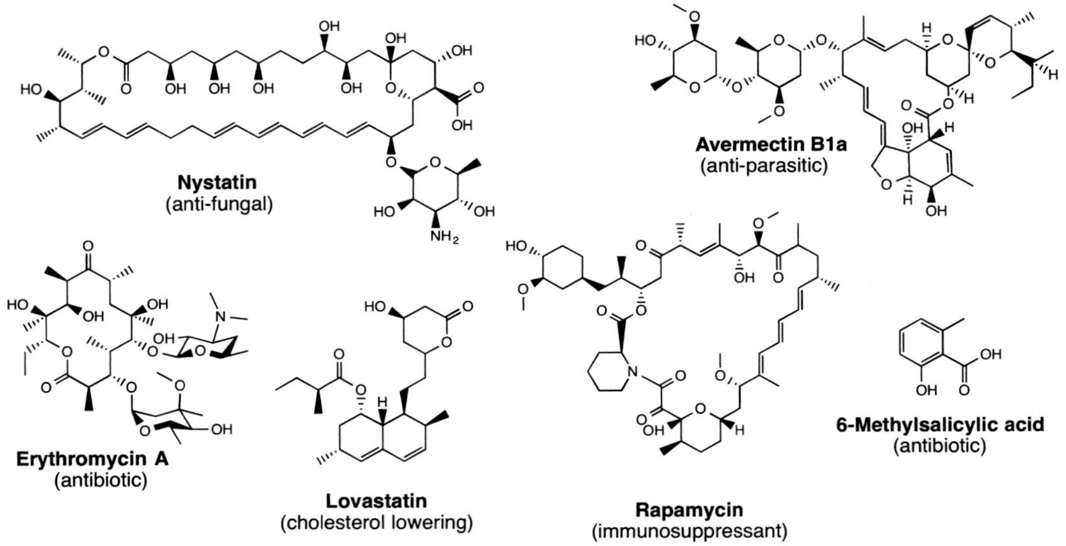

Fig. 1 Illustrative examples of polyketide natural products and their respective bioactivities

agents or signaling molecules [7, 8]. Commonly, they are biosynthesized through the action of giant multifunctional megaenzymes, termed modular polyketide synthases (PKSs), which utilize sequential condensation chemistry to construct elaborate product scaffolds from simple carboxylic acid substrates [9, 10]. This biosynthetic logic permits the assembly of a suite of molecules that are in principle of almost unlimited chemical diversity. In addition, the highly modular architecture of these systems raises intriguing possibilities for their modification and manipulation through the addition, removal, or substitution of synthase modules or domains therein [11, 12]. This approach has been exploited to generate an even larger portfolio of useful molecules, the so-called unnatural natural products [13]. With the advent of synthetic biology and the increasing adoption of the concepts and practices of rational design, it is inevitable that synthase reengineering will become more reliable and expedient in the future [14, 15]. Here we outline the basic principles of modular PKS enzymology, highlighting recent developments in the characterization of these systems, and discuss future prospects for harnessing their biosynthetic potential.

2   Modular PKSs and the Biosynthetic Process

Modular PKSs are amongst the largest and most sophisticated enzymes known. In extreme cases their proteinogenic mass can exceed 5 MDa. The distinguishing structural feature of these systems is their assembly line-like architecture, which comprises a series of linearly arranged, multi-domain extension modules,

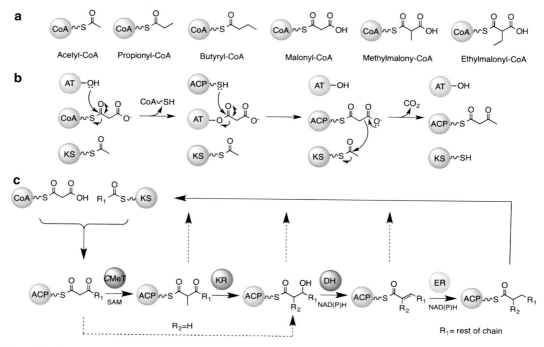

Fig. 2 (a) Chemical structures of example PKS starter and extender units. (b) Individual steps involved in extender unit selection and chain incorporation. (c) Illustrative chain modifications catalyzed by module embedded PKS domains

housed in sequence within giant polypeptide chains [1, 16]. Biosynthesis upon these systems proceeds in step-wise fashion, through the processive transfer and extension of the nascent polyketide chain, as it progresses from one module to the next. Each module within the synthase complex incorporates a single carboxylic acid substrate, derived from a pool of simple precursors, into the pathway product (Fig. 2) [1, 6, 17, 18]. The first module of the PKS initiates biosynthesis through the selection and loading of a starter unit. Each subsequent module within the PKS catalyzes the thiotemplated addition of a defined extender unit to this initial priming substrate. In some cases, however, phenomena including module 'skipping' or iterative 'stuttering' have been reported [19, 20]. The structures and properties of polyketides biosynthesized via this route are dictated by the number, identity, and incorporation order of the starter and extender units from which they are assembled. These factors are defined by the sequence of the modules that constitute the PKS assembly line. For this reason modular PKSs can be considered to possess an inherent biosynthetic programming, where synthase nucleotide sequence is colinearly related to product chemistry [21, 22]. This relationship provides a powerful tool for the prediction and analysis of synthase function, requiring knowledge only of PKS gene sequence. In addition, given that the open reading frames (ORFs) that encode PKS assembly lines and their associated tailoring enzymes are commonly clustered within the genomes of biosynthetically competent microorganisms, PKS nucleotide sequences provide a useful analytical tool by which synthase components can be readily identified and functionally assigned [23, 24].

To perform the required set of chemical reactions necessary to extend a polyketide chain by a single acyl unit each PKS extension module must minimally house three essential domains. These are; an acyltransferase (AT) domain, responsible for the selection and loading of carboxylic acid extender units derived from their coenzyme-A (CoA) thioesters; an acyl carrier protein (ACP), possessing a phosphopantetheine arm, which acts as a site of covalent tethering for both the growing product chain and selected extender units; and a ketosynthase (KS) domain, which catalyzes the carbon–carbon bond forming Claisen condensation of the downstream product chain with the upstream ACP tethered extender unit (Fig. 2). This minimal KS-AT-ACP module architecture may be elaborated to include additional auxiliary domains that further modify the incorporated acyl unit. Such modifications may take place at the reactive α-position or the β-carbonyl. An example of the former is the introduction of an α-methyl substituent via a reaction catalyzed by a C-methyltransferase (CMeT), which exploits S-adenosyl methionine as a methyl source (Fig. 2). Alternatively the β-carbonyl of the incorporated acyl unit may be reduced in a reaction catalyzed by a ketoreductase domain (KR) to form a β-hydroxy substituent. This may in turn undergo dehydration, catalyzed by a dehydratase domain (DH), yielding an α/β-unsaturated thioester. Finally an enoyl reductase (ER) may act to reduce this species yet further yielding a product fully reduced at the β-position (Fig. 2). Both the substrate selectivity of the module embedded AT domain and the module's compliment of auxiliary domains, offer complementary mechanisms for achieving diverse product chemistry [1, 6, 17, 18].

Upon reaching the final module of the PKS assembly line the immature polyketide product is liberated from the synthase via cleavage of its covalent tether. In the majority of modular PKSs this process is catalyzed by a thioesterase that forms either the C-terminal domain of the synthase, or functions as a stand-alone enzyme [25]. Frequently, following release from the PKS, the biosynthesized product is subjected to additional site-specific tailoring modifications catalyzed by free-standing enzymes. Tailoring modifications are in many cases implicitly required to establish the bioactivity of the synthase product. Examples include glycosylation, halogenation, methylation, and hydroxylation among others [1, 10, 26].

Modular PKS enzymology is readily distinguishable from that of other polyketide synthases. Iteratively acting type I PKSs (iPKSs) found in fungi are smaller but inherently more sophisticated systems, which use a single set of functional domains to accomplish the biosynthesis of their products, in a manner analogous to that of the type I fatty acid synthase [27–30]. Uniquely, the domains that comprise iPKSs often only catalyze their specific transformations during a subset of chain extension cycles, establishing a defined biosynthetic program that violates the colinearity of synthase sequence and product chemistry. Type II PKSs are transiently assembled iteratively acting complexes composed of discrete proteins [31, 32]. Consequently type II systems are significantly smaller and less intricate than both modular and iteratively acting type I PKSs. Biosynthesis upon type II PKSs proceeds in two discernable phases, initiation and chain extension, which are catalyzed by the minimal type II PKS assembly, comprising two KS like condensing enzymes and an ACP. The minimal KS-ACP machinery may be further elaborated through the incorporation of a selection of additional domains including ketoreductases, cyclases, and aromatases. Type III PKSs, which are widely distributed amongst bacteria, fungi, and plants, are in essence free standing KS domains that act to catalyze the sequential condensation of acetate units to a CoA derived starter unit [33–35]. Chemical diversity in the products of type III systems arises due to the choice of starter unit, the number of chain extension steps catalyzed, the mechanism of cyclization, and product modification through the action of allied tailoring enzymes.

3   The DEBS Paradigm

Early attempts to delineate the enzymology of modular PKSs focused on the deoxyerythronolide B synthase (DEBS). This system is responsible for the biosynthesis of 6-deoxyerythronolide (6-dEB), the macrolide nucleus of the clinically relevant antibiotic erythromycin A [4]. Studies of DEBS resulted in the establishment of many of the central tenets of modular PKS enzymology and as such this system serves as a useful test case for illustrating the basic principles of PKS function [36].

DEBS is encoded for by a series of ORFs clustered on the genome of the soil-dwelling actinomycete *Saccharopolyspora erythraea* [37]. This cluster comprises the genes *ery*AI-III that encode the core biosynthetic machinery of DEBS (DEBS 1-3), along with a number of associated tailoring and regulatory proteins (Fig. 3) [36, 38]. DEBS was the first modularPKS gene cluster to be sequenced, thus providing initial evidence for the distinctive modular assembly line-like architecture of bacterial PKSs [21, 37]. This analysis also proved instrumental in establishing the colinear relationship between synthase nucleotide sequence and product chemistry.

Each module within DEBS consists of a discrete set of domains that are responsible for the incorporation and reductive tailoring of

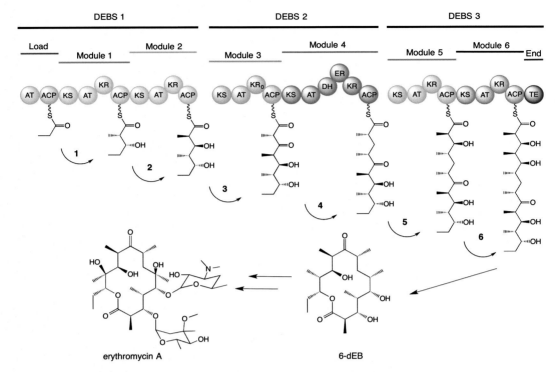

Fig. 3 Domain organization of the 6-deoxyerythronolide B synthase (DEBS) and the biosynthetic route to erythromycin A. Domains are colored based on their host module. *Numbered arrows* indicate the direction and order of product chain extension and transfer

a designated extender unit at a precise location within the growing product chain. Each domain is functionally specific and contributes to the incorporation of a single extender unit within its host module. Biosynthesis upon DEBS is initiated by a loading module, located at the *N*-terminus of DEBS 1, which consists of an AT, nominally selective for a propionate starter unit, and an ACP onto which this unit becomes tethered. It has been reported that this initial loading step is more promiscuous than was initially presumed and that in addition to propionate, non-native extender units can also be accepted by the starter module AT [39–41].

The remainder of DEBS comprises six extension modules distributed across DEBS 1-3 and a *C*-terminal TE domain that catalyzes the release and cyclization of the immature PKS product, the 14-membered 6-dEB macrolactone [42]. The modules within DEBS are numbered in the order in which they act during biosynthesis and each houses a single AT, ACP, and KS domain. Each of the extender module embedded AT domains of DEBS is selective for (2*S*)-methylmalonyl-CoA. In addition to their compliment of core domains, modules 1, 2, 5, and 6 also house active KRs. Consequently, extender units incorporated at these locations undergo reductive modification yielding a hydroxyl substituent at their β-position. The stereochemistries of the hydroxyl groups are dictated at the level of protein structure. The KR domain can attack the β-ketone from either side of the acyl chain, resulting in either a d- or l-configuration. The structure of the KR domain dictates the direction of substrate entry into the enzyme's active site, which in turn determines the face of the keto-group that is presented for hydride transfer from NADPH [43]. Module 4 of DEBS has a full complement of reductive domains (KR, DH, and ER). Extender units incorporated at this location undergo complete β-keto reduction, yielding an α/β alkane. By contrast, DEBS module 3 houses no functional reductive domains resulting in retention of the β-keto group. Notably, however, this module does contain an oft unreported 'broken' KR that controls epimerization of the incorporated methylmalonyl extender unit, a function that was previously ascribed to this module's KS domain [44, 45].

To generate the fully functional DEBS PKS there is a requirement to assemble DEBS 1-3 into a single complex. This is facilitated by protein–protein interaction interfaces at the N- and C-termini of the individual DEBS polypeptides, which serve to link these three proteins together to form the intact megaenzyme. DEBS docking domains comprise complimentary N- and C-terminal helical bundles of 30–50 and 80–130 residues respectively [46]. Complex formation between complimentary docking domains is implicitly required to permit transfer of the growing product chain between neighboring modules that are housed on different polypeptides.

Upon release from DEBS the cyclized 6-dEB macrolactone is subjected to post-PKS tailoring, catalyzed by a complement of stand-alone enzymes encoded within the *ery* gene cluster. 6-dEB is hydroxylated and glycosylated to form erythromycin D, which is then converted to erythromycin A through the action of the O-methyl transferase EryG and the hydroxylase EryK [47]. Interestingly these two transformations occur in a sequence independent fashion, forming the isolatable intermediates erythromycin B and erythromycin C respectively.

4 The Structural Enzymology of Modular PKSs

Significant insights into the enzymology of PKS chain extension and processing have been provided by structural studies of isolated synthase domains. Crystal structures have been reported for AT, ACP, KS, KR, DH, and ER domains amongst others, and NMR spectroscopy has been used extensively for the structural characterization of ACPs [38, 48–51]. These data have proven informative in establishing the roles and contributions of each domain during biosynthesis. Here we provide structural descriptions of the three core PKS domains (AT, ACP, and KS), but direct readers to references [49, 50] for more comprehensive descriptions of PKS domain structure.

Acyltransferases have been shown to possess a distinct two sub-domain architecture comprising a larger hydrolase like sub-domain fused to a smaller ferredoxin like sub-domain [52–55]. The AT active site sits at the interface of the two sub-domains, at the base of a solvent exposed channel, and houses an invariant His-Ser catalytic dyad. ATs are proposed to employ a ping-pong bi-bi mechanism that proceeds via an acyl-enzyme intermediate. The intermediate is stabilized during catalysis by an oxyanion hole formed by backbone amides from neighboring amino acids within the enzyme active site. Resolution of the acyl-enzyme intermediate occurs only in the presence of thiol nucleophiles rendering this intermediate sufficiently stable to permit its isolation and characterization in vitro [56]. The topology of the AT active site is in part dictated by amino acids that form defined substrate-selectivity motifs. Residues that occupy these positions play a role in dictating the specific acyl extender unit that the AT selects [57, 58]. Substitution of these motifs has been used to alter AT selectivity giving rise to new natural products, though the universality of this approach remains questionable [59–61]. Recently it has been proposed that extender unit selectivity is dictated more generally by a combination of structural features distributed throughout the enzyme fold [50].

ACPs are small negatively charged helical bundles that provide a site of anchorage for acyl intermediates during biosynthesis [51, 62]. The covalent attachment of intermediates to the ACP occurs via a post-translationally modified serine residue bearing a phosphopantetheine arm, which forms part of a conserved Asp-Ser-Leu motif. This motif is located at the N-terminus of helix 2 of the protein, which is considered the key portion of the ACP for mediating interactions with each of its respective binding partners [63–65]. In many instances ACPs exhibit a high degree of specificity for their cognate intra-module domains [63, 66]. Undoubtedly this property is dictated by a combination of ACP sequence and the identity of the substrate or intermediate to which the ACP is tethered. Although originally considered as somewhat as a passive component of the biosynthetic machinery, it is becoming increasingly clear that ACPs play a more active role. This includes for example shielding of tethered acyl units to allow their presentation at appropriate time points or locations within their host module [67–69].

Ketosynthase domains are dimeric proteins with a conserved thiolase fold [52, 53, 70]. Structurally KSs are composed of two α-β-α-β-α protomers arranged in the form of a five-layered core, within which three layers of α-helices are separated by two layers of β-sheets. Although there is some structural divergence between KSs, they all exhibit a small number of universal features. These include retention of the overall fold described above, extensive and highly hydrophobic dimer interfaces, and a conserved active site cysteine that acts as a site for the covalent attachment of substrates and intermediates. Variations in KS structure are largely confined to the enzyme's active site and associated regions, and have significant impact on substrate selectivity [52, 53, 71]. Examples of this include the identity and location of the key catalytic active site residues (excluding the universally conserved Cys) and the steric and electrostatic topology of the active site and its associated solvent exposed access channel. Such variations influence the ability of KSs to act upon intermediates of different chain length, saturation state and stereochemistry, and inadvertently provide a proofreading or gate-keeping function [66, 72, 73]. KS catalyzed Claisen condensation can be achieved using either a Cys-His-His or Cys-His-Asn catalytic triad, along with an intermediate stabilizing oxyanion hole.

Despite continued progress in the structural characterization of isolated PKS domains, it is only recently that structural techniques have been successfully applied to the study of intact PKS modules [74–77]. These analyses have afforded a step-change in the understanding of PKS enzymology and provided insight into the dynamic nature of chain extension and processing. Most significant amongst these studies has been the elucidation of the structure of module 5 from the pikromycin PKS (PikAIII) using cryo-electron microscopy (cryo-EM) [74, 75]. Uniquely, in this study, a number of structures were determined in a range of states that mimic all stages of the module's catalytic cycle. In addition, the resolution range within which these structures were obtained (7–11 Å) allowed the unambiguous placement of high resolution crystal structures and NMR models of domains homologous to those of PikAIII, yielding a series of pseudo-atomic models describing the process of chain extension in its entirety. In an elegant series of experiments, PikAIII, which possesses a KS-AT-KR-ACP domain architecture, is shown to adopt a distinctive symmetrical arch-like structure comprising a single PikAIII homodimer. Dimerization occurs via the KS domain, which sits at the top of the arch. The AT and KR domains from each monomer form descending struts, which together with the capping KS dimer form a single central reaction chamber accessible by the active sites of each of the KS, AT and KR domains that constitute the module (Fig. 4). The relative positions of each domain deviate significantly from those proposed in structural models of intact PKS modules. The location of the PikAIII module's two ACPs are also resolved and are shown, using targeted chemical modifications designed to mimic a range of reaction cycle intermediates, to occupy distinct locations within the chamber during catalysis (Fig. 4). These include; a *holo*-ACP form bearing a phosphopantetheine arm, where the ACPs are located adjacent to their respective AT or KR domains (state 1); a form in which the ACPs carry a pentaketide intermediate, locating

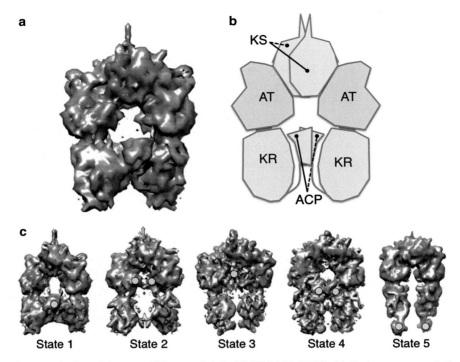

Fig. 4 (**a**) Solid rendering of the cryo-EM map of *holo* PikAIII (EMD-5647). (**b**) Cartoon representation of *holo* PikAIII showing the relative position of each domain within the module. (**c**) Conformational states of PikAIII observed by cryo-EM (state 1 EMD-5647, state 2 EMD-5663, state 3 EMD-5653, state 4 EMD-5664, state 5 EMD-5666). Each state equates to a defined step within the module's catalytic cycle. The location of each PikAIII ACP is indicated by a *yellow circle*. In states 1 and 3 the ACPs occupy overlapping positions at the front and rear of the module

them next to their respective AT's active sites (state 2); a form in which the ACPs are acylated with methylmalonate, and within which they occupy positions below their cognate KSs, primed for active site entry (state 3); a form within which the ACPs carries a β-ketohexaketide, mimicking the system following chain extension, where the ACPs locate next to their respective KRs ready for ketoreduction (state 4); and finally a form mimicking the culmination of the chain extension processes, where the ACPs, carrying β-hydroxyhexaketide groups, are expelled from the PKS reaction chamber, facilitating chain transfer to the down-stream module (state 5). In each of the structures reported the two ACPs occupy equivalent positions on either side of the module dimer, suggesting that they operate in synchronous fashion. This is a likely consequence of space constraints within the reaction chamber and steric hindrance imposed by the downstream KS dimer. Analysis of the reported PikAIII cryo-EM structures also hints at dynamic motions in other domains within the PKS module and is a powerful illustration of the value of analyzing the structures of individual proteins in the context of their interacting partners.

5   *Trans*-AT Synthases: A New Paradigm in Modular PKS Enzymology

A significant recent development in the study of PKSs was the identification of a second class of modular synthases whose domain and module architectures diverge significantly from those of the canonical DEBS like systems [78–80]. This new family of modular PKSs, the *trans*-AT synthases, are notable for their highly mosaic structures that incorporate disparate biosynthetic features within a single megaenzyme complex [81]. *Trans*-AT PKSs have been shown to have an evolutionary lineage distinct from that of the *cis*-AT DEBS like systems [46]. Although *trans*-AT PKSs make use of the same step-wise sequential condensation chemistry employed by *cis*-AT synthases, they exploit a much broader repertoire of functional domains, make use of *trans*-acting elements to modify synthase intermediates, and exhibit module architectures that diverge from the classical KS-AT-ACP paradigm [46, 81, 82]. Here we describe two examples of biosynthetic peculiarities common to *trans*-AT PKSs, but direct readers to more comprehensive reviews of this area [81, 82].

The defining feature of *trans*-AT synthases is the absence of module embedded AT domains throughout the PKS. Substrate loading in these systems is instead provided in *trans*, by freestanding *trans*-acting ATs encoded within the synthase gene cluster [83]. *Trans*-acting ATs are found either as stand-alone enzymes, or as di- or tri-domain fusions proteins partnered with decarboxylases (DCs), ERs, and/or proof-reading acyl-hydrolases (AH; Fig. 5) [78, 84–86]. The role of *trans*-ATs in substrate loading has been demonstrated both in vivo and in vitro, and initial structural and functional characterization of these enzymes suggests that they possess the same general structure and catalytic mechanism as module embedded ATs, but are able to furnish multiple ACPs throughout the PKS complex with extender units [87–90]. The mechanism of *trans*-AT recruitment to each extension module remains to be established, however, it has been suggested that there may be distinct acyltransferase docking domains located within *trans*-AT PKS modules that facilitate this process [91, 92].

Another distinctive biosynthetic feature common to *trans*-AT PKSs is their use of a multi-protein hydroxymethylglutaryl CoA synthase (HCS) enzyme cassette to catalyze the introduction of methyl groups at β-carbon positions within product incorporated acyl extender units [93]. To achieve this acetyl-ACP is condensed with the unreduced β-carbon of the nascent polyketide chain forming a β-hydroxy-β-carboxymethyl intermediate. This intermediate is then subjected to sequential dehydration and decarboxylation to form the β-methyl group (Fig. 5) [89, 94–97]. Structural studies of *trans*-AT PKS di-domain ACPs upon which β-branching takes place have identified a distinct amino acid signature that appears to

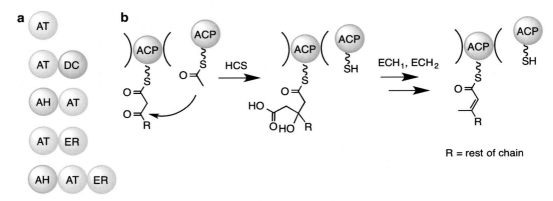

Fig 5 (a) Domain organization within polypeptides that house *trans*-acting acyltransferases. (b) Scheme for the introduction of β-methyl branches in to polyketides via the HCS cassette route. Only the ACP domains of the modules that support the product chain are shown. *ECH* enoyl CoA hydratase

target the branching machinery to these locations. This discovery raises the intriguing possibility of implementing β-branching chemistry at a range of locations throughout PKSs, by introducing appropriate amino acid signatures into ACPs that do not, in their native state, support branch formation.

6   Reengineering of PKS Assembly Lines

The highly modular architectures of PKSs make these systems attractive targets for reengineering as a mechanism for accessing unnatural natural products with novel or enhanced functionality. This approach gains credence from the notion that polyketide scaffolds have been evolutionarily selected for optimum performance within their producer's environmental niche, and as such they may be of limited utility in wider contexts. The targeted modification of polyketides through manipulation of the cellular machineries responsible for their biosynthesis may therefore yield derivatives of the parent compound with, for example, improved clinical efficacy. The tractability of PKS reengineering was initially demonstrated in DEBS [98]. These investigations resulted in the establishment of a set of rules for the purposeful manipulation of modular synthases [98]. Despite these initial successes, the reengineering of modular PKSs has, however, proved to be much more challenging than was initially anticipated. Though with recent advances in the delineation of synthase enzymology, the development of more robust experimental tools for targeted high-throughput genetic manipulation and DNA synthesis, and improvements in analytical techniques and computation, PKS reengineering appears poised for a renascence.

A number of strategies have been proposed for the successful manipulation of PKSs. These focus on targeted modifications at the module, domain, or amino acid level, or involve the refactoring of precursor biosynthesis or post-PKS tailoring. Approaches targeted at the module level have focused on the deletion, insertion or substitution of intact modules within PKS complexes. This approach has been successfully applied to DEBS, where substitution of the loading module of DEBS 1 with that from the tylosin PKS resulted in a hybrid system selective solely for a propionate starter unit [99]. Similarly, replacement of the DEBS 1 starter module with that from the oleandomycin PKS yielded a hybrid system within which acetate was exclusively incorporated as the starter unit [99]. Although pioneered in DEBS, loading module substitution has been successfully implemented in other systems. For example, replacement of the loading module initiating avermectin biosynthesis, which specifically incorporates isobutyryl-CoA, with that of the cyclohexanecarboxylic (CHC) phoslactomycin PKS loading module from *Streptomycesplatensis*, permitted the biosynthesis of the antiparasitic veterinary medicine doramectin [100]. To achieve this outcome it was also necessary to express the five proteins responsible for the biosynthesis of the CHC-CoA precursor in tandem with the hybrid PKS.

Targeted manipulation at the domain level requires less extensive interference with the PKS, however, the importance of intra-module protein–protein interactions should always be considered. The tolerance of DEBS to domain swapping has been probed extensively [101, 102]. Each extension module within this PKS contains an AT domain selective for (2*S*)-methylmalonyl-CoA. Substitution of these domains with ATs selective for malonyl, ethylmalonyl, or methoxymalonyl extender units has permitted the biosynthesis of regioselectively modified polyketides with the expected chemical composition [102–106]. Substitution of individual domains in DEBS modules 2, 5, and 6, with counterparts from the rapamycin PKS, possessing different substrate specificities and reductive capabilities, as well as the insertion of additional domains from the same system, has been used to produce an extensive range of 6-dEB analogs [102, 107, 108].

To minimize the deleterious effects of PKS reengineering, the targeted mutation of individual residues within modules or domains represents an attractive, less invasive approach. As the number of publically available genome sequences and protein structures increases exponentially, so increases the ability of researchers to make informed site-specific changes that confer or modulate protein function. This may include targeted changes that impact catalytic activity, substrate selectivity, cofactor binding, or stereoselectivity. Examples of the use of this approach in DEBS include deactivation of the enoyl reductase domain of module 4 permitting the biosynthesis of $\Delta^{6,7}$-anhydroerythromycin C [109], and targeted alteration of the substrate selectivities of extender module AT domains to allow malonyl-CoA and fluoromalonyl-CoA to be accepted as substrates [110]. This approach has been further informed by computational methods. For example the use of quantum mechanics/molecular mechanics (QM/MM) methods was instrumental in the design of a mutagenesis strategy to reengineer DEBS AT6 to accept a non-natural 2-propargylmalonyl extender unit [111]. The polyketide biosynthesized by this modified PKS possesses a synthetically functionalizable handle that can be exploited to generate an even greater number of useful derivatives.

In addition to reengineering strategies that focus explicitly on the PKS itself, complimentary approaches that target the allied starter and extender unit biosynthetic pathways, or focus on post synthase tailoring reactions have been investigated. The former have included the targeted mutation of acyl-CoA synthetases [112, 113], or the replacement of endogenous acyl-CoA pathways with those that generate alternative precursors [114]. For the latter, many successful approaches have involved the repurposing of glycosyltransferases to generate polyketide products with altered glycosylation patterns and consequently more favorable toxicities, solubilities, and bioavailabilities [115, 116].

7 Conclusions

In recent years significant progress has been made in the genetic, chemical, biochemical and structural characterization of modular PKSs and their constituent parts. Despite this, many questions still remain, and without doubt the study of modular synthases will remain a fertile area of research for many years to come. Studies of PKS structure and function are providing unexpected insights into the assembly, operation, and dynamics of these systems, and the frequency with which new PKS gene clusters are being identified and annotated continues to rise exponentially.

Notably, one area of modular PKS research that has progressed more slowly than was hoped is that of synthase reengineering. Methods developed to enable the rational redesign of modular PKSs and their products have proven nontrivial to implement successfully across multiple, often closely related systems, and consequently many approaches that showed initial promise have failed to deliver. That said, new fundamental insights into the enzymology of these systems will undoubtedly expedite the development of the necessary tools and technologies that are required for robust, broadly implementable synthase reengineering.

References

1. Staunton J, Weissman KJ (2001) Polyketide biosynthesis: a millennium review. Nat Prod Rep 18:380–416

2. Li JW-H, Vederas JC (2009) Drug discovery and natural products: end of an era or an endless frontier? Science 325:161–166

3. Demain AL, Vaishnav P (2011) Natural products for cancer chemotherapy. Microb Biotechnol 4:687–699

4. Washington JA, Wilson WR (1985) Erythromycin: a microbial and clinical perspective after 30 years of clinical use (1). Mayo Clin Proc 60:189–203

5. Campbell WC (2012) History of avermectin and ivermectin, with notes on the history of other macrocyclic lactone antiparasitic agents. Curr Pharm Biotechnol 13:9–11

6. Weissman KJ (2009) Introduction to polyketide biosynthesis. Methods Enzymol 459:3–16

7. Challis GL, Hopwood DA (2003) Synergy and contingency as driving forces for the evolution of multiple secondary metabolite production by *Streptomyces* species. Proc Natl Acad Sci U S A 100(Suppl 2):14555–14561

8. Firn RD, Jones CG (2003) Natural products - a simple model to explain chemical diversity. Nat Prod Rep 20:382

9. Walsh CT (2004) Polyketide and nonribosomal peptide antibiotics: modularity and versatility. Science 199:1805–1811

10. Walsh CT (2008) The chemical versatility of natural-product assembly lines. Acc Chem Res 41:4–10

11. Williams G (2013) Engineering polyketide synthases and nonribosomal peptide synthetases. Curr Opin Struct Biol 23:603–612

12. Wong FT, Khosla C (2012) Combinatorial biosynthesis of polyketides - a perspective. Curr Opin Chem Biol 16:117–123

13. Hertweck C (2015) Decoding and reprogramming complex polyketide assembly lines: prospects for synthetic biology. Trends Biochem Sci 40:189–199

14. Cummings M, Breitling R, Takano E (2014) Steps towards the synthetic biology of polyketide biosynthesis. FEMS Microbiol Lett 351:116–125

15. Goss RJM, Shankar S, Fayad AA (2012) The generation of "unnatural" products: synthetic biology meets synthetic chemistry. Nat Prod Rep 29:870–889

16. Weissman KJ (2004) Polyketide biosynthesis: understanding and exploiting modularity. Philos Trans A Math Phys Eng Sci 362:2671–2690

17. Fischbach MA, Walsh CT (2006) Assembly-line enzymology for polyketide and nonribosomal Peptide antibiotics: logic, machinery, and mechanisms. Chem Rev 106:3468–3496

18. Hertweck C (2009) The biosynthetic logic of polyketide diversity. Angew Chem Int Ed Engl 48:4688–4716

19. Xue Y, Sherman DH (2000) Alternative modular polyketide synthase expression controls macrolactone structure. Nature 403:571–575

20. Wilkinson B et al (2000) Novel octaketide macrolides related to 6-deoxyerythronolide B provide evidence for iterative operation of the erythromycin polyketide synthase. Chem Biol 7:111–117

21. Donadio S, Staver MJ, Mcalpine JB et al (1991) Modular organization of genes required for complex polyketide biosynthesis. Science 252:675–679

22. Callahan B, Thattai M, Shraiman BI (2009) Emergent gene order in a model of modular polyketide synthases. Proc Natl Acad Sci U S A 106:19410–19415

23. Fedorov O, Niesen FH (2012) Kinase inhibitor selectivity profiling using differential scanning fluorimetry. Methods Mol Biol 795:109–118

24. Helfrich EJN, Reiter S, Piel J (2014) Recent advances in genome-based polyketide discovery. Curr Opin Biotechnol 29:107–115

25. Horsman ME, Hari TPA, Boddy CN (2015) Polyketide synthase and non-ribosomal peptide synthetase thioesterase selectivity: logic gate or a victim of fate? Nat Prod Rep (in Press)

26. Lin S, Huang T, Shen B (2012) Tailoring enzymes acting on carrier protein-tethered substrates in natural product biosynthesis. Methods Enzymol 516:321–343

27. Maier T, Leibundgut M, Ban N (2008) The crystal structure of a mammalian fatty acid synthase. Science 321:1315–1323

28. Townsend CA (2014) Aflatoxin and deconstruction of type I, iterative polyketide synthase function. Nat Prod Rep 31:1260–1265

29. Vederas JC (2014) Explorations of fungal biosynthesis of reduced polyketides - a personal viewpoint. Nat Prod Rep 31:1253–1259

30. Simpson TJ (2014) Fungal polyketide biosynthesis - a personal perspective. Nat Prod Rep 31:1247–1252

31. Hertweck C, Luzhetskyy A, Rebets Y et al (2007) Type II polyketide synthases: gaining a deeper insight into enzymatic teamwork. Nat Prod Rep 24:162–190

32. Das A, Khosla C (2009) Biosynthesis of aromatic polyketides in bacteria. Acc Chem Res 42:631–639

33. Yu D, Xu F, Zeng J et al (2012) Type III polyketide synthases in natural product biosynthesis. IUBMB Life 64:285–295

34. Hashimoto M, Nonaka T, Fujii I (2014) Fungal type III polyketide synthases. Nat Prod Rep 31:1306–1317

35. Austin MB, Noel JP (2003) The chalcone synthase superfamily of type III polyketide synthases. Nat Prod Rep 20:79–110

36. Katz L (2009) The DEBS paradigm for type I modular polyketide synthases and beyond, 1st edn. Elsevier Inc., Amsterdam

37. Cortes J, Haydock SF, Roberts GA et al (1990) An unusually large multifunctional polypeptide in the erythromycin-producing polyketide synthase of *Saccharopolyspora erythraea*. Nature 348:176–178

38. Weissman KJ (2015) Uncovering the structures of modular polyketide synthases. Nat Prod Rep 32:436–453

39. Kao CM, Katz L, Khosia C (1994) Engineered biosynthesis of a complete macrolactone in a heterologous host. Science 265:509–512

40. Wiesmann KEH et al (1995) Polyketide synthesis in vitro on a modular polyketide synthase. Chem Biol 2:582–589

41. Rowe CJ, Gaisser S, Staunton J et al (1998) Construction of new vectors for high-level expression in actinomycetes. Gene 216: 215–223

42. Pinto A, Wang M, Horsman M et al (2012) 6-Deoxyerythronolide B synthase thioesterase-catalyzed macrocyclization is highly stereoselective. Org Lett 14:2278–2281

43. Siskos AP et al (2005) Molecular basis of Celmer's rules: stereochemistry of catalysis by isolated ketoreductase domains from modular polyketide synthases. Chem Biol 12:1145–1153

44. Weissman KJ et al (1997) The molecular basis of Celmer's rules: the stereochemistry of the condensation step in chain extension on the erythromycin polyketide synthase. Biochemistry 36:13849–13855

45. Keatinge-Clay AT, Stroud RM (2006) The structure of a ketoreductase determines the organization of the beta-carbon processing enzymes of modular polyketide synthases. Structure 14:737–748

46. Nguyen T et al (2008) Exploiting the mosaic structure of trans-acyltransferase polyketide synthases for natural product discovery and pathway dissection. Nat Biotechnol 26: 225–233

47. Zhang Q et al (2011) Knocking out of tailoring genes eryK and eryG in an industrial erythromycin-producing strain of Saccharopolyspora erythraea leading to overproduction of erythromycin B, C and D at different conversion ratios. Lett Appl Microbiol 52:129–137

48. Khosla C (2009) Structures and mechanisms of polyketide synthases. J Org Chem 74: 6416–6420

49. Keatinge-Clay AT (2012) The structures of type I polyketide synthases. Nat Prod Rep 29:1050–1073

50. Tsai S-CS, Ames BD (2009) Structural enzymology of polyketide synthases, 1st edn. Elsevier Inc., Amsterdam

51. Crosby J, Crump MP (2012) The structural role of the carrier protein—active controller or passive carrier. Nat Prod Rep 29: 1111–1137

52. Tang Y, Kim C, Mathews II et al (2006) The 2.7-Å crystal structure of a 194-kDa homodimeric fragment of the 6-deoxyerythronolide B synthase. Proc Natl Acad Sci U S A 103:11124–11129

53. Tang Y, Chen AY, Kim C-Y et al (2007) Structural and mechanistic analysis of protein interactions in module 3 of the 6-deoxyerythronolide B synthase. Chem Biol 14:931–943

54. Bergeret F et al (2012) Biochemical and structural study of the atypical acyltransferase domain from the mycobacterial polyketide synthase Pks13. J Biol Chem 287:33675–33690

55. Park H, Kevany BM, Dyer DH et al (2014) A polyketide synthase acyltransferase domain structure suggests a recognition mechanism for its hydroxymalonyl-acyl carrier protein substrate. PLoS One 9:e110965

56. Lau J, Cane DE, Khosla C (2000) Substrate specificity of the loading didomain of the erythromycin polyketide synthase. Biochemistry 39:10514–10520

57. Haydock SF et al (1995) Divergent sequence motifs correlated with the substrate specificity of (methyl) malonyl-CoA: acyl carrier protein transacylase domains in modular polyketide synthases. FEBS Lett 374:246–248

58. Yadav G, Gokhale RS, Mohanty D (2003) Computational approach for prediction of domain organization and substrate specificity

of modular polyketide synthases. J Mol Biol 328:335–363

59. Reeves CD et al (2001) Alteration of the substrate specificity of a modular polyketide synthase acyltransferase domain through site-specific mutations. Biochemistry 40: 15464–15470

60. Lau J, Fu H, Cane DE et al (1999) Dissecting the role of acyltransferase domains of modular polyketide synthases in the choice and stereochemical fate of extender units. Biochemistry 38:1643–1651

61. Del Vecchio F et al (2003) Active-site residue, domain and module swaps in modular polyketide synthases. J Ind Microbiol Biotechnol 30:489–494

62. Lai JR, Koglin A, Walsh CT (2006) Carrier protein structure and recognition in polyketide and nonribosomal peptide biosynthesis. Biochemistry 45:14869–14879

63. Crump MP et al (1997) Solution structure of the actinorhodin polyketide synthase acyl carrier protein from *Streptomyces coelicolor* A3 (2). Biochemistry 36:6000–6008

64. Findlow SC, Winsor C, Simpson TJ et al (2003) Solution structure and dynamics of oxytetracycline polyketide synthase acyl carrier protein from *Streptomyces rimosus*. Biochemistry 42:8423–8433

65. Li Q, Khosla C, Puglisi JD, Liu CW (2003) Solution structure and backbone dynamics of the holo form of the frenolicin acyl carrier protein. Biochemistry 42:4648–4657

66. Chen AY, Schnarr NA, Kim C et al (2006) Extender unit and acyl carrier protein specificity of ketosynthase domains of the 6-Deoxyerythronolide B synthase. J Am Chem Soc 128:3067–3074

67. Płoskoń E et al (2010) Recognition of intermediate functionality by acyl carrier protein over a complete cycle of fatty acid biosynthesis. Chem Biol 17:776–785

68. Evans SE et al (2009) Probing the interactions of early polyketide intermediates with the actinorhodin ACP from *S. coelicolor* A3(2). J Mol Biol 389:511–528

69. Johnson MNR, Londergan CH, Charkoudian LK (2014) Probing the phosphopantetheine arm conformations of acyl carrier proteins using vibrational spectroscopy. J Am Chem Soc 136:11240–11243

70. Gay DC et al (2014) A close look at a ketosynthase from a trans-acyltransferase modular polyketide synthase. Structure 22:444–451

71. Khosla C, Gokhale RS, Jacobsen JR et al (1999) Tolerance and specificity of polyketide synthases. Annu Rev Biochem 68:219–253

72. Chen AY, Cane DE, Khosla C (2007) Structure-based dissociation of a type I polyketide synthase module. Chem Biol 14:784–792

73. Watanabe K, Wang CCC, Boddy CN et al (2003) Understanding substrate specificity of polyketide synthase modules by generating hybrid multimodular synthases. J Biol Chem 278:42020–42026

74. Dutta S et al (2014) Structure of a modular polyketide synthase. Nature 510:512–517

75. Whicher JR et al (2014) Structural rearrangements of a polyketide synthase module during its catalytic cycle. Nature 510:560–564

76. Edwards AL, Matsui T, Weiss TM et al (2014) Architectures of whole-module and bimodular proteins from the 6-deoxyerythronolide B synthase. J Mol Biol 426:2229–2245

77. Davison J et al (2014) Insights into the function of *trans*-acyl transferase polyketide synthases from the SAXS structure of a complete module. Chem Sci 5:3081

78. Cheng Y, Tang G, Shen B (2003) Type I polyketide synthase requiring a discrete acyltransferase for polyketide biosynthesis. Proc Natl Acad Sci U S A 100:3149–3154

79. El-sayed AK et al (2003) Characterization of the mupirocin biosynthesis gene cluster from *Pseudomonas fluorescens* NCIMB 10586. Chem Biol 10:419–430

80. Moldenhauer J, Chen X-H, Borriss R et al (2007) Biosynthesis of the antibiotic bacillaene, the product of a giant polyketide synthase complex of the *trans*-AT family. Angew Chem Int Ed Engl 46:8195–8197

81. Piel J (2010) Biosynthesis of polyketides by *trans*-AT polyketide synthases. Nat Prod Rep 27:996–1047

82. Till M, Race PR (2014) Progress challenges and opportunities for the re-engineering of *trans*-AT polyketide synthases. Biotechnol Lett 36:877–888

83. Cheng Y-Q, Coughlin JM, Lim S-K et al (2009) Type I polyketide synthases that require discrete acyltransferases. Methods Enzymol 459:165–186

84. Liu T, Huang Y, Shen B (2009) The bifunctional acyltransferase/decarboxylase LnmK as the missing link for-alkylation in polyketide biosynthesis. J Am Chem Soc 131:6900–6901

85. Thomas CM, Hothersall J, Willis CL et al (2010) Resistance to and synthesis of the antibiotic mupirocin. Nat Rev Microbiol 8:281–289

86. Jensen K et al (2012) Polyketide proofreading by an acyltransferase-like enzyme. Chem Biol 19:329–339

87. Wong FT, Jin X, Mathews II et al (2011) Structure and mechanism of the *trans*-acting acyltransferase from the disorazole synthase. Biochemistry 50:6539–6548

88. Musiol EM et al (2011) Supramolecular templating in kirromycin biosynthesis: the acyltransferase KirCII loads ethylmalonyl-CoA extender onto a specific ACP of the *trans*-AT PKS. Chem Biol 18:438–444

89. Calderone CT, Kowtoniuk WE, Kelleher NL et al (2006) Convergence of isoprene and polyketide biosynthetic machinery: isoprenyl-S-carrier proteins in the pksX pathway of Bacillus subtilis. Proc Natl Acad Sci U S A 103:8977–8982

90. Lopanik NB et al (2010) *In vivo* and *in vitro* trans-acylation by BryP, the putative bryostatin pathway acyltransferase derived from an uncultured marine symbiont. Chem Biol 15:1175–1186

91. Chan YA, Thomas MG (2010) Recognition of (2S)-aminomalonyl-acyl carrier protein (ACP) and (2R)-hydroxymalonyl-ACP by acyltransferases in zwittermicin A biosynthesis. Biochemistry 49:3667–3677

92. Musiol EM, Weber T (2012) Discrete acyltransferases involved in polyketide biosynthesis. Med Chem Commun 3:871

93. Calderone CT (2008) Isoprenoid-like alkylations in polyketide biosynthesis. Nat Prod Rep 25:845–853

94. Calderone CT, Iwig DF, Dorrestein PC et al (2007) Incorporation of nonmethyl branches by isoprenoid-like logic: multiple beta-alkylation events in the biosynthesis of myxovirescin A1. Chem Biol 14:835–846

95. Gu L et al (2006) Metabolic coupling of dehydration and decarboxylation in the curacin A pathway: functional identification of a mechanistically diverse enzyme pair. J Am Chem Soc 128:9014–9015

96. Simunovic V, Müller R (2007) 3-hydroxy-3-methylglutaryl-CoA-like synthases direct the formation of methyl and ethyl side groups in the biosynthesis of the antibiotic myxovirescin A. Chembiochem 8(5):497–500

97. Simunovic V, Müller R (2007) Mutational analysis of the myxovirescin biosynthetic gene cluster reveals novel insights into the functional elaboration of polyketide backbones. Chembiochem 8:1273–1280

98. McDaniel R, Ebert-Khosla S, Hopwood DA et al (1995) Rational design of aromatic polyketide natural products by recombinant assembly of enzymatic subunits. Nature 375:549–554

99. Long PF et al (2002) Engineering specificity of starter unit selection by the erythromycin-producing polyketide synthase. Mol Microbiol 43:1215–1225

100. Wang J-B, Pan H-X, Tang G-L (2011) Production of doramectin by rational engineering of the avermectin biosynthetic pathway. Bioorg Med Chem Lett 21:3320–3323

101. Rodriguez E, McDaniel R (2001) Combinatorial biosynthesis of antimicrobials and other natural products. Curr Opin Microbiol 4:526–534

102. McDaniel R et al (1999) Multiple genetic modifications of the erythromycin polyketide synthase to produce a library of novel 'unnatural' natural products. Proc Natl Acad Sci U S A 96:1846–1851

103. Oliynykl M, Brown MJB, Cort J et al (1996) A hybrid modular polyketide synthase obtained by domain swapping. Chem Biol 3:833–839

104. Ruan X et al (1997) Acyltransferase domain substitutions in erythromycin polyketide synthase yield novel erythromycin derivatives. J Bacteriol 179:6416–6425

105. Stassi DL et al (1998) Ethyl-substituted erythromycin derivatives produced by directed metabolic engineering. Proc Natl Acad Sci U S A 95:7305–7309

106. Kato Y et al (2002) Functional expression of genes involved in the biosynthesis of the novel polyketide chain extension unit, methoxymalonyl-acyl carrier protein, and engineered biosynthesis of 2-desmethyl-2-methoxy-6-deoxyerythronolide B. J Am Chem Soc 124:5268–5269

107. McDaniel R et al (1997) Gain-of-function mutagenesis of a modular polyketide synthase. J Am Chem Soc 119:4309–4310

108. Bedford D, Jacobsen JR, Luo G et al (1996) A functional chimeric modular polyketide synthase generated via domain replacement. Chem Biol 3:827–831

109. Donadio S, Mcalpine JB, Sheldont PJ et al (1993) An erythromycin analog produced by reprogramming of polyketide synthesis. Proc Natl Acad Sci U S A 90:7119–7123

110. Walker MC et al (2013) Expanding the fluorine chemistry of living systems using engineered polyketide synthase pathways. Science 341:1089–1094

111. Sundermann U et al (2013) Enzyme-directed mutasynthesis: a combined experimental and theoretical approach to substrate recognition of a polyketide synthase. ACS Chem Biol 8:443–450

112. Koryakina I, Williams GJ (2011) Mutant malonyl-CoA synthetases with altered specificity for polyketide synthase extender unit generation. Chembiochem 12:2289–2293

113. Koryakina I et al (2013) Poly specific trans-acyltransferase machinery revealed via engineered acyl-CoA synthetases. ACS Chem Biol 8:200–208

114. Lechner A et al (2013) Designed biosynthesis of 36-methyl-FK506 by polyketide precursor pathway engineering. ACS Synth Biol 2:379–383

115. Rix U, Fischer C, Remsing LL et al (2002) Modification of post-PKS tailoring steps through combinatorial biosynthesis. Nat Prod Rep 19:542–580

116. Olano C, Méndez C, Salas JA (2010) Post-PKS tailoring steps in natural product-producing actinomycetes from the perspective of combinatorial biosynthesis. Nat Prod Rep 27:571

Part II

In Vitro Methods

Chapter 3

Measurement of Nonribosomal Peptide Synthetase Adenylation Domain Activity Using a Continuous Hydroxylamine Release Assay

Benjamin P. Duckworth, Daniel J. Wilson, and Courtney C. Aldrich

Abstract

Adenylation is a crucial enzymatic process in the biosynthesis of nonribosomal peptide synthetase (NRPS) derived natural products. Adenylation domains are considered the gatekeepers of NRPSs since they select, activate, and load the carboxylic acid substrate onto a downstream peptidyl carrier protein (PCP) domain of the NRPS. We describe a coupled continuous kinetic assay for NRPS adenylation domains that substitutes the PCP domain with hydroxylamine as the acceptor molecule. The pyrophosphate released from the first-half reaction is then measured using a two-enzyme coupling system, which detects conversion of the chromogenic substrate 7-methylthioguanosine (MesG) to 7-methylthioguanine. From profiling substrate specificity of unknown or engineered adenylation domains to studying chemical inhibition of adenylating enzymes, this robust assay will be of widespread utility in the broad field NRPS enzymology.

Key words Adenylation, Adenylate-forming, Hydroxamate, MesG, Enzyme assay

1 Introduction

Adenylation domains prime nonribosomal peptide synthetase (NRPS) biosynthetic pathways by catalyzing a two-step reaction (Fig. 1) [1, 2]. In the first step, the adenylation domain recognizes and binds ATP and its cognate carboxylic acid substrate **1**, which is usually an amino acid but can also be an α-hydroxy acid, aryl acid, or fatty acid (Fig. 1). The enzyme then catalyzes the nucleophilic attack (step **a**) of the substrate carboxylic acid on the α-phosphate of ATP to form an acyl-adenylate intermediate **2** and pyrophosphate. In the second-half reaction (step **b**), the adenylation domain binds a peptidyl carrier protein (PCP) domain and transfers the activated acid onto the phosphopantetheinyl arm of the PCP domain to provide the thioester tethered carboxylic acid building block (**3**).

NRPS adenylation domains (A-domains) are usually located in *cis* to the PCP domain as part of a multifunctional NRPS protein; however, A-domains can also be located in *trans* as separate pro-

Bradley S. Evans (ed.), *Nonribosomal Peptide and Polyketide Biosynthesis: Methods and Protocols*, Methods in Molecular Biology, vol. 1401, DOI 10.1007/978-1-4939-3375-4_3, © Springer Science+Business Media New York 2016

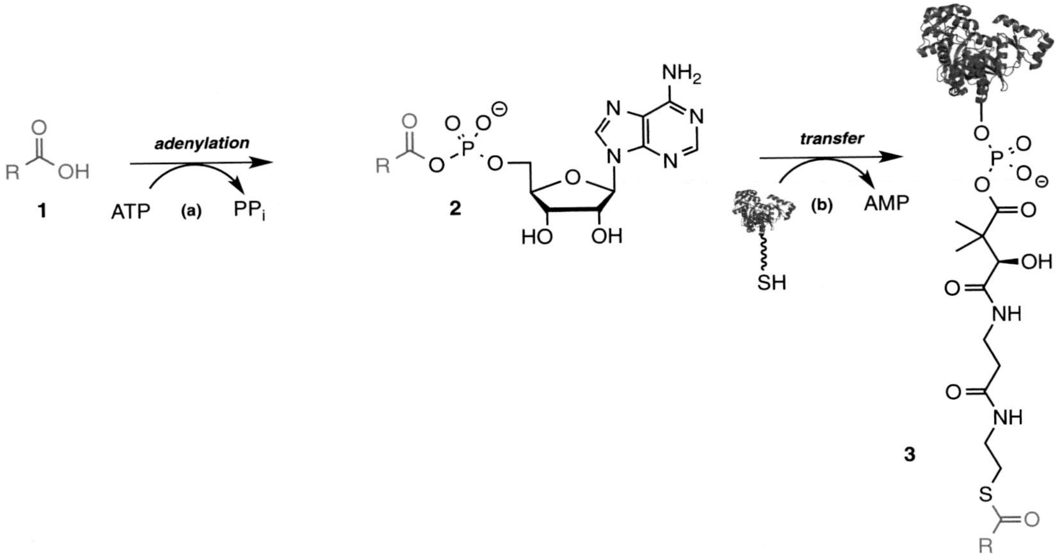

Fig. 1 Two-step mechanism of adenylating enzymes

teins [3]. In the former case, the A-domain cannot be analyzed by steady-state kinetic techniques since only a single turnover can be performed whereas in the latter case stoichiometric amounts of the cognate PCP domain are required. The most common method to measure A-domain activity that obviates the need for a PCP domain is the pyrophosphate exchange radioassay wherein one measures incorporation of [^{32}P]PPi into ATP. Bachmann and coworkers have also reported an innovative mass spectrometric-based pyrophosphate exchange assay employing [^{18}O]ATP that supplants the requirement for radioisotopes [4]. These exchange assays are useful for most A-domains, but not all A-domains undergo pyrophosphate exchange [5, 6]. Moreover, the pyrophosphate exchange assay only evaluates the adenylation partial reaction in the reverse direction and therefore may not provide physiologically meaningful kinetic parameters. In the absence of an acceptor molecule such as a PCP domain, the acyl-adenylate slowly leaks out of the active site, enabling slow turnover (the leak rates are typically 100-fold slower than the overall rates using the cognate PCP domain) [6, 7]. Garneau-Tsodikova and coworkers have exploited this phenomenon to develop a nonradioactive assay employing pyrophosphatase that cleaves the liberated pyrophosphate to inorganic phosphate, which is detected by malachite green [8]. All of these aforementioned assays are end-point assays. Ideally, it would be useful to develop a continuous assay, which measures the overall A-domain catalyzed reaction in the kinetically relevant forward direction and that employs a reactive surrogate for the PCP domain in order to provide fast enzyme turnover.

Herein we report a simple continuous coupled assay based on an amalgamation of several reported assays, which proceeds in the forward direction and employs hydroxylamine as an alternative and highly reactive acceptor molecule for the PCP domain [5, 9, 10]. The assay is rendered continuous by monitoring the pyrophosphate produced from the first-half reaction. The pyrophosphate is first cleaved to inorganic phosphate by inorganic pyrophosphatase (IP); the resulting phosphate is a substrate of purine nucleoside phosphorylase (PNP), which converts 7-methylthioguanosine (MesG, **5**) to 7-methylthioguanine (**6**), whose formation can be continuously monitored at 360 nm. Our lab has validated this hydroxylamine-MesG coupled adenylation assay against several stand-alone A-domains as well as A-domains from multifunctional NRPS proteins. Moreover, the specificity constants (k_{cat}/K_M) obtained with this new assay are virtually identical to values obtained employing the standard, radioactive pyrophosphate exchange radioassay [6].

The following protocol details the hydroxylamine-MesG assay performed with the A-domain of the canonical NRPS known as GrsA that activates D-phenylalanine. GrsA is involved in the biosynthesis of the prototypical nonribosomal peptide gramicidin [11] and its adenylation domain (GrsA-A_{Phe}) has been successfully studied using the hydroxylamine-MesG continuous assay [6]. Below, we describe two assays that are run in separate wells (*see* **Note 1**); the first contains all assay components (GrsA-A_{Phe} enzyme, inorganic pyrophosphatase (IP), purine nucleoside phosphorylase (PNP), 7-methylthioguanosine (MesG), TCEP (tris(2-carboxyethyl)phosphine), hydroxylamine, ATP, and D-phenylalanine (D-Phe)), while the second reaction does not contain D-Phe and is used to measure background activity. Our lab uses a plate reader (Molecular Devices Spectramax M5e) that reads both absorbance in the kinetic mode and the pathlength of each individual well following the completion/end point of the assay. Additionally, this assay may be conveniently run on a UV-spectrophotometer using a micro-cuvette. As noted below, the slope of the absorbance progress curve (in mAU/min) for each well is normalized to its own path length before being converted to concentration units (μM/min).

2 Materials

2.1 Expression of Adenylating Enzyme

Adenylating enzymes in our lab are routinely cloned into pET vectors and transformed into *E. coli* BL21 (DE3) for overexpression. His-tagged proteins are successively purified by Ni-NTA and gel filtration chromatography and aliquoted and stored at –80 °C in buffer containing 5 % glycerol [11].

2.2 Reagent Stock Solution Preparation

1. Water for this assay: Prepare ultrapure water for this assay by passing water through a Millipore-Q system or equivalent.

2. 2× Adenylation Assay Buffer: 100 mM Tris–HCl, pH 8.0, 10 mM $MgCl_2$. Weigh and transfer 12.1 g of Tris (Sigma) and 2 g of $MgCl_2 \cdot 6H_2O$ (Sigma) to a 1 L graduated beaker. Add water to a volume of 900 mL, stir to dissolve, and then adjust the pH to 8.0 with 6 N HCl. Dilute up to 1 L with water and store at room temperature.

3. 100 mM TCEP: Dissolve 29 mg of TCEP (HCl salt, Fisher Scientific) in 1 mL of water and store at −20 °C.

4. 4 M hydroxylamine solution: Dissolve 2.78 g of hydroxylamine (HCl salt, Sigma) in 10 mL of water and store at 4 °C.

5. 7 M NaOH solution: Dissolve 2.8 g of sodium hydroxide (Sigma) in 10 mL of water on ice and store at 4 °C.

6. 40 U/mL inorganic pyrophosphatase (IP): add 2.5 mL of water to one vial of IP (Sigma I1643-100UN) and store at 4 °C.

7. 100 U/mL purine nucleoside phosphorylase (PNP): add 1 mL of water to one vial of PNP (Sigma N8264-100UN) and store at 4 °C.

8. 1 mM MesG: dissolve 3.0 mg of MesG (Berry & Associates) in 10 mL of water, aliquot into 1 mL aliquots, and store at −80 °C for up to 1 month.

9. 100 mM ATP: dissolve 55 mg of ATP (disodium salt hydrate, Sigma) in 1 mL of 1 M Tris–HCl pH 8 and stored at −80 °C (*see* **Note 2**).

10. 100 mM acid substrate in 100 mM Tris–HCl, pH 8.0: (*see* **Note 3**).

11. Concentrated adenylating enzyme: Make intermediate enzyme stock in 2× adenylation buffer (*see* Subheading 2.1).

12. Preparation of working solution of hydroxylamine: On the day that the assay is to be run, prepare a 2 M solution of hydroxylamine, pH 7.0. To a 1.5 mL centrifuge tube on ice, add 400 µL of the 4 M hydroxylamine stock solution. To this, add 175 µL of water and 225 µL of 7 M NaOH stock solution dropwise. Confirm that the pH of this solution is 7–8 using pH paper.

13. Prepare Master Mix: The assay will be initiated by adding 95 µL of master mix (GrsA-A_{Phe} enzyme, IP, PNP, MesG, TCEP, hydroxylamine, and ATP in Tris–HCl buffer pH 8.0) to 5 µL D-Phe (or buffer alone) in a well of a UV clear half-area 96-well plate. Therefore, all of the components in the master mix are made at 1.05× the concentration, so that they are at a final concentration of 1× when diluted to a final volume of 100 µL in the assay plate.

    (a) Make stock concentrations of assay components (*see* Subheading 2.1 and Column A in Table 1).

Table 1
Master mix preparation

	Column A	Column B	Column C	Column D	Column E
	[Stock]	[Final] in 100 µL well	[Final] in 95 µL	Volume needed for 1 well (µL)	Total volume needed in master mix (µL)
2× Adenylation assay buffer	–	1×	1×	50	125
Water	–	–	–	22.9	70.3
GrsA-A	1 µM	5 nM	5.25 nM	0.53	1.3
IP	40 U/mL	0.04 U	0.042 U	1.05	2.6
PNP	100 U/mL	0.1 U	0.105 U	1.05	2.6
MesG	1 mM	100 µM	105 µM	5.25	13.2
TCEP	100 mM	1 mM	1.05 mM	1.05	2.6
Hydroxylamine	2 M	150 mM	157.5 mM	7.7	19.3
ATP	100 mM	5 mM	5.25 mM	5.3	13.1

(b) Determine final concentrations of enzyme and substrates (Column B, Table 1) that will provide initial velocity conditions (*see* **Note 5**).

(c) Multiply values in Column B by a factor of 1.05 to obtain values in Column C.

(d) Calculate the volume of each assay component for one assay well (Column D).

(e) Multiply the volumes in Column D by 2.5 to obtain the total volume needed for all of the assay wells (Column E). In the assay example described here, two assays are run in separate wells (with and without D-Phe). The multiplication factor is increased from 2 to 2.5 in order to account for small volume losses due to pipetting.

2.3 Additional Supplies and Equipment

(a) 96-well UV clear half-area plates (Corning #3679).

(b) Plate reader capable of reading UV absorbance at 360 nm and calculating the path length of the wells (*see* **Note 4**).

3   Methods

3.1   Initiate Assay

1. To one well of a 96-well UV clear half-area plate add 5 µL of a 20 mM solution of D-Phe (made in 100 mM Tris, pH 8.0) and to another well add 5 µL of 100 mM Tris–HCl, pH 8.0. Add 95 µL of the Master Mix to both wells and ensure that no bubbles form.

3.2   Read Plate and Path Length

1. Read the absorbance at 360 nm for 5 min in the kinetic read mode (Fig. 2).
2. Obtain slope (in mAU/min) using the plate reader software.
3. Read the pathlength of each well.

3.3   Calculate Initial Velocity

1. The slopes measured by the plate reader for the two assays in Fig. 2 are:

   +D-Phe = 21.972 mAU/min

   −D-Phe = 2.195 mAU/min

2. The path lengths at the end point of the assays are:

   +D-Phe = 0.512 cm

   −D-Phe = 0.505 cm

3. Calculate the path length normalized slope by dividing the slopes by the pathlength of each well:

   +D-Phe: 21.972 mAU/min ÷ 0.512 cm = 42.914 mAU/min/cm

   −D-Phe: 2.195 mAU/min ÷ 0.505 cm = 4.347 mAU/min/cm

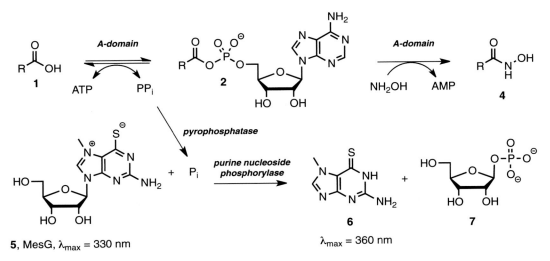

5, MesG, λ_{max} = 330 nm λ_{max} = 360 nm

Fig. 2 Time course of 7-methylguanine production. Each reaction contained 50 mM Tris–HCl, pH 8.0, 5 mM MgCl$_2$, 5 nM GrsA-A$_{Phe}$, 0.04 U IP, 0.1 U PNP, 200 µM MesG, 1 mM TCEP, 150 mM hydroxylamine, 5 mM ATP. The control reaction contained no D-Phe while the positive reaction contained 1 mM D-Phe

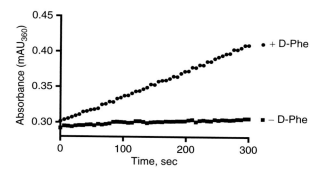

Fig. 3 Continuous hydroxamate-MesG assay for measuring adenylation activity

4. Convert units from mAU to AU (*see* **Note 7**):

+D-Phe: 42.914 mAU/min/cm ÷ 1000 = 0.0429 min^{-1} cm^{-1}

−D-Phe: 4.347 mAU/min/cm ÷ 1000 = 0.0043 min^{-1} cm^{-1}

5. Divide by the extinction coefficient of 7-methylthioguanine (**6**, Fig. 3):

+D-Phe: 0.0429 min^{-1} cm^{-1} ÷ 11,000 M^{-1} cm^{-1} = 3.9 × 10^{-6} M/min

−D-Phe: 0.0043 min^{-1} cm^{-1} ÷ 11,000 M^{-1} cm^{-1} = 3.9 × 10^{-7} M/min

6. Convert to µM:

+D-Phe: 3.9 × 10^{-6} M/min × 10^{6} µM/M = 3.9 µM/min

+D-Phe: 3.9 × 10^{-7} M/min × 10^{6} µM/M = 0.39 µM/min

7. Divide by 2 since there are two molecules of 7-methylguanine formed per turnover:

+D-Phe: 3.9 µM/min ÷ 2 = 1.95 µM/min

+D-Phe: 0.39 µM/min ÷ 2 = 0.195 µM/min

8. Subtract the background rate (−D-Phe) from the full enzymatic rate (+D-Phe):

Rate: 1.95 µM/min − 0.195 µM/min = 1.76 µM/min

9. The rate value obtained in **step 8** can now be used to obtain Michaelis–Menten kinetic parameters of all substrates (by varying substrate concentrations) or study enzyme inhibition (by varying inhibitor concentrations) (*see* **Note 6**).

4   Notes

1. Assays should be run in triplicate. The examples shown here are run in singlicate for clarity of presentation.

2. To dissolve ATP, a high concentration of Tris–HCl Buffer (1 M, pH 8.0) must be used in order to neutralize ATP, which is acidic. ATP should not be stored for more than 1 week at −80 °C.

3. Our lab routinely makes 10–100 mM solutions of the acid substrate in 100 mM Tris–HCl, pH 8.0. The concentration of the acid stock will depend on the acid's aqueous solubility.

4. The amount of adenylating enzyme (in units) should always be much lower (<50×) than the amount of coupling enzymes. This will ensure that the coupling enzymes are not the rate limiting enzymes.

5. Our lab uses a Spectramax M5e plate reader, which reads and calculates path length using the PathCheck® function.

6. If inhibition of adenylating enzymes is to be studied, add the Master Mix to the inhibitor (final DMSO ≤1 %) and incubate for 10 min prior to initiating the reaction by adding to the well containing the substrate.

7. Absorbance is unitless. Therefore, the unit AU is not shown after converting from mAU to AU.

Acknowledgement

C.C.A. acknowledges membership in and support from the Region V "Great Lakes" Regional Center of Excellence in Biodefense and Emerging Infectious Diseases Consortium (National Institutes of Health award 1-U54-AI-057153) and support from R01-AI-070219 for this work.

References

1. Schmelz S, Naismith JH (2009) Adenylate-forming enzymes. Curr Opin Struct Biol 19:666–671

2. Gulick AM (2009) Conformational dynamics in the acyl-CoA synthetases, adenylation domains of non-ribosomal peptide synthetases, and firefly luciferase. ACS Chem Biol 4:811–827

3. Hur GH, Vickery CR, Burkart MD (2012) Explorations of catalytic domains in non-ribosomal peptide synthetase enzymology. Nat Prod Rep 29:1074–1098

4. Phelan VV, Du Y, McLean JA, Bachmann BO (2009) Adenylation enzyme characterization using γ-^{18}O$_4$-ATP pyrophosphate exchange. Chem Biol 16:473–478

5. Kadi N, Challis GL (2009) Chapter 17 siderophore biosynthesis: a substrate specificity assay for nonribosomal peptide synthetase-independent siderophore synthetases involving trapping of acyl-adenylate intermediates with hydroxylamine. Methods Enzymol 458:431–457

6. Wilson DJ, Aldrich CC (2010) A continuous kinetic assay for adenylation enzyme activity and inhibition. Anal Biochem 404:56–63

7. Ehmann DE, Shaw-Reid CA, Losey HC et al (2000) The EntF and EntE adenylation domains of *Escherichia coli* enterobactin synthetase: sequestration and selectivity in acyl-AMP transfers to thiolation domain cosubstrates. Proc Natl Acad Sci U S A 97:2509–2514

8. McQuade TJ, Shallop AD, Sheoran A et al (2009) A nonradioactive high-throughput assay for screening and characterization of adenylation domains for nonribosomal peptide combinatorial biosynthesis. Anal Biochem 386:244–250

9. Webb MR (1992) A continuous spectrophotometric assay for inorganic phosphate and for measuring phosphate release kinetics in biological systems. Proc Natl Acad Sci U S A 89:4884–4887

10. Upson RH, Huagland RP, Malekzadeh MN et al (1996) A spectrophotometric method to measure enzymatic activity in reactions that generate inorganic pyrophosphate. Anal Biochem 243:41–45

11. Stachelhaus T, Marahiel M (1995) Modular structure of peptide synthetases revealed by dissection of the multifunctional enzyme GrsA. J Biol Chem 270:6163–6169

Chapter 4

Affinity Purification Method for the Identification of Nonribosomal Peptide Biosynthetic Enzymes Using a Synthetic Probe for Adenylation Domains

Fumihiro Ishikawa and Hideaki Kakeya

Abstract

A series of inhibitors have been designed based on 5′-O-sulfamoyl adenosine (AMS) that display tight binding characteristics towards the inhibition of adenylation (A) domains in nonribosomal peptide synthetases (NRPSs). We recently developed an affinity probe for A domains that could be used to facilitate the specific isolation and identification of NRPS modules. Our synthetic probe, which is a biotinylated variant of L-Phe-AMS (L-Phe-AMS-biotin), selectively targets the A domains in NRPS modules that recognize and convert L-Phe to an aminoacyl adenylate in whole proteomes. In this chapter, we describe the design and synthesis of L-Phe-AMS-biotin and provide a summary of our work towards the development of a series of protocols for the specific enrichment of NRPS modules using this probe.

Key words Nonribosomal peptide synthetase (NRPS), Adenylation domain, Small-molecule probe, Affinity purification, *Aneurinibacillus migulanus* ATCC 9999, Gramicidin S synthase 1 (GrsA)

1 Introduction

Microorganisms produce a large number of natural products that display a broad range of biologically interesting properties, including antimicrobial, immunosuppressant, and anticancer activities [1]. These microbial metabolites belong to a large class of natural products known as polyketide (PK), nonribosomal peptide (NRP), and PK-NRP hybrid molecules. As well as the 20 proteinogenic amino acids, peptide-based microbial metabolites also consist of non-proteinogenic amino acids, aryl acids, fatty acids, and other acids, with different combinations of these building blocks leading to complex natural chemical libraries [2]. A wide range of peptide-based natural products are biosynthesized by large, highly versatile multifunctional enzymes known as polyketide synthase (PKS)-nonribosomal peptide synthetase (NRPS) hybrids and NRPSs. The results of extensive periods of biochemical investigation have provided a detailed understanding of the functional characteristics

Bradley S. Evans (ed.), *Nonribosomal Peptide and Polyketide Biosynthesis: Methods and Protocols*, Methods in Molecular Biology, vol. 1401, DOI 10.1007/978-1-4939-3375-4_4, © Springer Science+Business Media New York 2016

and molecular basis underpinning the specific substrate recognition of the A domains in NRPS modules [3, 4]. Progress in this area has also made it possible to predict the enzymatic processes of newly isolated peptide-based compounds produced by the NRPS enzymes uncovered by genomic information. The amino acid building blocks these natural products consist of often correlate quite readily with the amino acid specificity of the A domains assembled on their associated biosynthetic enzymes [5, 6]. By taking advantage of the strict substrate recognition characteristics of A domains, the aim of this study was to selectively isolate NRPS modules from bacterial proteomes using synthetic probes for A domains and directly link the chemotypes of the expressed peptide-based compounds with their associated NRPS biosynthetic enzymes. Methods for the selective detection of biosynthetic enzymes have allowed us to develop a deeper understanding of the expression dynamics of biosynthetic enzymes in the producers and have therefore facilitated the discovery of the expressed PKS, NRPS, and PKS-NRPS gene clusters expressed in unsequenced organisms using a combination with mass spectrometry-based approach [7, 8].

We recently reported the design, synthesis, and application of a biotinylated L-Phe-AMS (L-Phe-AMS-biotin) and highlighted its use as a proteomic tool for the specific isolation and identification of A domain-containing NRPS modules (Fig. 1) [9]. The A domains can selectively recognize cognate amino, acyl, and

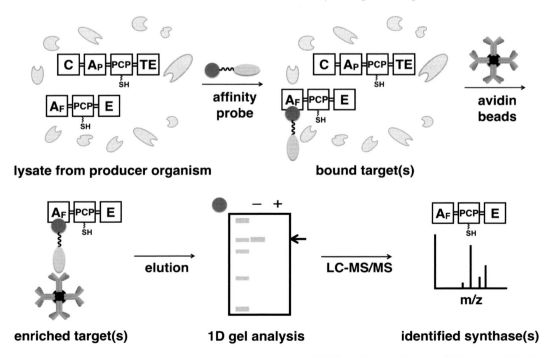

Fig. 1 Schematic summary for the proteomic analysis of the NRPS modules using an affinity probe [9]. The modules are composed of peptidyl carrier protein (PCP), adenylation (A) [A_F: L-Phe specific; A_P: L-Pro specific], epimerase (E), condensation (C), and thioesterase (TE) domains

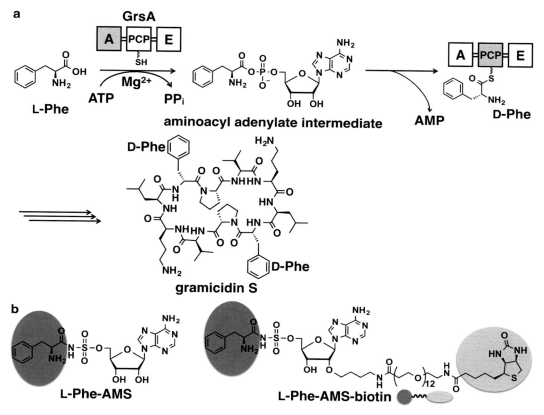

Fig. 2 Biosynthetic pathway and small molecules described in this study [9]. (**a**) Biosynthetic pathway of the gramicidin S. (**b**) Structures of L-Phe-AMS and L-Phe-AMS-biotin

other acid groups from the cellular pool, and catalyze the formation of acyl adenylate intermediates during NRP biosynthesis using ATP. The A domains can therefore act as the gatekeeper of the NRPS enzymatic processes. The adenylated substrate can then transfer the acyl group onto a 4′-phosphopantetheine prosthetic arm via the nucleophilic attack of a thiol on a downstream peptidyl carrier protein (PCP) of the NRPS assembly line (Fig. 2a) [10, 11]. 5′-O-sulfamoyl adenosine (AMS) has been used extensively in the design of inhibitors requiring tight binding characteristics to the A domains of NRPS enzymes [12, 13]. Marahiel et al. [12] reported the synthesis of the simple substrate-based mimetic 5′-O-[N-(phenylalanyl)-sulfamoyl] adenosine (L-Phe-AMS, Fig. 2b), where the highly reactive acylphosphate linkage was replaced by a bioisosteric and chemically stable non-hydrolyzable acylsulfamate group [12].

The current study began with the design of a suitably tagged affinity probe that could be used to target A domains (Fig. 2b). Tridomain GrsA was selected as the target protein in this study because GrsA is a single NRPS module with the domain organization A (L-Phe)-PCP-epimerase (E) that incorporates D-Phe by selectively recognizing, activating and epimerizing L-Phe in the biosynthesis of gramicidin S (Fig. 2a). The design of the probe used in the current study was based on the tight-binding bisubstrate inhibitor, L-Phe-AMS, and a long pegylated spacer unit (56.0 Å) was selected to reduce steric hindrance between the biosynthetic enzymes and the streptavidin molecules during the affinity purification step. The linker molecule was attached to the 2′-OH group of adenosine based on an X-ray crystal structure of the GrsA A domain in complex with AMP and L-Phe [3].

In this chapter, we have described the method used for the preparation of L-Phe-AMS-biotin, as well as a protocol for specific isolation of NRPS modules using L-Phe-AMS-biotin. We have also described the selective isolation and identification of GrsA from the proteome of the gramicidin S producer, *Aneurinibacillus migulanus* ATCC 9999, as a specific example of the overall utility of this technique. Combinations of this probe with other domain-specific reactive probes, such as thioesterase [14] and dehydratase [15], could potentially be used to further improve the isolation and detection of biosynthetic enzymes containing NRPS modules in proteomic environments.

2   Materials

2.1   Synthetic Procedure of L-Phe-AMS-Biotin

All of the materials and reagents described below can be purchased from commercial suppliers and used without further purification.

1. Obtain the organic solvents used in following experiments, including ethyl acetate, hexanes, acetone, methanol, *N,N-dimethylformamide*, dichloromethane, tetrahydrofuran, and dimethoxyethane from Nacalai Tesque, Wako, and Sigma-Aldrich. Acetonitrile should be purchased as the high-performance liquid chromatography (HPLC) grade for the HPLC purification of the adenosine derivative.

2. Prepare L-Phe-AMS according to the literature procedure, which will also be used to afford an analogue of L-Phe-AMS-biotin for comparative activity studies [9, 12].

2.2   Bacterial Culture of A. migulanus ATCC 9999

1. Nutrient agar: Add 3.0 g of beef extract, 5.0 g of peptone, and 15 g of agar to 1 L of distilled water, and sterilize the resulting mixture at 121 °C for 20 min before being poured into petri dishes and allow to cool to room temperature.

2. Nutrient broth: Add 3.0 g of beef extract and 5.0 g of peptone to 1 L of distilled water, and sterilize the resulting mixture at 121 °C for 20 min.

3. YPG medium: Add 50 g of yeast extract, 50 g of Bacto-peptone, and 5 g of glucose to 1 L of distilled water and sterilize the resulting mixture at 121 °C for 20 min.

4. PBS buffer: Add 9.6 g of Dulbecco's PBS to 1 L of distilled water and sterilize the resulting mixture at 121 °C for 20 min.

5. 80 % (v/v) glycerol solution: Add 80 mL of glycerol to 20 mL of distilled water, and sterilize the resulting mixture at 121 °C for 20 min.

2.3 Preparations of the Samples for the Proteomic Studies

1. Buffer A: 20 mM Tris–HCl (pH 8.0), 1 mM MgCl$_2$, 1 mM TCEP, 0.05 % Igepal CA-630, and the protease inhibitor cocktail: For example, add 10 µL of 1 M MgCl$_2$, 10 µL of 1 M TCEP, and 2.5 µL of Igepal CA-630 to 10 mL of 20 mM Tris (pH 8.0) and 100 µL of protease inhibitor cocktail (100×), which, to prevent the decomposition of the protease inhibitors, is cooled on ice prior to being used.

2. 10 mg/mL lysozyme solution: Add 10 mg of lysozyme from egg white to 1 mL of 20 mM Tris–HCl (pH 8.0).

3. CBB solution for the Bradford Assay: Add 2 mL of protein assay CBB solution (5×) to 8 mL of distilled water.

2.4 Affinity Purification Procedure

1. Buffer A: 20 mM Tris–HCl (pH 8.0), 1 mM MgCl$_2$, 1 mM TCEP, and 0.05 % Igepal CA-630: For example, mix 50 µL of 1 M MgCl$_2$, 50 µL of 1 M TCEP, and 25 µL of Igepal CA-630 were mixed in 50 mL of 20 mM Tris (pH 8.0).

2. 9.5 mM DMSO solution of L-Phe-AMS-biotin.

3. 10 mM DMSO solution of L-Phe-AMS.

4. RIPA buffer composed of 50 mM Tris–HCl (pH 7.5), 150 mM NaCl, 1 % NP-40, 1 % sodium deoxycholate, and 0.1 % SDS.

5. SDS-PAGE gel loading buffer (5×): 0.25 M Tris–HCl (pH 6.8) is mixed with 10 % sodium dodecyl sulfate (SDS), 0.5 M dithiothreitol (DTT), 50 % glycerol, and 0.1 % bromophenol blue (BPB). Keep one aliquot of the resulting mixture and store it at 4 °C for immediate use. Store the remaining aliquots at −20 °C for later use.

6. SDS-PAGE running buffer (10×): Add 30.2 g of tris(hydroxymethyl)aminomethane, 144 g of glycine, and 10 g of SDS to 1 L of distilled water, and dilute the resulting solution to 1× for running the gels.

3 Methods

3.1 Synthetic Procedure for the Preparation of L-Phe-AMS-Biotin

L-Phe-AMS-biotin was synthesized from adenosine, 4-dibromobutane, Boc-Phe-OSu, and EZ-Link NHS-PEG12-Biotin using the eight step route shown in Fig. 3 [9], which provided access to the desired material in an overall yield of 1.8 % from adenosine.

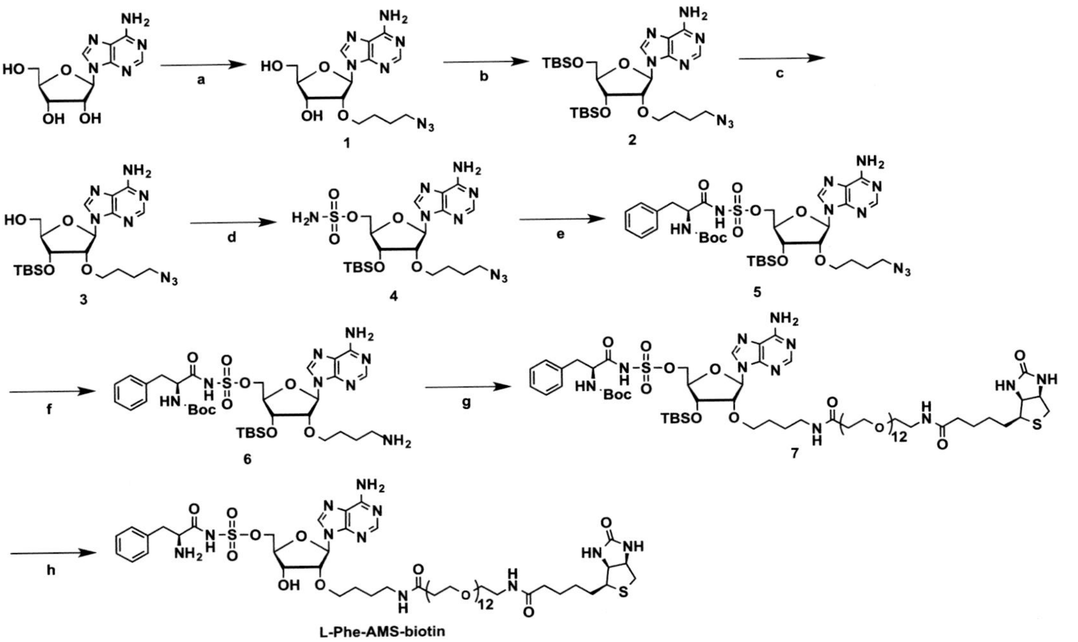

Fig. 3 Synthetic route to L-Phe-AMS-biotin [9]. *Reagents and conditions*: [*a*] NaH, 4-azide-1-bromobutane, TBAI, DMF, 55 °C, 45 %. [*b*] TBSCl, imidazole, DMAP, DMF, 0 °C → rt, 65 %. [*c*] 50 % aqueous TFA, THF, 0 °C, 75 %. [*d*] NH$_2$SO$_2$Cl, NaH, DME, 0 °C → rt, 62 %. [*e*] Boc-Phe-OSu, Cs$_2$CO$_3$, DMF, rt, 67 %. [*f*] Pd/C, EtOH, rt, 25 %. [*g*] EZ-link NHS-PEG$_{12}$-Biotin, DIEA, DMF, rt, 81 %. [*h*] 80 % aqueous TFA, rt, 99 %

3.1.1 Synthesis of 1-Azido-4-bromobutane [16]

1. Dissolve 1,4-dibromobutane in *N*,*N*-dimethylformamide under an atmosphere of nitrogen at room temperature.

2. Add 0.49 eq. of sodium azide to the solution, and stir the resulting mixture at 50 °C for 16 h.

3. Cool the mixture to ambient temperature and partition between a 1:1 (v/v) mixture of ethyl acetate/hexane and water.

4. Collect the organic layer and wash it sequentially with water (three times) and brine (one time).

5. Dry the organic layer over anhydrous sodium sulfate and remove the solvent under vacuum to give the crude product.

6. Purify the residue by flash column chromatography.

3.1.2 Synthesis of Sulfamoyl Chloride [17]

1. Dissolve chlorosulfonyl isocyanate in dichloromethane in a three-necked round-bottom flask equipped with an addition funnel, nitrogen gas inlet, and glass stopper, and cool the resulting solution to 0 °C.

2. Add a solution of 1.0 eq. of formic acid solution in dichloromethane to the reaction in a drop-wise manner over a period of 5–10 min (*see* **Note 1**), and stir the resulting mixture at 0 °C for 10 min.

3. Remove the reaction mixture from the ice bath and allowed the mixture to warm to room temperature.

4. Stir the resulting white suspension at room temperature for 12 h.

5. Place the suspension in a freezer at −20 °C freezer for 6 h.

6. Dry the resulting white solid under vacuum to give the title compound as a white solid. This material can be used in the next step without further purification.

3.1.3   Synthesis of 1

1. Dissolve adenosine in *N,N*-dimethylformamide at 50 °C under an atmosphere of nitrogen, and cool the resulting solution to 5 °C.

2. Add 1.3 eq. of sodium hydride (60 % dispersion in mineral oil) to the solution followed by 0.22 eq. of tetrabutylammonium iodide, and allow the resulting mixture to warm to room temperature with stirring over 30 min.

3. Add 1.1 eq. of 4-azide-1-bromobutane to the reaction, and stir the resulting mixture at 55 °C for 2 days.

4. Cool the reaction mixture to ambient and remove the solvent under vacuum to give the crude product.

5. Purify the residue by HPLC (*see* **Note 2**) to give the desired product.

3.1.4   Synthesis of 2

1. Dissolve compound **1**, 6.0 eq. of imidazole and 0.15 eq. of *N,N*-dimethyl-4-aminopyridine in *N,N*-dimethylformamide under an atmosphere of nitrogen, and cool to the resulting solution to 0 °C.

2. Add a solution of 2.5 eq. of *tert*-butyldimethylsilyl chloride in *N,N*-dimethylformamide to the reaction mixture, and stir the resulting solution at room temperature for 16 h.

3. Dilute the reaction mixture with ethyl acetate and wash it with brine (three times).

4. Collect and dry the organic layer over anhydrous sodium sulfate and then remove the solvent under vacuum to give the crude product.

5. Purify the residue by flash column chromatography.

3.1.5   Synthesis of 3

1. Add a 50 % aqueous solution of trifluoroacetic acid (4 mL) to a solution of compound **2** in tetrahydrofuran (4 mL) and stir the resulting mixture at 0 °C for 3 h (*see* **Note 3**).

2. Remove the solvent under vacuum to give the crude product.

3. Purify the residue by flash column chromatography.

3.1.6   Synthesis of 4

1. Add 1.5 eq. of sodium hydride (60 % dispersion in mineral oil) to a solution of compound **3** in dimethoxyethane at 0 °C under an atmosphere of nitrogen.

2. Stir the mixture at 0 °C for 30 min.

3. Add 1.5 eq. of sulfamoyl chloride to the mixture at 0 °C (*see* **Note 4**), and stir the resulting mixture at 0 °C for 1 h.

4. Warm the reaction mixture to room temperature and stir the mixture at this temperature for 10 h.

5. Quench the reaction with methanol and then remove the solvent under vacuum to give the crude product.

6. Purify the residue by flash column chromatography.

3.1.7 Synthesis of 5

1. Dissolve compound **4** in *N,N*-dimethylformamide under an atmosphere of nitrogen at room temperature.

2. Add 1.5 eq. of Boc-Phe-OSu and 3.0 eq. of cesium carbonate to the reaction, and stir the resulting mixture at room temperature for 12 h.

3. Filter the reaction mixture through a pad of celite to remove the cesium carbonate, and wash the pad with additional *N,N*-dimethylformamide.

4. Combine the filtrates and remove the solvent under vacuum to give the crude product.

5. Purify the residue by flash column chromatography.

3.1.8 Synthesis of 6

1. Dissolve compound **5** in ethanol.

2. Add 10 % Pd/C, and hydrogenate the resulting suspension under an atmosphere of hydrogen at room temperature for 12 h.

3. Filter the reaction mixture through a pad of celite to remove Pd/C, and wash the pad with additional ethanol.

4. Evaporate the combined filtrates under vacuum to give the crude product.

5. Purify the residue by flash column chromatography.

3.1.9 Synthesis of 7

1. Dissolve compound **6** in *N,N*-dimethylformamide under an atmosphere of nitrogen at room temperature.

2. Add 1.0 eq. of EZ-link NHS-PEG$_{12}$-Biotin and 1.2 eq. of *N,N-diisopropylethylamine*, and stir the resulting solution at room temperature for 7 h.

3. Evaporate the reaction solvent under vacuum to give the crude product.

4. Purify the residue by flash column chromatography.

3.1.10 Synthesis of ʟ-Phe-AMS-Biotin

1. Dissolve compound **7** in a mixture of 4:1 (v/v) mixture of trifluoroacetic acid and water, and stir the resulting solution at room temperature for 5 h.

2. Evaporate the solution under vacuum to give the crude product.

3. Purify the residue by flash column chromatography (*see* **Note 5**).

3.2 Bacterial Culture

This section describes the propagation and culture procedures used for the gramicidin S producer *A. migulanus* ATCC 9999.

3.2.1 Bacterial Propagation Procedure

1. Open vial containing culture lyophilisate of *A. migulanus* ATCC 9999.

2. Start with five culture tubes containing 5 mL of nutrient broth.

3. Remove 1 mL of nutrient broth using a 1 mL pipette and rehydrate the pellet.

4. Aseptically transfer this aliquot back into a culture tube, and mix well.

5. Use several drops of this mixture to inoculate the remaining tubes of nutrient broth and nutrient agar plates.

6. Incubate all of the tubes and plates at 37 °C for 24 h.

7. Add 100 µL of 80 % sterile glycerol to each of the 2 mL screw cap vials.

8. Add 700 µL of liquid bacterial culture to each of the vials, mix well, and store at –80 °C.

3.2.2 Bacterial Culture Procedure

1. Prepare nutrient agar plates (*see* Subheading 2.2).

2. Prepare a rich, complex medium (YPG) according to the procedure developed by Matteo et al. (*see* Subheading 2.2) [18].

3. Streak *A. migulanus* ATCC 9999 for single colonies on nutrient agar plates, and incubate the resulting plates at 37 °C for 24 h.

4. Pick a single colony.

5. Prepare a seed culture of 2–5 mL YPG medium in a culture tube, and incubate the culture at 37 °C for 24 h on a reciprocating shaker at 220 rpm.

6. Inoculate 2 mL of the seed culture into 250 mL of YPG in a 1 L baffled flask, and incubate the resulting mixture at 37 °C for 10 h ($OD_{660} = 7.6$) on a reciprocating shaker at 220 rpm.

7. Determine growth at specific time points by measuring the absorbance at 660 nm on a spectrophotometer.

8. Harvest the cells into 50 mL plastic tubes and centrifuge them for 15 min at $15,000 \times g$ at 4 °C.

9. Discard the supernatant and suspend the cell pellets in PBS buffer.

10. Divide the cells into 1.5 mL eppendorf tubes and centrifuge them for 15 min at $15,000 \times g$ at 4 °C.

11. Discard the supernatant and store the cell pellets at –80 °C until they are required.

3.3 Sample Preparations for Proteomic Studies

This section describes the procedures used for the preparation of proteomic samples from *A. migulanus* ATCC 9999.

1. Prepare buffer A (i.e., mix 20 mM Tris–HCl, pH 8.0, 1 mM MgCl$_2$, 1 mM TCEP, 0.05 % Igepal CA-630, and the protease inhibitor cocktail).

2. Prepare a 10 mg/mL lysozyme solution using 20 mM Tris–HCl (pH 8.0) (*see* **Note 6**).

3. Resuspend the frozen cell pellets in 0.5–1 mL of buffer A on ice.

4. Add 10–20 µL of the 10 mg/mL lysozyme solution to the resuspended cells, and incubate the resulting mixture over ice for 30 min with gentle shaking.

5. Transfer the mixture to a water bath at 30 °C and incubate for 30 min with gentle shaking.

6. Centrifuge the mixture for 30 min at 15,000×*g* at 4 °C and collect the supernatant.

7. Measure the total protein concentration on a plate reader according to the Bradford Assay using BSA as a standard with an absorbance of 595 nm [19].

3.4 Affinity Purification Procedure

This section provides a description of the procedure used for the specific enrichment of the endogeneous GrsA obtained from the *A. migulanus* ATCC 9999 proteome using L-Phe-AMS-biotin.

3.4.1 Preparation of Streptavidin-Agarose Resin

1. Pour 50–100 µL of the streptavidin-agarose slurry into a 1.5 mL tube and wash with 10 column volumes of buffer A (i.e., 20 mM Tris–HCl, pH 8.0, 1 mM MgCl$_2$, 1 mM TCEP, and 0.05 % Igepal CA-630).

2. Centrifuge the resulting mixture for 5 min at 15,000×*g* at 4 °C and discard the supernatant.

3. Repeat this cycle three times.

3.4.2 Isolation of the Cellular Target of L-Phe-AMS-Biotin (See Fig. 4a)

1. Prepare a 1.0–2.0 mg/mL solution of *A. migulanus* ATCC 9999 cellular lysates using buffer A on ice (*see* Subheading 3.3 for the composition of the buffer solution).

2. Add 1 µL of L-Phe-AMS-biotin (9.5 µM from a 9.5 mM stock in DMSO to give a final DMSO concentration of 0.1 %) into 1 mL of the 1.0–2.0 mg/mL lysate solution, and incubate the resulting solution for 12 h at 4 °C on tube rotator.

3. Add the solution to a the streptavidin-agarose resin and incubate the resulting mixture for 1 h at 4 °C on a tube rotator.

4. Collect the resins by centrifugation at 15,000×*g* for 5 min at 4 °C, and wash them three times with 1 mL of ice-cold RIPA buffer (*see* Subheading 3.4.1 for the composition of the buffer solution).

5. Elute the bound proteins by competitive elution in the presence of L-Phe-AMS (*see* Subheading 3.4.4 for competitive elution).

6. Mix the eluent with 10 µL of 5× gel-loading buffer, and incubate the resulting mixture at 95 °C for 5 min.

7. Analyze the eluted proteins by 8 % SDS-PAGE and stain the gel with a Silver Stain Kit (Nacalai Tesque).

*3.4.3 Competitive Inhibition Assay (See Fig. 4b and **Note 7**)*

1. Prepare a 1.0–2.0 mg/mL solution of *A. migulanus* ATCC 9999 cellular lysates using buffer A on ice (*see* Subheading 3.3 for the composition of the buffer solution).

2. Preincubate 1 mL of the 1.0–2.0 mg/mL lysate solution with 1 µL of L-Phe-AMS (100 µM from a 10 mM stock in DMSO), and incubate the resulting mixture for 1 h at 4 °C on a tube rotator.

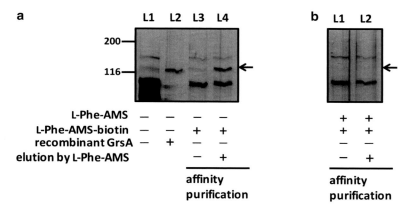

Fig. 4 Affinity purification of endogenous GrsA from *A. migulanus* ATCC 9999 proteome [9]. The cells were grown at OD_{660} of 7.6 in this study. (**a**) Affinity purification of endogenous GrsA by L-Phe-AMS-biotin. *A. migulanus* ATCC 9999 lysate (1.2 mg/mL) was incubated with 9.5 µM of L-Phe-AMS-biotin for 12 h at 4 °C. Bound proteins were eluted by incubation with L-Phe-AMS [DMSO alone (*Lane 3*) and 0.5 mM (*Lane 4*)], subjected to SDS-PAGE with purified GrsA standard, and subsequently visualized with silver stain. Gel lanes are depicted as follows: *L1* = *A. migulanus* ATCC 9999 cellular lysate, *L2* = recombinant GrsA, *L3* = eluate by treatment of DMSO alone, and *L4* = eluate by treatment with 0.5 mM of L-PHE-AMS. (**b**) Affinity purification of endogenous GrsA in the presence of L-PHE-AMS. The lysate (1.2 mg/mL) was treated with 100 µM of L-Phe-AMS prior to the administration of 9.5 µM of L-Phe-AMS-biotin. Bound proteins were eluted by incubation with L-Phe-AMS [DMSO alone (*Lane 1*) and 0.5 mM (*Lane 2*)]. Endogenous GrsA marked with *arrowhead* at ~120 kDa in (**a**) specifically eluted from the resin by the treatment with not vehicle (DMSO) but L-Phe-AMS. In contrast, the affinity purification of endogenous GrsA by L-Phe-AMS-biotin was competitively inhibited in pretreatment with L-Phe-AMS (marked with *arrowhead* **b**)

3. Add 1 µL of L-Phe-AMS-biotin (9.5 µM from a 9.5 mM stock in DMSO) to the mixture, and incubate the resulting mixture for 12 h at 4 °C on tube rotator. It is important to mention that the total DMSO concentration in this experiment should be kept at 0.2 %.

4. Add the solution to a streptavidin-agarose resin, and incubate the resulting mixture for 1 h at 4 °C on tube rotator.

5. Collect the resins by centrifugation at $15,000 \times g$ for 5 min at 4 °C and wash them three times with 1 mL of ice-cold RIPA buffer (see Subheading 3.4.1 for the composition of the buffer solution).

6. Elute the bound proteins by competitive elusion with L-PHE-AMS (see Subheading 3.4.4 for details of the competitive elution method).

7. Mix the eluents with 10 µL of 5× gel-loading buffer, and incubate the resulting mixture at 95 °C for 5 min.

8. Analyze eluted proteins by 8 % SDS-PAGE and stain the gel with a Silver Stain Kit.

*3.4.4 Competitive Elution Procedure (See **Note 8**)*

1. Prepare a 500 µM solution of L-Phe-AMS in buffer A on ice (see Subheading 3.3 for the composition of the buffer solution). It is noteworthy that the total DMSO concentration was kept at 5 % in all of the elution experiments.

2. Add 40 µL of the 500 µM L-Phe-AMS solution to the washed resins (see Subheadings 3.4.2 and 3.4.3 and repeat **steps 4** and **5**, respectively), and incubate the resulting mixtures at room temperature for 2 h. Controls should also be prepared by eluting with 40 µL of buffer A (i.e., in the absence of L-Phe-AMS with only DMSO).

3. Centrifuge the mixtures at $15,000 \times g$ for 5 min at 4 °C and collect the supernatant for analysis (40 µL) (see Subheadings 3.4.2 and 3.4.3 to repeat **steps 6** and **7**, respectively).

3.4.5 In-Gel Digestion and Mass Spectroscopic Analysis

The details of the experimental procedures used for the identification of the protein can be found in refs. 20, 21.

4 Notes

1. The apparatus should be vented because gaseous byproducts can form during the addition of the formic acid solution.

2. HPLC purification was performed under the following conditions: Senshu pak PEGASIL ODS SP100 column (250×20 mm) using a 20:80 (v/v) mixture of acetonitrile and 0.1 % aqueous trifluoroacetic acid as the mobile phase at a flow rate of 8.0

mL/min. Detection was conducted at 210 nm. The retention times of the materials were as follows: t_R: 2′-isomer 1 = 13.5 min, 3′-isomer of 1 = 22.0 min. The acetonitrile and trifluoroacetic acid were removed under vacuum, whereas the water was removed by lyophilization to give the 2′-isomer 1.

3. A 1:1 (v/v) mixture of tetrahydrofuran and 50 % aqueous trifluoroacetic acid was used for the deprotection step. It is noteworthy that the use of an extended reaction time resulted in the deprotection of both the 5′- and 3′-TBS groups. In these cases, the reaction time should be shortened. The progress of the deprotection of 5′-TBS group should therefore be carefully monitored by thin-layer chromatography (TLC) every 30 min.

4. The sulfamoyl chloride should be used immediately after being weighed because it is a deliquescent reagent.

5. L-Phe-AMS-biotin was purified by flash column chromatography eluting with a 1:4 (v/v) mixture of MeOH and CHCl$_3$. The polarity of the eluent was then increased to 100 % MeOH because of the high polarity of the biotinylated material.

6. The synthetase was found to be particularly sensitive to the mechanical cell disruption processes, and it was therefore necessary to treat the cells gently to obtain the intracellular proteins [22].

7. To confirm the nature of the proteins bound to L-Phe-AMS-biotin, we recommend the use of a competitive inhibition study using L-Phe-AMS as a competitive inhibitor during the affinity purification processes. Typically, the purification processes were completely inhibited by the pretreatment of the lysate with L-Phe-AMS, which resulted in the loss of the target protein bands (Fig. 4b).

8. The elution of the affinity-purified proteins should not be conducted by boiling the resin (i.e., 95 °C for 5 min), but should be achieved by the treatment of L-Phe-AMS according to the competitive elution method (Fig. 4a), because the elution by boiling samples can also provide many of the nonspecifically bound proteins, resulting in a low-resolution for the detection of target proteins by SDS-PAGE analysis.

Acknowledgement

This work was supported in part by a Grant-in Aid for Young Scientists (B) 26750370 (F.I.) and research grants from the Japan Society of the Promotion of Science and the Ministry of Education, Culture, Sports, Science and Technology, Japan (H.K.).

References

1. Finking R, Marahiel MA (2004) Biosynthesis of nonribosomal peptides. Annu Rev Microbiol 58:453–488
2. Hur GH, Vickery CR, Burkart MD (2012) Explorations of catalytic domains in nonribosomal peptide synthetase enzymology. Nat Prod Rep 10:1074–1098
3. Conti E, Stachelhause T, Marahiel MA et al (1997) Structure basis for the activation of phenylalanine in the non-ribosomal biosynthesis of gramicidin S. EMBO J 16:4174–4183
4. Stachelhaus T, Mootz HD, Marahiel MA (1999) The specificity-conferring code of adenylation domains in nonribosomal peptide synthetases. Chem Biol 6:493–505
5. Mootz HD, Marahiel MA (1999) The tyrocidine biosynthesis operon of *Bacillus brevis*: complete nucleotide sequence and biochemical characterization of functional internal adenylation domains. J Bacteriol 179:6843–6850
6. Boettger D, Hertweck C (2013) Molecular diversity sculpted by fungal PKS-NRPS hybrids. ChemBioChem 14:28–42
7. Bumpus SB, Evans BS, Thomas PM et al (2009) A proteomics approach to discovering natural products and their biosynthetic pathways. Nat Biotechnol 10:951–956
8. Meier JL, Niessen S, Hoover HS et al (2009) An orthogonal active site identification system (OASIS) for proteomic profiling of natural product biosynthesis. ACS Chem Biol 4:948–957
9. Ishikawa F, Kakeya H (2014) Specific enrichment of nonribosomal peptide synthetase module by an affinity probe for adenylation domains. Bioorg Med Chem Lett 24:865–869
10. Kleinkauf H, Gevers W, Lipmann F (1969) Interrelation between activation and polymerization in gramicidin S biosynthesis. Proc Natl Acad Sci U S A 62:226–233
11. Lee SG, Lipmann F (1977) Isolation of amino acid activating subunit-pantetheine protein complexes: their role in chain elongation in tyrocidine synthesis. Proc Natl Acad Sci U S A 74:2343–2347
12. Finking R, Neumüller A, Solsbacher J et al (2003) Aminoacyl adenylate substrate analogues for the inhibition of adenylation domains of nonribosomal peptide synthetases. ChemBioChem 4:903–906
13. Qiao C, Gupte A, Boshoff HI et al (2007) 5′-O-[(N-Acyl)sulfamoyl]adenosines as antitubercular agents that inhibit MbtA: an adenylation enzymes required for siderophore biosynthesis of the mycobactins. J Med Chem 50:6080–6094
14. Ishikawa F, Haushalter RW, Burkart MD (2012) Dehydratase-specific probes for fatty acid and polyketide synthases. J Am Chem Soc 134:769–772
15. Meier JL, Mercer AC, Burkart MD (2008) Fluorescent profiling of modular biosynthetic enzymes by complementary metabolic and activity based probes. J Am Chem Soc 130:5443–5445
16. Andrew HD, Rohde RD, Millward SW et al (2009) Iterative in situ click chemistry creates antibody-like protein-capture agents. Angew Chem Int Ed 48:4944–4948
17. Brodsky BH, Bois JD (2005) Oxaziridine-mediated catalytic hydroxylation of unactivated 3° C-H bonds using hydrogen peroxide. J Am Chem Soc 127:15391–15393
18. Matteo CC, Glade M, Tanaka A et al (1975) Microbiological studies on the formation of gramicidin S synthetases. Biotechnol Bioeng 17:129–142
19. Bradford MM (1976) A rapid and sensitive method for the quantitation of microgram quantities of protein utilizing the principle of protein-dye binding. Anal Biochem 72:248–254
20. Shevchenko A, Wilm M, Vorm O et al (1996) Mass spectrometric sequencing of proteins silver-stained polyacrylamide gels. Anal Chem 68:850–858
21. Gharahdaghi F, Weinberg CR, Meagher DA et al (1999) Mass spectrometric identification of proteins from silver-stained polyacrylamide gel: a method for the removal of silver ions to enhance sensitivity. Electrophoresis 20:601–605
22. Augenstein DC, Thrasher KD, Sinskey AJ et al (1974) Optimization in the recovery of a labile intracellular enzyme. Biotechnol Bioeng 16:1433–1447

Chapter 5

Colorimetric Detection of the Adenylation Activity in Nonribosomal Peptide Synthetases

Chitose Maruyama, Haruka Niikura, Masahiro Takakuwa, Hajime Katano, and Yoshimitsu Hamano

Abstract

Nonribosomal peptide synthetases (NRPSs) are multifunctional enzymes consisting of catalytic domains. The substrate specificities of adenylation (A) domains determine the amino-acid building blocks to be incorporated during nonribosomal peptide biosynthesis. The A-domains mediate ATP-dependent activation of amino-acid substrates as aminoacyl-O-AMP with pyrophosphate (PPi) release. Traditionally, the enzymatic activity of the A-domains has been measured by radioactive ATP–[^{32}P]-PPi exchange assays with the detection of ^{32}P-labeled ATP. Recently, we developed a colorimetric assay for the direct detection of PPi as a yellow 18-molybdopyrophosphate anion ($[(P_2O_7)Mo_{18}O_{54}]^{4-}$). $[(P_2O_7)Mo_{18}O_{54}]^{4-}$ was further reduced by ascorbic acid to give a more readily distinguishable blue coloration. Here we demonstrate the lab protocols for the colorimetric assay of PPi released in A-domain reactions.

Key words Nonribosomal peptide synthetase, Adenylation domain, Colorimetric assay, Poly anion, ATP–[^{32}P]-PPi exchange assay

1 Introduction

Nonribosomal peptides constitute a major class of secondary metabolites produced in microorganisms and are synthesized by nonribosomal peptide synthetases (NRPSs). Unlike post-ribosomal peptide synthesis, NRPSs can accept nonproteinogenic amino-acid building blocks as substrates, thereby offering greater structural diversity. NRPSs are multifunctional enzymes consisting of catalytic domains [1–3]. The amino-acid substrate is activated as an aminoacyl-O-AMP by an adenylation (A) domain and subsequently loaded onto the 4'-phosphopantetheine (4'-PP) arm of the adjacent thiolation (T) domain with AMP and pyrophosphate (PPi) releases, resulting in the formation of an aminoacyl-S-enzyme (Fig. 1a). A condensation (C) domain catalyzes a peptide-bond formation between two amino-acid substrates activated as the aminoacyl-S-enzyme. The substrate specificities of A-domains determine

Bradley S. Evans (ed.), *Nonribosomal Peptide and Polyketide Biosynthesis: Methods and Protocols*, Methods in Molecular Biology, vol. 1401, DOI 10.1007/978-1-4939-3375-4_5, © Springer Science+Business Media New York 2016

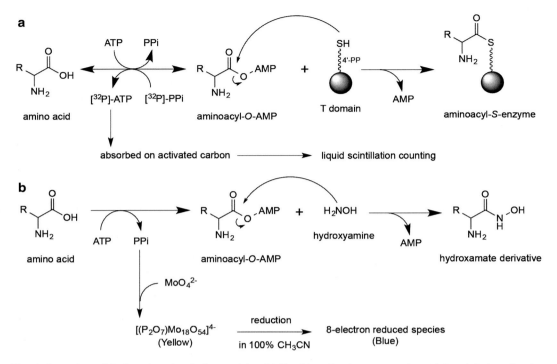

Fig. 1 Detection of A-domain adenylation activity. (**a**) Traditionally, the enzymatic activity of A-domains has been measured by radioactive ATP–[^{32}P]-PPi exchange assays with the detection of ^{32}P-labeled ATP. (**b**) Addition of hydroxylamine into the A-domain reaction mixture enhances PPi release. PPi is directly detected as a yellow 18-molybdopyrophosphate anion ([(P$_2$O$_7$)Mo$_{18}$O$_{54}$]$^{4-}$). [(P$_2$O$_7$)Mo$_{18}$O$_{54}$]$^{4-}$ is further reduced by ascorbic acid to give a more readily distinguishable blue coloration

the amino-acid building blocks to be incorporated during nonribosomal peptide biosynthesis. Traditionally, the enzymatic activity of A-domains has been measured by radioactive ATP–[^{32}P]-PPi exchange assays through the detection of ^{32}P-labeled ATP produced by a reversible reaction of the A-domain [4, 5]. In 2009, McQuade et al. reported a nonradioactive high-throughput assay for the screening and characterization of A-domains [6]. Their assay uses malachite green to measure orthophosphate (Pi) concentrations after degradation by inorganic pyrophosphatase of the PPi released during aminoacyl-*O*-AMP formation. However, this method seems to be inadequate for A-domains that have high catalytic rates of the reverse reaction, because the released PPi should be immediately converted to ATP, particularly in the reaction mixture without a T-domain (Fig. 1a).

Recently, we developed a colorimetric assay for the direct detection of PPi as a yellow 18-molybdopyrophosphate anion ([(P$_2$O$_7$)Mo$_{18}$O$_{54}$]$^{4-}$) [7, 8]. [(P$_2$O$_7$)Mo$_{18}$O$_{54}$]$^{4-}$ was further reduced by ascorbic acid to give an eight-electron reduced species, which shows a more readily distinguishable blue coloration. Using this assay, the enzymatic activity was successfully measured in acetyl-CoA synthetase that forms AMP + PPi. However, we were unable to detect the enzymatic activities in A-domains, probably due to these enzymes' PPi-consuming reverse reaction. Although the addition of a T-domain to the reaction mixture should facilitate PPi release, a large amount of T-domain is needed to achieve this. To address this problem, we explored the use of nucleophilic reagents instead of T-domains. Our recent study demonstrated that the aminoacyl-*O*-AMPs produced by A-domains are converted to hydroxamate derivatives in an enzyme reaction containing hydroxylamine [8]. In addition, the resulting PPi was detected by our colorimetric assay (Fig. 1b). Here we demonstrate the lab protocols for the colorimetric assay of A-domains.

2 Materials

Prepare all solutions using analytical-grade reagents. Prepare and store all reagents at room temperature (unless otherwise described).

2.1 A Domain Reaction Mixture

1. Tris buffer: 1 M Tris (pH 9.0) (*see* **Note 1**) in water.
2. Magnesium solution: 100 mM $MgCl_2$ in water.
3. ATP (pH 7.0): Weigh 551 mg adenosine 5′-triphosphate (ATP) disodium salt anhydrate and transfer it to a test tube. Add water to a volume of 7 mL and mix. Adjust pH with NaOH. Make up to 10 mL with water. Store in suitable aliquots at –70 °C. Final concentration 100 mM.
4. Hydroxylamine (pH 7.2): Weigh 1.4 g hydroxylamine hydrochloride and transfer it to a glass beaker. Add water to a volume of 80 mL and mix. Adjust pH with KOH. Make up to 100 mL with water. Store at 4 °C (*see* **Note 2**). Final concentration 200 mM.
5. Amino-acid substrates: 20–100 mM amino-acid solutions are prepared and used for A-domain reaction mixtures (*see* **Note 3**).
6. A-domains (enzymes): The recombinant enzyme of an A-domain, which is purified to homogeneity by affinity chromatography, is required (*see* **Note 4**).

2.2 Colorimetric Assay of PPi

1. Concentrated hydrochloric acid: 5 M HCl in water.
2. Acetonitrile (anhydrous, 99.8 %).
3. 1 M Na_2MoO_4: Weigh 24.2 g disodium molybdate(VI) dihydrate and transfer it to a glass beaker. Make up to 100 mL with water.
4. Mo(VI) solution: Mix 6 mL of concentrated hydrochloric acid and 30 mL of acetonitrile. Add water to a volume of 45 mL. Add 1 mL of 1 M Na_2MoO_4 slowly to the solution while

stirring. Make up to 50 mL with water to give a working solution containing 20 mM Na_2MoO_4, 0.6 M HCl, and 60 % acetonitrile. This solution should be freshly prepared for use.

5. 50 mM bis(triphenylphosphoranylidene)ammonium chloride (BTPPACl): Weigh 1.44 g BTPPACl and transfer it to a glass beaker. Add acetonitrile (not water) to a volume of 50 mL.

6. Ascorbic acid solution: Mix 2 mL of 5 M HCl and 3 mL of acetonitrile. Add 0.44 g L-ascorbic acid to the solution. This solution should be freshly prepared for use.

3   Methods

3.1   A Domain Reaction

1. In a 1.5-mL microfuge tube, mix the solution components using the values given in Table 1 (*see* **Note 5**). Mix in the order shown.

2. Incubate the reaction mixture for 10–60 min at 30 °C.

3.2   PPi Detection by Colorimetric Assay

1. Transfer 50 μL of the reaction mixture to a fresh 1.5-mL microfuge tube containing 500 μL of Mo(VI) solution. Mix thoroughly and incubate at room temperature for 2–5 min (*see* **Note 6**).

2. Add 10 μL of 50 mM BTPPACl, mix thoroughly, and incubate at room temperature for 5 min (*see* **Note 7**).

3. Centrifuge the resulting cloudy solution at 20,000×*g* for 15 min and remove all liquid.

4. Dissolve the precipitant in 100 μL of acetonitrile by mixing vigorously.

Table 1
Solutions for preparing A domain assay mixture

Solution component	Component volume (μL)	Final concentration	Reference
H_2O	47		
Tris Buffer	5	50 mM	
ATP (pH 7.0)	5	5 mM	
Magnesium solution	5	5 mM	
Hydroxylamine (pH 7.2)	16	32 mM	Note 5
500 μg/mL A domain	20	100 μg/mL	
Amino-acid substrate	2	2 mM	Note 5
Reaction volume	100		

5. Spin briefly to collect the contents at the bottom of the tube.

6. Transfer 100 μL of the solution to a 96-well plate and add 10 μL of ascorbic acid solution.

7. After mixing the solution by pipetting, incubate at room temperature for 10 min.

8. Measure the absorbance of the well at 620 nm on a plate reader (Fig. 2).

3.3 Standard Curves of PPi Concentration

1. Instead of the enzyme reaction mixture, 50 μL of 0–1000 μM $Na_4P_2O_7$ solution is used for the PPi colorimetric assay.

2. The PPi colorimetric assay is carried out by the same method described above (**steps 1–8** in Subheading 3.2).

3. Obtain a standard curve from the results (Fig. 3).

3.4 Determination of A-Domain–Specific Activity

1. In a 1.5-mL microfuge tube, mix the solution components (except for the amino-acid substrate) using the values given in Table 1. Mix in the order shown (*see* **Note 8**).

2. Incubate the mixture for 3 min at 30 °C (preincubation).

3. Add 2 μL of the amino-acid solution or water (control) to start the enzyme reaction (*see* **Note 9**).

4. Incubate the mixture for 5–30 min at 30 °C. Terminate the enzyme reaction by adding 500 μL of Mo(VI) solution.

5. Carry out the PPi colorimetric assay using the method described above (**steps 1–8** in Subheading 3.2). Determine the concentration of PPi using the PPi standard curve (Subheading 3.3).

6. Determine the specific activity based on the PPi production (Fig. 4).

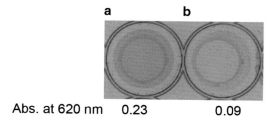

Abs. at 620 nm   0.23   0.09

Fig. 2 Colorimetric assay of PPi released in an A-domain reaction. The purified rORF 19 (100 μg/mL) is incubated with (**a**) and without (**b**) 2 mM β-lysine in a reaction mixture containing 32 mM hydroxylamine for 30 min at 30 °C (*see* **Notes 5** and **8**). The PPi released in the reaction is detected by the colorimetric assay

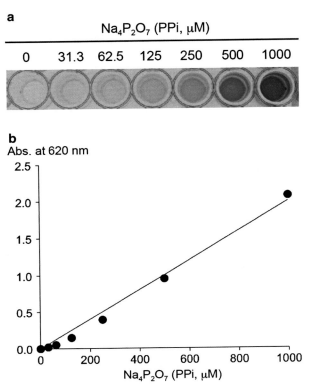

Fig. 3 Standard curve of PPi obtained from various concentrations of $Na_4P_2O_7$. Instead of the enzyme reaction mixture, 50 μL of 0–1000 μM $Na_4P_2O_7$ solution is used for the PPi colorimetric assay (**a**). The absorbances of the samples are measured at 620 nm on a plate reader (**b**)

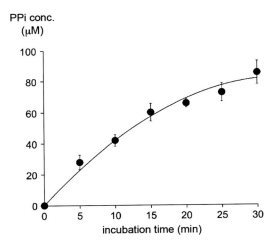

Fig. 4 Determination of A-domain-specific activity. The purified rORF 19 (100 μg/mL) is incubated with 2 mM β-lysine in a reaction mixture containing 32 mM hydroxylamine for 0–30 min at 30 °C. The reactions are terminated by the addition of Mo(VI) solution

4   Notes

1. Buffers and their pH should be optimized for A-domains. Tris(hydroxymethyl)aminomethane(Tris),3-morpholinopropanesulfonic acid (MOPS), and N-tris(hydroxymethyl) methyl-3-aminopropanesulfonic acid (TAPS) seem to give good results at pH 7–9. Phosphate buffer is not recommended, because it inhibits the activity of the A domains and also gives a high background in the following PPi assay.

2. Hydroxylamine solution can be stored for up to 3 days at 4 °C. However, the pH should be checked before use.

3. When an amino-acid substrate dissolved in a diluted HCl or NaOH is used for the enzyme reaction, the pH of the reaction mixture should be checked. Water-insoluble amino acids can be dissolved in dimethyl sulfoxide (DMSO).

4. Cell-free extract will give a high background in the colorimetric assay of PPi, probably due to the hydrolysis of the ATP by phosphatases. Therefore, a highly purified A-domain should be used for the enzyme reaction. His-tagged recombinant enzymes give good results in our laboratory. NaCl and imidazole, which are used for the purification steps in Ni-affinity chromatography, do not interfere with the colorimetric assay of PPi.

5. In the enzyme reaction, the concentration of hydroxylamine should be optimized. 10–60 mM hydroxylamine gives good results in a large number of A domains. For example, 32 mM is the optimum concentration for a recombinant enzyme of ORF 19 (rORF 19), which is a stand-alone A-domain involved in the biosynthesis of streptothricin antibiotics [9]. In addition, 50 mM Tris–HCl (pH 9.0) and 2 mM β-lysine are used for the rORF 19 reaction as the buffer and substrate, respectively. As a control reaction, the enzyme reaction should be performed without an amino-acid substrate (Fig. 2).

6. The addition of Mo(VI) solution terminates the enzyme reaction and forms a yellow 18-molybdopyrophosphate anion ($[(P_2O_7)Mo_{18}O_{54}]^{4-}$). Prolonged incubation (more than 5 min) increases the background in the PPi colorimetric assay.

7. The $[(P_2O_7)Mo_{18}O_{54}]^{4-}$ anion is precipitated with the BTPPA$^+$ cation.

8. The enzyme concentration should be optimized. For example, in rORF 19 (see **Note 5**), 100 µg/mL enzyme is good for determining the specific activity (Fig. 4).

9. The substrate concentration should be optimized.

Acknowledgments

This work was supported in part by KAKENHI (25108720), the Asahi Glass Foundation, and the Japan Foundation for Applied Enzymology.

References

1. Marahiel MA, Stachelhaus T, Mootz HD (1997) Modular peptide synthetases involved in nonribosomal peptide synthesis. Chem Rev 97:2651–2674
2. Mootz HD, Schwarzer D, Marahiel MA (2002) Ways of assembling complex natural products on modular nonribosomal peptide synthetases. ChemBioChem 3:490–504
3. Schwarzer D, Finking R, Marahiel MA (2003) Nonribosomal peptides: from genes to products. Nat Prod Rep 20:275–287
4. Bryce GF, Brot N (1972) Studies on the enzymatic synthesis of the cyclic trimer of 2,3-dihydroxy-N-benzoyl-L-serine in Escherichia coli. Biochemistry 11:1708–1715
5. Rusnak F, Faraci WS, Walsh CT (1989) Subcloning, expression, and purification of the enterobactin biosynthetic enzyme 2,3-dihydroxybenzoate-AMP ligase: demonstration of enzyme-bound (2,3-dihydroxy-benzoyl)adenylate product. Biochemistry 28:6827–6835
6. McQuade TJ, Shallop AD, Sheoran A et al (2009) A nonradioactive high-throughput assay for screening and characterization of adenylation domains for nonribosomal peptide combinatorial biosynthesis. Anal Biochem 386:244–250
7. Katano H, Tanaka R, Maruyama C et al (2012) Assay of enzymes forming AMP + PPi by the pyrophosphate determination based on the formation of 18-molybdopyrophosphate. Anal Biochem 421:308–312
8. Katano H, Watanabe H, Takakuwa M et al (2013) Colorimetric determination of pyrophosphate anion and its application to adenylation enzyme assay. Anal Sci 29:1095–1098
9. Maruyama C, Toyoda J, Kato Y et al (2012) A stand-alone adenylation domain forms amide bonds in streptothricin biosynthesis. Nat Chem Biol 8:791–797

Chapter 6

Facile Synthetic Access to Glycopeptide Antibiotic Precursor Peptides for the Investigation of Cytochrome P450 Action in Glycopeptide Antibiotic Biosynthesis

Clara Brieke, Veronika Kratzig, Madeleine Peschke, and Max J. Cryle

Abstract

The glycopeptide antibiotics are an important class of complex, medically relevant peptide natural products. Given that the production of such compounds all stems from in vivo biosynthesis, understanding the mechanisms of the natural assembly system—consisting of a nonribosomal-peptide synthetase machinery (NRPS) and further modifying enzymes—is vital. In order to address the later steps of peptide biosynthesis, which are catalyzed by Cytochrome P450s that interact with the peptide-producing nonribosomal peptide synthetase, peptide substrates are required: these peptides must also be in a form that can be conjugated to carrier protein domains of the nonribosomal peptide synthetase machinery. Here, we describe a practical and effective route for the solid phase synthesis of glycopeptide antibiotic precursor peptides as their Coenzyme A (CoA) conjugates to allow enzymatic conjugation to carrier protein domains. This route utilizes Fmoc-chemistry suppressing epimerization of racemization-prone aryl glycine derivatives and affords high yields and excellent purities, requiring only a single step of simple solid phase extraction for chromatographic purification. With this, comprehensive investigations of interactions between various NRPS-bound substrates and Cytochrome P450s are enabled.

Key words Glycopeptide antibiotics, Solid phase peptide synthesis, Coenzyme A, Bio-conjugation, Nonribosomal peptide synthetase, Cytochrome P450

1 Introduction

Nonribosomal peptide synthetases (NRPSs) are large, multi-enzyme complexes acting as modular peptide assembly lines that are involved in the biosynthesis of numerous peptide natural products independent of the ribosome [1]. NRPSs are organized into functional modules, in which each module enables the incorporation of one amino acid and consists of several domains: these domains are responsible for the selection/activation of amino acids and the elongation of the growing peptide chain (*see* Fig. 1). During all steps following amino acid activation the peptide is conjugated as a thioester to peptidyl carrier protein domains (PCP- or

Bradley S. Evans (ed.), *Nonribosomal Peptide and Polyketide Biosynthesis: Methods and Protocols*, Methods in Molecular Biology, vol. 1401, DOI 10.1007/978-1-4939-3375-4_6, © Springer Science+Business Media New York 2016

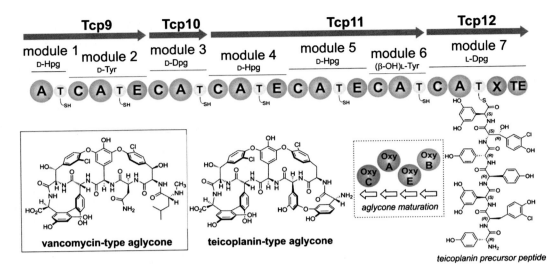

Fig. 1 Schematic pathway of teicoplanin biosynthesis by nonribosomal peptide synthesis and structures of vancomycin and teicoplanin aglycones (type I- and type IV-GPAs). A: adenylation; C: condensation; E: epimerization; T: thiolation/peptidyl carrier protein; TE: thioesterase; X: oxygenase recruitment domain

thiolation (T)-domains) via a posttranslationally introduced 4′-phosphopantetheine linker. Thus, assembly of a peptide requires multiple interactions between catalytic domains within the NRPS-machinery and is further complemented by additional interactions with enzymes acting in *trans*, which are often focused on the final peptide maturation process.

A prominent class of NRPS-metabolites that require extensive external modification of the NRPS-bound peptide is the glycopeptide antibiotics (GPAs) such as vancomycin and teicoplanin [2, 3]: a group of peptide natural products that are clinically relevant for the effective treatment of gram-positive bacterial infections that includes multidrug-resistant strains. These compounds possess a complex structure consisting of a heptapeptide backbone with a high content of non-proteinogenic amino acids that is modified by glycosylation and further decorations. Moreover, the amino acid side chains are highly cross-linked through several phenolic and aryl cross-links leading to the rigid 3D structure of GPAs that is crucial for their antibiotic activity [2]. During biosynthesis, these cross-links are introduced by several Cytochrome P450 enzymes (P450s) that act on the peptide while it remains bound to the NRPS machinery (Fig. 1).

Cytochromes P450 (P450s) are heme-dependent monooxygenases with a highly conserved overall protein fold found in prokaryotes and eukaryotes. These enzymes are capable of catalyzing a broad range of oxidative transformations, often in a stereo- as well as regioselective manner. A special group of P450s, which is involved in the biosynthesis of secondary metabolites, oxidizes their substrates that include amino acids, fatty acids or peptides only when these are bound to a carrier protein domain from an NRPS, polyketide, or fatty acid synthase [3, 4]. The interaction between the P450 and the NRPS systems in GPA biosynthesis is transient in nature and thus challenging to investigate [5].

In the biosynthesis of GPAs, three (in the case of vancomycin, type I) or four (in the case of teicoplanin, type IV) P450s named Oxys are necessary for aglycone maturation. Several of these enzymes have been characterized and the order of oxidation and individual catalytic roles have been assigned for type I and type IV GPAs [3, 6–10]. Recently, Haslinger et al. revealed that a special NRPS-domain in the final module of the teicoplanin NRPS machinery—the so-called X-domain—is crucial for the recruitment of these Oxy enzymes to the NRPS, thus providing new insight into the mode of interaction of these enzymes in teicoplanin biosynthesis [11]. In this study it was also shown that this recruitment event is essential for in vitro activity of the both P450s catalyzing the C-O-D (OxyB$_{tei}$) and D-O-E (OxyA$_{tei}$) cyclization of a teicoplanin precursor peptide bound to a PCP-X di-domain construct. In contrast to this, the C-O-D ring catalyzing P450 from the vancomycin gene cluster (OxyB$_{van}$) exhibits not only in vitro oxidation activity on a precursor peptide bound to a single PCP domain in the absence of an X-domain [12, 13], but it also accepts peptide/PCP-pairs from alternate GPA gene clusters, thus showing a degree of promiscuity regarding the peptide substrate as well as the PCP domain [14].

The reasons for these differences between highly related P450 enzymes remain to be clarified. However, in order to perform such studies the straightforward access to a variety of substrates—linear precursor peptides covalently linked to PCP domains—is crucial: such a method is discussed in this chapter.

The production of such peptide–PCP conjugates requires the initial synthesis of the linear precursor peptide and, in a second step, the conjugation of the peptide to the desired PCP domain. Tethering a cargo to *apo*-PCP domains can efficiently be performed by using the promiscuous phosphopantetheinyl transferase Sfp [15, 16]. This enzyme transfers the 4′-phosphopantetheinyl moiety of Coenzyme A (CoA) onto a conserved serine residue of PCPs while tolerating a wide range of small molecules attached to the CoA molecule via thioester bonds. The production of linear precursor peptides is practically performed using solid phase peptide synthesis methodology, which allows the assembly of peptides also containing unnatural or D-amino acids very efficiently and in high yields. For GPA precursor peptides, however, such syntheses are challenging due to the high sensitivity to epimerization of the arylglycine derivatives 4-hydroxyphenylglycine (Hpg) and 3,5-dihydroxyphenylglycine (Dpg) found in the aglycones of GPAs [17]: three such residues appear in the case of vancomycin and teicoplanin even includes five such residues. Due to this sensitivity, special care has to be taken to prepare such peptides in good yields and purities. The Robinson group has developed a mild solid phase peptide synthesis (SPPS) route based on Alloc-chemistry to access vancomycin precursor peptides [18, 19]. As the respective Alloc-protected amino acid buildings blocks are not commercially available, this route required extensive effort for the synthesis and optimization of the SPPS conditions, which limits the range of accessible substrate peptides possibly suited for biochemical investigations.

Driven to overcome these obstacles our group decided to perform a careful investigation of conditions suitable for a standard SPPS strategy based on Fmoc-chemistry, which has the advantage of allowing the utilization of the broad range of commercially available reagents for such syntheses. By optimizing the conditions for amino acid coupling and for cleavage of the temporary Fmoc-protecting group on the amine functional group, racemization of arylglycine residues could be suppressed on the solid phase during synthesis [20].

This enabled us to move forward and establish a rapid route to the respective peptide-CoA conjugates (*see* Fig. 2) appropriate for the enzymatic carrier protein loading with a significant reduced purification workload during the entire synthesis. The synthesis strategy we adopted is based on the SPPS of precursor peptides using commercially available Dawson resin [21]: this resin can be activated to release the synthesized peptide directly as an activated thioester, which in a subsequent step can be converted with CoA to the respective peptide-CoA conjugate. By optimizing reaction conditions in each step, purification is reduced to a single step employing solid-phase extraction to isolate these peptide-CoA

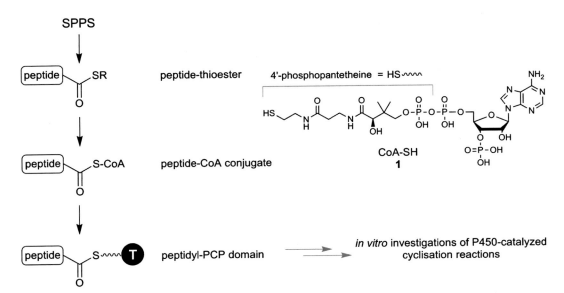

Fig. 2 Synthetic outline affording peptidyl-PCP thioester conjugates for in vitro investigations of Cytochrome P450 catalyzed cyclization reactions in the biosynthesis of GPAs

conjugates [14]. As not only GPAs but also other NRPS-derived natural products such as feglymycin [22] or β-lactam antibiotics [23] contain arylglycine derived residues, we believe that this methodology could be of general relevance for the exploration of other biosynthetic pathways.

2 Materials

2.1 Synthesis of Protected Amino Acids

1. Acetonitrile (MeCN).
2. Sodium carbonate solution: 10 % sodium carbonate (Na$_2$CO$_3$)-solution in Milli-Q water.
3. Fmoc- N-hydroxysuccinimide   ester   (Fmoc-OSu, Sigma-Aldrich).
4. (d)-4-hydroxyphenylglycine/(l)-4-hydroxyphenylglycine (Sigma-Aldrich).
5. Hydrochloric acid: 2 M aqueous hydrochloric acid.
6. Ethyl acetate.
7. Hexane.
8. Brine: saturated aqueous NaCl; prepare a solution in water by using an excess of NaCl (>360 g/L) in order to guarantee saturation of solution.
9. Sodium sulfate (Na$_2$SO$_4$).
10. 1,4-Dioxane.
11. Sodium hydroxide solution: 1 M aqueous sodium hydroxide.
12. Di-*tert*-butyl dicarbonate (Di-Boc, Sigma-Aldrich).
13. Sodium hydrogen carbonate (NaHCO$_3$).
14. Concentrated phosphoric acid (H$_3$PO$_4$, 85 %).
15. Acetone.

2.2 Solid Phase Peptide Synthesis (SPPS)

1. Commercially available amino acid building blocks: Fmoc-(l)-Tyr(tBu)-OH, Fmoc-(d)-Tyr(tBu)-OH, Fmoc-(l)-Asn(Trt)-OH, Boc-(d)-Leu-OH (Merck Novabiochem).
2. Dawson Dbz AM resin (Merck Novabiochem).
3. (1-cyano-2-ethoxy-2-oxoethylidenaminooxy)dimethylamino-morpholino-carbenium hexafluorophosphate, COMU (Merck Novabiochem).
4. Triethylamine (NEt$_3$, >99 %).
5. 1,8-Diazabicyclo[5.4.0]undec-7-ene (DBU, >99 %).
6. N,N-Dimethylformamide (DMF, >99 %).
7. Dichloromethane (DCM, reagent grade).

2.3   Resin Activation and Peptide Displacement Using MPAA

1. *p*-nitrophenylchloroformate (pNPCF) solution in DCM (50 mM), immediately prepared before use.

2. *N,N*-diisopropylethylamine (DIPEA) solution in *N,N*-dimethylformamide (86 mM), immediately prepared before use.

3. 4-mercaptophenylacetic acid (MPAA), tri-*n*-butylphosphine (PBu$_3$, 97 %, Sigma-Aldrich), *N,N*-dimethylformamide (DMF, anhydrous).

4. TFA cleavage mixture: 95 % trifluoroacetic acid (TFA), 2.5 % triisopropylsilane, 2.5 % water, ice-cold diethyl ether.

2.4   Activated Peptide Thioester Displacement with CoA

1. Reaction buffer: 50 mM potassium phosphate buffer pH 8.3 water-MeCN 2:1.

2. TCEP stock solution: 500 mM TCEP in Milli-Q water pH 7 (the solution was stored in aliquots at -24 °C).

3. Coenzyme A (CoA) (Affymetrix).

4. Strata-X solid phase extraction columns (200 mg resin/3 mL tubes) (Phenomenex).

5. SPE equilibration solvent: 50 mM KPi pH 7.0.

6. SPE wash solvent: methanol, 25 % solution in Milli-Q water.

7. SPE elution solvent: 35 % MeCN solution in Milli-Q water.

2.5   PCP Loading

1. Loading buffer: 1 M Hepes buffer pH 7.0, 5 M NaCl, 1 M MgCl$_2$.

2. Peptide-CoA conjugate (*see* Subheading 3.4).

3. Expressed PCP-domain [11, 14, 24].

4. 4′-phosphopantetheinyl-transferase Sfp R4-4 mutant (Sfp) [25].

5. Centrifugal filters with a molecular weight cutoff MW = 3000 (Amicon).

6. Wash buffer: 50 mM Hepes pH 7.0, 50 mM NaCl.

2.6   OxyB-Catalyzed Turnover and Turnover Workup

1. NAD(P)H stock solution: 50 mM NADH or NADPH in Milli-Q water, make freshly every time before use.

2. Glucose oxidase stock solution: 4 mg/mL in Milli-Q water, can be stored for 1 week at 4 °C.

3. D-glucose stock solution: 20 % in Milli-Q water, stored at 4 °C.

4. Reaction buffer: 50 mM Hepes, 50 mM NaCl pH 7.0.

5. PCP loading reaction: 60 μM *peptidyl*-PCP in 50 mM Hepes, 50 mM NaCl pH 7.0 (*see* Subheading 3.5).

6. Redox system: ferredoxin HaPuX and corresponding ferredoxin reductase HaPuR (*see* **Note 28**) [26].

7. NADH-regeneration system: glucose oxidase (0.033 mg/mL) and 0.33 % *w/v* glucose.

8. Cytochrome P450 OxyB$_{van}$ [27].

9. SPE-columns: Strata-X 33u polymeric reversed phase solid phase extraction columns 30 mg/mL (Phenomenex).

10. SPE elution solvent: methanol including 0.1 % formic acid.

11. SPE wash solvent: 5 % MeCN in Milli-Q water, 50 % MeCN in Milli-Q water.

12. 50 % Methylhydrazine (98 %, Sigma-Aldrich) in Milli-Q water.

13. Formic acid.

14. Syringe filters, PVDV, pore size 0.45 µM.

2.7 Equipment

1. Automated solid phase peptide synthesizer (Protein Technologies).

2. SpeedVac concentrator (Eppendorf).

3. Freeze-dryer (Christ).

4. High performance liquid chromatography (HPLC) system (Shimadzu) equipped with analytical C18 columns (Waters XBridge™ BEH 300 Prep C18 column, particle size: 5 and 10 µm, 4.6 × 250 mm).

5. Mass spectrometer (Shimadzu).

6. Nuclear magnetic resonance (NMR) spectrometer (Bruker).

7. Thermomixer (Eppendorf).

8. Amicon Ultra-0.5 mL Centrifugal Filters (Merck Millipore).

9. Centrifuge (Eppendorf).

3   Methods

3.1 Synthesis of Protected Amino Acids (Fig. 3)

1. Dissolve (d)-4-hydroxyphenylglycine or (l)-4-hydroxyphenyl-glycine 2 (5.00 g, 29.9 mmol) in a 1:1 solution of MeCN/sodium carbonate solution (10 %, 100 mL) with stirring at room temperature. Add Fmoc-OSu (10.1 g, 29.9 mmol) in portions over 30 min to this solution (*see* **Note 1**). Leave the reaction mixture stirring overnight, and then dilute it with water (100 mL). Extract this solution two times with diethyl ether (100 mL) and separate the organic layers. Acidify the aqueous phase to pH 3 using hydrochloric acid solution (2 M) and again extract the aqueous phase with ethyl acetate (3 × 100 mL). Extract the combined ethyl acetate organic layers with brine (2 × 100 mL), separate the organic layer and dry it over Na$_2$SO$_4$. Remove the solid by filtration and concentrate the filtrate in vacuo. Recrystallize the crude product in a mixture of ethyl acetate and hexane (*see* **Note 2**). For proper crystallization, keep the flask at 4 °C overnight, then separate the

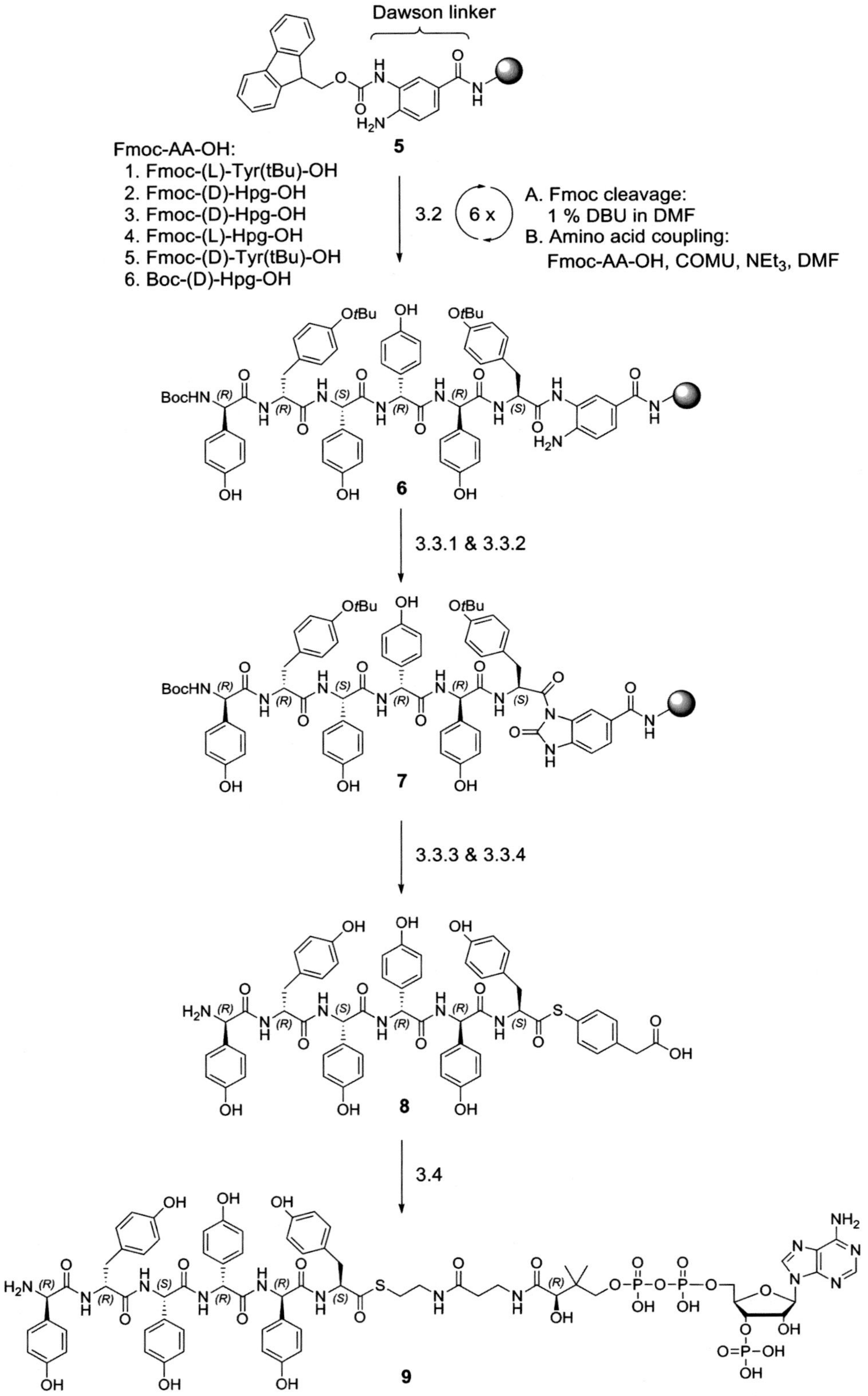

Fig. 3 Synthesis of the protected arylglycine building blocks needed for SPPS. The *numbering* refers to the corresponding paragraphs in the text

precipitated solid and dry in vacuo to obtain Fmoc-Hpg-OH **3** as a white solid (9.27 g, 23.8 mmol, 80 % yield). ^{1}H-NMR (300 MHz, DMSO-d$_6$): $\delta = 9.49$ (bs, 1H, COOH), 8.04 (d, 1H, NH), 7.89 (d, 2H, H4-, H5-Fmoc), 7.76 (d, 2H, H1-, H8-Fmoc), 7.44–7.38 (m, 2H, H3-, H6-Fmoc), 7.36–7.27 (m, 2H, H2-, H7-Fmoc), 7.21 (d, 2H, $^3J = 8.5$ Hz, Har-2,6), 6.74 (d, 2H, H$_{ar}$-3,5), 5.02 (d, 1H, Hα), 4.32–4.18 (m, 3H, H9-Fmoc, CH$_2$-Fmoc) ppm.

2. To a mixture of (d)-4-hydroxyphenylglycine **2** (2.00 g, 12.0 mmol) in 2:1 dioxane/water (30 mL), stirred on an ice bath, add sodium hydroxide solution (1 M, 10 mL) and 10 min later Di-Boc (3.9 g, 18.0 mmol) and sodium bicarbonate (1.01 g, 12.0 mmol). Leave the solution stirring overnight, allowing the mixture to warm up to room temperature. Remove the organic solvent by concentrating the solution in vacuo. Add ethyl acetate (40 mL) to the remaining mixture and acidify the organic layer to pH 3 with concentrated phosphoric acid. Separate the layers and extract the aqueous layer with ethyl acetate (2×40 mL). Wash the combined organic layers with brine (2×50 mL) and dry them over Na$_2$SO$_4$. Remove the solid by filtration and concentrate the filtrate in vacuo. Triturate the remaining oil with hexane (6×30 mL) until the oil becomes a paste (*see* **Note 3**). Dissolve this in acetone (100 mL) and concentrate the solution in vacuo, which affords Boc-(d)-Hpg-OH **4** (2.3 g, 8.6 mmol, 72 %) as an easily powdered white foam. ^{1}H-NMR (300 MHz, DMSO-d$_6$): $\delta = 7.36$ (d, 1H, NH), 7.17 (d, 2H, H$_{ar}$), 6.70 (d, 2H, H$_{ar}$), 4.95 (d, 1H, Hα), 1.38 (s, 9H, C(CH$_3$)$_3$) ppm.

3.2 Solid Phase Peptide Synthesis (Fig. 4)

The solid phase peptide synthesis (SPPS) of a teicoplanin precursor peptide is performed using an automated peptide synthesizer (*see* **Note 4**) on a 50 µM scale using the following conditions:

1. Resin swelling: Shake the Dawson resin for 10 min in 3 mL DMF; repeat this procedure three times (*see* **Notes 5** and **6**).

2. Fmoc deprotection (A, Fig. 4): Treat the resin with 3 mL of 1 % DBU in DMF for 30 s, filter the cleavage solution and check photometrically for the Fmoc cleavage product ($\lambda = 301$ nm). Repeat this procedure until the absorbance stays constant (*see* **Note 7**); subsequently wash the resin with DMF (five times with 3 mL).

3. Amino acid activation and coupling (B, Fig. 4): Solubilize 4 eq. of amino acid and 4 eq. of COMU activator in 2 mL of a 0.1 M Et$_3$N solution in DMF (4 eq.), incubate for 2 min and then add this solution to the resin. Let this resin shake for 30 min at room temperature (*see* **Note 8**). Following this, remove the soluti on and wash the resin with DMF (5×3 mL).

4. Repeat **steps 2** and **3** of Subheading 3.2, for each amino acid.

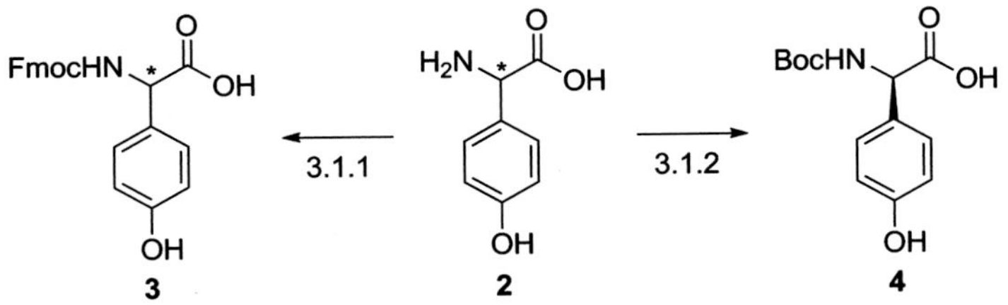

Fig. 4 Assembly of the teicoplanin-like precursor hexapeptide **6** on solid phase resin using Fmoc-chemistry followed by conversion into the activated MPAA thioester conjugate **8** and subsequently the respective CoA-conjugate **9**. *Numbering* refers to the corresponding paragraphs in the main text

3.3 Resin Activation and Peptide Displacement Using MPAA (Fig. 4)

1. Wash the resin thoroughly with DCM (4×3 mL) before adding *p*-nitrophenylchloroformate (50 mg, 0.25 mmol, 5 eq.) dissolved in 5 mL of DCM. Agitate the resin with this solution for 40 min, then remove the solution and wash the resin with DCM (four times) and DMF (three times).

2. Add a solution of DIPEA (44 μL, 5 eq.) in 3 mL of DMF to the resin and agitate the reaction for 15 min. Wash the resin with DMF (three times) and DCM (three times) and dry it under a stream of nitrogen (*see* **Notes 9** and **10**).

3. Transfer the resin into a 50 mL falcon tube, add 3 mL of dry DMF and agitate 15 min for swelling. Saturate the mixture with an inert gas (argon or nitrogen) by bubbling it through the solution for 10 min. Add 4-mercaptophenylacetic acid (84 mg, 0.5 mmol, 10 eq.) and PBu$_3$ (200 μL, 1.2 mmol) to the resin (*see* **Note 11**) and leave this mixture agitating in a closed tube for 24 h at room temperature. Separate the DMF solution from the resin; wash the resin beads with DMF (2 mL) and remove the solvent in vacuo (*see* **Note 12**).

4. Suspend the remaining residue in 5 mL of freshly prepared TFA cleavage mixture and incubate it under gentle agitation for 90 min (*see* **Note 13**). Concentrate the solution under a stream of nitrogen to ~1 mL and precipitate the peptide-MPAA thioester **8** via addition of ice cold diethyl ether (~10 mL). To ensure complete precipitation, store the mixture at −24 °C overnight. After collecting the peptide-MPAA thioester via centrifugation and decanting of the ether layer, dissolve the residue in 50 % acetonitrile–50 % water, analyze the raw mixture using analytical HPLC together (*see* **Note 14**) with mass spectrometry and lyophilize the raw peptide-MPAA thioester (*see* Fig. 5a).

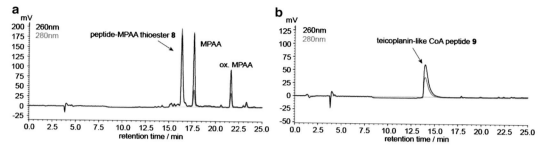

Fig. 5 Analytical reversed-phase HPLC trace of crude peptide-MPAA thioester **8** (**a**) and SPE-purified teicoplanin-like precursor CoA peptide **9** (**b**)

3.4 Activated Peptide Thioester Displacement with CoA (Fig. 4)

1. Dissolve the initial peptide-MPAA thioester **8** from 50 μmol scale synthesis in reaction buffer (~6 mL) (*see* **Note 15**) and 20 mM TCEP (*see* **Note 16**). Add CoA (35 mg, 0.875 equivalents) and let the reaction gently agitate at room temperature for 1–2 h, until—according to HPLC reaction monitoring—the peptide-MPAA thioester **8** is consumed (*see* **Note 17**).

2. In the meantime whilst the displacement reaction is underway prepare the SPE columns (four for one peptide-CoA synthesis 50 μmol scale) for purification: (1) Add methanol (5 mL) to each column for activation and let this run through the column under gravity, (2) Equilibrate the columns with SPE equilibration solvent (pH 7.0, 3 mL).

3. Dilute the reaction mixture with SPE equilibration solvent until the MeCN concentration is 5 % (*see* **Note 18**) and load it onto the equilibrated SPE columns (*see* **Note 19**). After complete loading wash the columns with SPE equilibration solvent (pH 7.0, 3 mL) and in a second wash step with SPE wash solvent in water (3 mL). To elute the peptide-CoA conjugate from the SPE columns add SPE elution solvent (pH 5–6, 3 mL) (*see* **Note 20**). Collect the elution fraction and lyophilize the isolated CoA peptide **9**. The peptides are sufficiently pure to allow for PCP loading (*see* **Note 21** and Figs. 5b and 6).

3.5 PCP-Loading (Fig. 5)

In order to study Cytochrome P450-catalyzed oxidative cross-linking reactions, PCP-domains from module 7 of several GPAs NRPS systems (balhimycin (type I), teicoplanin (type IV), complestatin (type V)) can be exploited (*see* **Note 23**).

1. Prior to the loading reaction, dilute the lyophilized peptide-CoA conjugates in Milli-Q water to a concentration of 1.5 mM (*see* **Note 24**).

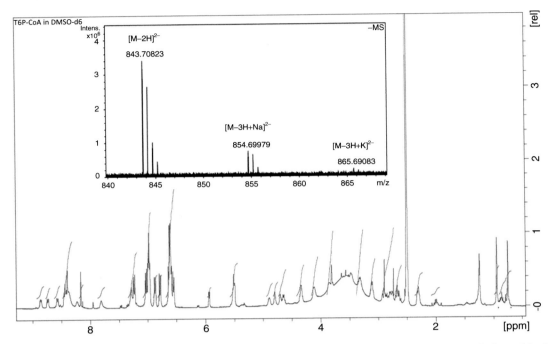

Fig. 6 ¹H-NMR in DMSO-d₆ (400 MHz) and HRMS spectra of HPLC-purified of teicoplanin-like CoA peptide **9** (ions are doubly charged and correspond to the protonated, sodiated, and potassiated species, *see* **Note 22**)

2. In an Eppendorf tube (*see* **Note 25**), prepare loading buffer (50 mM Hepes buffer, pH 7.0, 50 mM NaCl, 10 mM MgCl₂) (*see* **Note 26**) and add the expressed PCP7-construct to a concentration of 60 μM together with a fourfold excess of peptide-CoA conjugate [9].

3. Start the loading reaction by adjusting the Sfp concentration in the reaction to 6 μM (PCP:Sfp ratio 10:1). Gently mix the reaction mixture and incubate at 30 °C for 1 h (*see* **Note 27**).

4. Remove the excess of unloaded peptide by centrifugation at 4 °C using centrifugal filters with a molecular weight cutoff that allows the unloaded peptide-CoA conjugate pass through (MW 3000). Concentrate the loading reactions by centrifugation to approximately 100 μL, then add wash buffer (in a dilution of 1:5). Repeat this step three additional times. After concentrating the reaction for the fifth time, add wash buffer to reach the volume of the initial loading reaction (here: 175 μL). Keep the samples on ice until further use.

3.6 OxyB-Catalyzed Turnover and Turnover Workup (Fig. 7)

1. Prepare sufficient volumes of the NADH/NADPH, glucose oxidase and glucose stock solutions (*see* Subheading 2.6, **Note 28**).

2. Set up the turnover reaction in a total volume of 210 μL in reaction buffer. Thus, mix peptidyl-PCP from the PCP loading

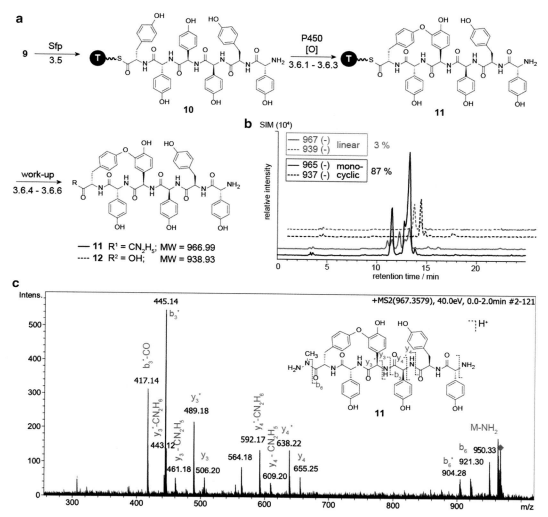

Fig. 7 (**a**) Enzymatic loading of teicoplanin-like CoA peptide **9** onto a PCP-domain (T) using the promiscuous transferase Sfp and oxidation of *peptidyl*-PCP **10** by the Cytochrome P450 OxyB$_{van}$ followed by workup. (**b**) HPLC-MS analysis of turnover reaction using single ion monitoring in negative mode; the masses correspond to the mono-cross-linked methylhydrazine (**11**, 965) and hydrolysis (**12**, 937) products and to the linear methylhydrazine (967) and hydrolysis (939) peptides. (**c**) ESI-MS/MS analysis of the C-*O*-D cross-linked product **11** from OxyB$_{van}$-catalyzed turnover. *Numbering* refers to the corresponding paragraphs in the main text

reaction (*see* Subheading 3.5, **Note 29**), ferredoxin, ferredoxin reductase (*see* **Note 28**) and OxyB in a micromolar ratio of 50:5:1:2. Add the NADH-regeneration system to the reaction. Start the reaction by adding 2 mM NADH/NADPH and incubate the mixture for 1 h at 30 °C with gentle shaking (*see* **Note 30**).

3. In the meantime prepare the solid phase extraction cartridge for workup of the turnover reaction: wash the cartridge with 1.5 mL of methanol followed by an equilibration step with water (1.5 mL). Do not allow the cartridge to dry.

4. Stop the turnover reaction by adding 30 μL of the methylhydrazine solution and incubate the mixture for 15 min at room
temperature.

5. After dilution and neutralization of the turnover mixture using
11 μL of formic acid (*see* **Note 31**), load the solution onto the
equilibrated SPE-column and allow the solution enter the column bed by gravity flow. Wash the column with a 5 % methanol solution (1 mL) and elute the peptides from the cartridge
with 500 μL methanol (incl. 0.1 % formic acid). Concentrate
the elution fraction in vacuo (*see* **Note 32**).

6. For HPLC-MS analysis, dilute the elution fraction to a final
volume of 100 μL using 50 % MeCN in water and filter the
sample. Analyze the samples by analytical reversed-phase
HPLC-MS using single ion monitoring (*see* **Note 33**).

4   Notes

1. Slow dissolution of the amino acid as well as Fmoc-OSu can
occur, in which case more MeCN can be added (up to 50 mL).
Overnight, the reaction mixture becomes a clear solution.

2. Dissolve the crude product in as little as possible volume of
ethyl acetate (~50 mL) and add hexane drop wise until a solid
precipitates and does not dissolve anymore with gentle
shaking.

3. Add hexane to the remaining oil and sonicate or mix this mixture (~5 min) for dissolving impurities into the hexane layer.
After separation of the phases, remove the hexane layer by
careful decanting or pipetting.

4. The solid phase peptide synthesis was performed on a Tribute
UV peptide synthesizer from Protein Technologies, but it can
also be performed manually.

5. Do not use a magnetic stir bar to mix the resin as this destroys
the beads.

6. It is mandatory that the resin is swollen well to guarantee efficient solvent access to the resin.

7. The Tribute UV peptide synthesizer features feedback monitoring allowing to measure the amount of Fmoc cleaved after
DBU treatment and to decide automatically if further cleavage
cycles are necessary for complete deprotection (typically three
to four cycles). Best results are obtained by incubating the
resin more often with a fresh DBU solution than incubating it
over a longer period.

8. Effective mixing of the resin is important to ensure proper
coupling.

9. Resin activation can also be performed on an automated peptide synthesizer.

10. The dried resin can be stored in a closed tube at −24 °C until further conversion.

11. PBu$_3$ is oxidized by oxygen and is therefore unstable at air; use syringes for transfer.

12. This can be done by filtering the resin through a glass pipette equipped with a piece of cotton wool. The filtrate can be collected in 2 mL tubes and concentrated in vacuo using a SpeedVac concentrator.

13. By adding the TFA cleavage mixture a white solid can be formed due to oxidation of MPAA. Its formation is reduced by proper saturating the reaction mixture with inert gas and by adding PBu$_3$ (*see* Subheading 3.3, **step 3**). This solid should be kept in the following reaction steps.

14. For analytical HPLC a Waters XBridge BEH300 Prep C18 column (10 μm, 4.6×250 mm) was used. The solvents used were water + 0.1 % formic acid (solvent A) and HPLC-grade acetonitrile + 0.1 % formic acid (solvent B); gradient: 0–2 min 95 % solvent A, 2–25 min up to 25 % solvent B, flow rate 1 mL/min.

15. Check the pH of the solution and adjust it carefully to pH 8.3 using 1 M NaOH if necessary.

16. TCEP is added to reduce oxidation of excess MPAA in the reaction mixture. Oxidized MPAA has proven to be hard to separate during SPE purification and to have a negative influence the reaction yield due to solubility problems.

17. The pH can drop during the reaction slowing down the reaction and therefore needs to be checked. If necessary, readjust it to pH 8.3. For HPLC *see* **Note 14**.

18. MeCN can interfere with the SPE purification: at excessively high MeCN concentrations the product may not bind to the solid phase, which results in reduced yields.

19. For proper product adsorption to the solid phase use gravity flow in the loading step. For washing and elution steps some pressure can be generated by using a pipette bulb.

20. The amount of MeCN varies depending on the polarity of the peptide: for vancomycin-like precursor peptides 25 % MeCN has proven to be sufficient for peptide elution, for teicoplanin-like precursor peptides a higher content of acetonitrile is needed.

21. After SPE purification no other CoA species, which could interfere with PCP loading, can be detected in the samples according to HPLC-MS analysis (*see* **Note 14**).

22. ^{1}H-NMR (400 MHz, 1H-1H-COSY, DMSO-d$_6$): $\delta = 8.86$ (d, 1H, $^3J = 7.8$ Hz, Tyr$_6^{NH}$), 8.74 (d, 1H, $^3J = 8.3$ Hz, Hpg^{NH}), 8.57 (d, 1H, $^3J = 7.5$ Hz, Hpg^{NH}), 8.47–8.32 (m, 5H, {8.44} Tyr$_2^{NH}$, {8.40} H8-CoA, Hpg^{NH}, NH$_2$), 8.23 (m, 1H, N**H**CH$_2$CH$_2$S-CoA), 8.17 (s, 1H, H2-CoA), 7.81 (m, 1H, N**H**CH$_2$CH$_2$CO-CoA), 7.34–7.20 (m, 4H, H$_{ar}$), 7.07–6.94 (m, 7H, H$_{ar}$), 6.87 (m, 2H, H$_{ar}$), 6.79 (m, 2H, H$_{ar}$), 6.66–6.53 (m, 9H, H$_{ar}$), 5.94 (d, 1H, $^3J = 5.3$ Hz, H1′-CoA), 5.56–5.48 (m, 3H, $3 \times$ Hpg$^\alpha$), 4.89 (m, 1H, H3′-CoA), 4.70 (m, 1H, H2′-CoA), 4.65 (m, Tyr$_2^\alpha$), 4.37–4.29 (m, 2H, H4′-CoA, Tyr$_6^\alpha$), 4.11 (m, 2H, H5′-CoA), 3.87–3.78 (m, under water peak, {3.85} OC**H$_2^a$**C(CH$_3$)$_2$-CoA, Hpg$_1^\alpha$), 3.59 (m, under water peak, C**H**(OH)CO-CoA), 3.46 (m, under water peak, OC**H$_2^b$**C(CH$_3$)$_2$-CoA), 3.35–3.28 (m, under water peak, NHC**H$_2$**CH$_2$CO-CoA), 3.15–3.07 (m, 2H, NHC**H$_2$**CH$_2$S-CoA), 2.94–2.62 (m, 6H, NHCH$_2$C**H$_2$**S-CoA, Tyr$_6\beta$, Tyr$_2\beta$), 2.34–2.26 (m, 2H, NHCH$_2$C**H$_2$**CO-CoA), 0.94, 0.75 (s, 2×3H, $2 \times$ gem-CH$_3$-CoA) ppm. Numbering of Coenzyme A accords to [28].

23. To increase the yield and in vitro-stability of isolated PCP7-domains it is recommended to express them as fusion proteins with thioredoxin or GB1 (IgG binding B1 domain) [14, 24, 29].

24. Always dilute the peptide-CoA conjugate freshly prior to each experiment due to instability of the peptide-CoA conjugate dissolved in water.

25. The total volume of a loading reaction for one turnover reaction (*see* Subheading 3.6) is 175 µL. For scaling up increase the total volume of the reaction, not the concentration of the components.

26. The peptide loading reaction is best carried out at pH 7.0 as higher pH leads to a faster hydrolysis of peptide-CoA minimizing the yield of the loading reaction.

27. The reaction conditions for the loading reaction described here might not be optimal for every system and can be changed if necessary (e.g., incubation temperature, incubation time).

28. The correct coenzyme (NADH or NADPH) needs to be selected depending on the redox system used. It has been shown that several redox systems such as HaPuR/HaPuX [26], PuR/PuxB [30] (NADH-dependent) or spinach ferredoxin together with *E. coli* flavodoxin reductase [12] (NADPH-dependent) are compatible with OxyB$_{van}$.

29. Always use the *peptidyl*-PCP directly after the PCP loading reaction due to the hydrolysis of the peptide from the PCP.

30. To minimize peptide hydrolysis from the PCP prior to the turnover reaction the reaction can be prepared during the PCP loading reaction. Therefore mix the required volumes of reaction buffer and 0.33 % glucose in the bottom of a 1.5 mL Eppendorf tube. Add the remaining components of the reaction except *peptidyl*-PCP separated from each other to the wall of the Eppendorf tube. Start the reaction after addition of *peptidyl*-PCP through centrifugation.

31. Depending on the peptide-PCP combination used neutralization of the solution can cause protein precipitation, which leads to a reduction of the observed peptide signal during HPLC-MS analysis. In this case another acid or a different neutralization procedure can be used.

32. Do not remove the complete solvent during peptide concentration as it can be difficult to properly redissolve the peptide fraction.

33. Waters XBridge BEH 300 Prep C18 column (particle size: 5 µm, 4.6×250 mm); gradient: 0–4 min 95 % solvent A, 4–4.5 min up to 15 % solvent B, 4.5–25 min up to 50 % solvent B; flow rate 1 mL/min (*see* **Note 14**).

References

1. Hur GH, Vickery CR, Burkart MD (2012) Explorations of catalytic domains in non-ribosomal peptide synthetase enzymology. Nat Prod Rep 29:1074–1098

2. Yim G, Thaker MN, Koteva K et al (2014) Glycopeptide antibiotic biosynthesis. J Antibiot 67:31–41

3. Cryle MJ, Brieke C, Haslinger K (2014) Oxidative transformations of amino acids and peptides catalysed by Cytochromes P450. In: Farkas E, Ryadnov M (eds) Amino acids, peptides and proteins, vol 38. Royal Society of Chemistry, Cambridge, pp 1–36

4. Cryle MJ, Schlichting I (2008) Structural insights from a P450 carrier protein complex reveal how specificity is achieved in the P450$_{Biol}$-ACP complex. Proc Natl Acad Sci U S A 105:15696–15701

5. Haslinger K, Brieke C, Uhlmann S et al (2014) The structure of a transient complex of a nonribosomal peptide synthetase and a cytochrome P450 monooxygenase. Angew Chem Int Ed 53:8518–8522

6. Süssmuth RD, Pelzer S, Nicholson G et al (1999) New advances in the biosynthesis of glycopeptide antibiotics of the vancomycin type from *Amycolatopsis mediterranei*. Angew Chem Int Ed 38:1976–1979

7. Bischoff D, Pelzer S, Holtzel A et al (2001) The biosynthesis of vancomycin-type glycopeptide antibiotics—new insights into the cyclization steps. Angew Chem Int Ed 40:1693–1696

8. Bischoff D, Pelzer S, Bister B et al (2001) The biosynthesis of vancomycin-type glycopeptide antibiotics—the order of the cyclization steps. Angew Chem Int Ed 40:4688–4691

9. Hadatsch B, Butz D, Schmiederer T et al (2007) The biosynthesis of teicoplanin-type glycopeptide antibiotics: assignment of P450 mono-oxygenases to side chain cyclizations of glycopeptide A47934. Chem Biol 14:1078–1089

10. Stegmann E, Pelzer S, Bischoff D et al (2006) Genetic analysis of the balhimycin (vancomycin-type) oxygenase genes. J Biotechnol 124:640–653

11. Haslinger K, Peschke M, Brieke C et al (2015) X-domain of peptide synthetases recruits oxygenases crucial for glycopeptide biosynthesis. Nature. 521:105–109

12. Woithe K, Geib N, Zerbe K et al (2007) Oxidative phenol coupling reactions catalyzed by OxyB: a cytochrome P450 from the vancomycin producing organism. Implications for vancomycin biosynthesis. J Am Chem Soc 129:6887–6895

13. Schmartz PC, Wölfel K, Zerbe K et al (2012) Substituent effects on the phenol coupling reaction catalyzed by the vancomycin biosynthetic P450 enzyme OxyB. Angew Chem Int Ed 51:11468–11472

14. Brieke C, Kratzig V, Haslinger K et al (2015) Rapid access to glycopeptide antibiotic precursor peptides coupled with cytochrome P450-mediated catalysis: towards a biomimetic synthesis of glycopeptide antibiotics. Org Biomol Chem 13:2012–2021

15. Quadri LEN, Weinreb PH, Lei M et al (1998) Characterization of Sfp, a *Bacillus subtilis* phosphopantetheinyl transferase for peptidyl carrier protein domains in peptide synthetases. Biochemistry 37:1585–1595

16. Vitali F, Zerbe K, Robinson JA (2003) Production of vancomycin aglycone conjugated to a peptide carrier domain derived from a biosynthetic non-ribosomal peptide synthetase. Chem Commun 21:2718–2719

17. Nicolaou KC, Boddy CNC, Bräse S et al (1999) Chemistry, biology, and medicine of the glycopeptide antibiotics. Angew Chem Int Ed 38:2096–2152

18. Freund E, Robinson JA (1999) Solid-phase synthesis of a putative heptapeptide intermediate in vancomycin biosynthesis. Chem Commun 24:2509–2510

19. Bo Li D, Robinson JA (2005) An improved solid-phase methodology for the synthesis of putative hexa- and heptapeptide intermediates in vancomycin biosynthesis. Org Biomol Chem 3:1233–1239

20. Brieke C, Cryle MJ (2014) A facile Fmoc solid phase synthesis strategy to access epimerization-prone biosynthetic intermediates of glycopeptide antibiotics. Org Lett 16:2454–2457

21. Blanco-Canosa JB, Dawson PE (2008) An efficient Fmoc-SPPS approach for the generation of thioester peptide precursors for use in native chemical ligation. Angew Chem Int Ed 47:6851–6855

22. Dettner F, Hänchen A, Schols D et al (2009) Total synthesis of the antiviral peptide antibiotic feglymycin. Angew Chem Int Ed 48:1856–1861

23. Davidsen JM, Bartley DM, Townsend CA (2013) Non-ribosomal propeptide precursor in nocardicin A biosynthesis predicted from adenylation domain specificity dependent on the MbtH family protein NocI. J Am Chem Soc 135:1749–1759

24. Haslinger K, Maximowitsch E, Brieke C et al (2014) Cytochrome P450 OxyB$_{tei}$ catalyzes the first phenolic coupling step in teicoplanin biosynthesis. ChemBioChem 15:2719–2728

25. Sunbul M, Marshall NJ, Zou Y et al (2009) Catalytic turnover-based phage selection for engineering the substrate specificity of Sfp phosphopantetheinyl transferase. J Mol Biol 387:883–898

26. Bell SG, Tan ABH, Johnson EOD et al (2010) Selective oxidative demethylation of veratric acid to vanillic acid by CYP199A4 from *Rhodopseudomonas palustris* HaA2. Mol Biosyst 6:206–214

27. Zerbe K, Pylypenko O, Vitali F et al (2002) Crystal structure of OxyB, a cytochrome P450 implicated in an oxidative phenol coupling reaction during vancomycin biosynthesis. J Mol Biol 277:47476–47485

28. Dordine RL, Paneth P, Anderson VE (1995) 13C NMR and 1H-1H NOEs of coenzyme-A: conformation of the pantoic acid moiety. Bioorg Chem 23:169–181

29. Bogomolovas J, Simon B, Sattler M et al (2009) Screening of fusion partners for high yield expression and purification of bioactive viscotoxins. Protein Expr Purif 64:16–23

30. Bell SG, Xu F, Johnson EOD et al (2010) Protein recognition in ferredoxin-P450 electron transfer in the class I CYP199A2 system from *Rhodopseudomonas palustris*. J Biol Inorg Chem 15:315–328

Chapter 7

Reconstitution of Fungal Nonribosomal Peptide Synthetases in Yeast and In Vitro

Ralph A. Cacho and Yi Tang

Abstract

The emergence of next-generation sequencing has provided new opportunities in the discovery of new nonribosomal peptides (NRPs) and NRP synthethases (NRPSs). However, there remain challenges for the characterization of these megasynthases. While genetic methods in native hosts are critical in elucidation of the function of fungal NRPS, in vitro assays of intact heterologously expressed proteins provide deeper mechanistic insights in NRPS enzymology. Our previous work in the study of NRPS takes advantage of *Saccharomyces cerevisiae* strain BJ5464-*npgA* as a robust and versatile platform for characterization of fungal NRPSs. Here we describe the use of yeast recombination strategies in *S. cerevisiae* for cloning of the NRPS coding sequence in 2μ-based expression vector; the use of affinity chromatography for purification of NRPS from the total *S. cerevisiae* soluble protein fraction; and strategies for reconstitution of NRPSs activities in vitro.

Key words Fungal secondary metabolite, Nonribosomal peptide synthetase, Transformation-assisted recombination, Yeast protein expression

1 Introduction

Nonribosomal peptides are an important class of natural compounds that have wide spectrum of biological and medical utility as antibiotics, immunosuppressants, and antitumor drugs [1, 2]. Unlike ribosomal proteins, nonribosomal peptides do not go through the ribosomal machinery in their biogenesis but are made by assembly-line-like megasynthases known as nonribosomal peptide synthetases (NRPSs). NRPSs are organized in functional units known as modules, with each module incorporating one amino acid to the nonribosomal peptide scaffold. The NRPS modules, in turn, are made up of domains. Three domains, adenylation, thiolation and condensation, are minimally required for selection and condensation of one amino acid building block. The adenylation domain selects and activates the building blocks of the peptide scaffold as amino-acyl-AMP (adenosyl-monophosphate).

Bradley S. Evans (ed.), *Nonribosomal Peptide and Polyketide Biosynthesis: Methods and Protocols*, Methods in Molecular Biology, vol. 1401, DOI 10.1007/978-1-4939-3375-4_7, © Springer Science+Business Media New York 2016

The activated substrate is then transferred to the thiol moiety of the phosphopantetheine prosthetic group of the thiolation domain. Thereafter, the condensation domain catalyzes the peptide bond formation where the T-domain bound aminoacyl unit performs a nucleophilic attack on the thioester group of the growing peptide attached to the T domain of the preceding module [2–4].

Much of the understanding of the underlying mechanisms of the NRPS machinery were brought forth from structural and biochemical studies of heterologously expressed NRPS and the dissected modules and domains thereof [5]. The logical choice for the heterologous expression system for these protein constructs is *Escherichia coli* due to its robust features as well as the diverse genetic and synthetic biology toolkits available. And while considerable successes were obtained from using *E. coli* as a heterologous host for bacterial NRPS systems, the reconstitution of fungal NRPS systems in *E. coli* has been limited to mono- and dimodule NRPSs [6, 7].

Saccharomyces cerevisiae is an excellent alternative heterologous expression system to *E. coli*. Like *E. coli*, *S. cerevisiae* has a myriad of flexible and well-characterized genetic and synthetic biology tools such as strong and controllable promoters and terminators, plasmid systems with a high-copy (2μ) or a single-copy origin of replication (CEN/ARS), as well as a host of strains with variety of auxotrophic markers and gene deletions [8]. *S. cerevisiae* is also evolutionarily much closer to filamentous fungi that are rich in NRPSs, thereby providing a more compatible environment for expression of foreign proteins of fungal origin.

However, due to the presence of vacuolar proteases, protein overexpression system for megasynthases in yeast has been difficult. Previous examples of expression of the megasynthase in yeast, such as 6-methylsalicylic acid synthase (MSAS), resulted in an active protein in vivo but the protein expression is insufficient for viable purification of the megasynthase [9]. Among the major contributor to the yeast protease activity are endoproteinases A and B, which are the two most abundant lumenal vacuolar proteases. To circumvent this problem, the genes *pep4* and *prb1*, encoding for endoproteinase A and B respectively, were deleted in *S. cerevisiae* (MATα *ura3-52 trp1 leu2Δ1 his3Δ200 can1* GAL) to construct the strain BJ5464 [10]. This strain is successfully used for heterologous overexpression of protein such as the human DNA polymerase η [11], and recombinant single-chain antibody fragment [12], which would otherwise be degraded by the proteases.

Furthermore, NRPSs as well as polyketide synthases (PKSs) requires a posttranslational installation of a phosphopantetheine prosthetic group to the active site serine of peptidyl carrier (also known as thiolation) or acyl carrier domain, respectively, in order for the megasynthases to be functional [13]. As such, the yeast strain BJ5464 was modified in order to integrate the *Aspergillus*

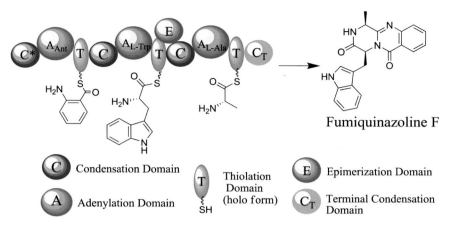

Fig. 1 Assembly of fumiquinazoline F by the trimodular NRPS TqaA showing the assembly-line arrangement of the domains in NRPS

terreus phosphopantetheine transferase gene *npgA*, which is under the control of the late-growth phase promoter ADH2.

Using this approach, our group as well as several others has successfully reconstituted the activity of fungal NRPSs such as the fumiquinazoline F synthetase (Fig. 1) TqaA/Afu6g12080 [14, 15], benzodiazepinone synthetase AnaPS [14], ardeemin synthetase ArdA [16], asperlicin synthetase AspA [17], and fungal cyclooligomer depsipeptide synthetases [18]. These works, in particular in vitro biochemical studies, brought about significant insights into the fungal NRPS machinery such as the mechanism of cyclization and release of fungal cyclic NRPs, and characterization of the iterative module of AspA in the biosynthesis of asperlicins [14, 17].

Furthermore, the approach described in these works is not limited to cloning and expression of fungal NRPS but also to other megasynthases such as non-reducing polyketide synthases (NR-PKSs) [15, 19–23], highly reducing PKSs (HR-PKSs) [19, 23–27], and PKS-NRPS hybrid megasynthases [28].

In addition to its utility as an expression system for megasynthases, the *S. cerevisiae* strain can also serve as a tool for the assembly and cloning of expression constructs for fungal megasynthases by taking advantage of the endogenous yeast homologous recombination system. Assembly of yeast plasmids via transformation-associated-recombination (TAR) has been successfully performed for plasmids up to 100 kbp, albeit with the more stable CEN-ARS origin of replication [29, 30]. However, in our hands, we have successfully assembled and cloned yeast plasmids with 2μ origin of replication in excess of 50 kbp.

In this protocol, we describe the use of yeast strain BJ5464-*npgA* for the assembly and cloning of the NRPS expression plasmid for yeast. In addition, we will describe the method for purification of megasynthases from the total soluble yeast protein fraction via FLAG-tag affinity chromatography and methods to characterize the products of the heterologously expressed NRPS.

2 Materials

2.1 Cloning of NRPS Coding Sequence (CDS)

1. High-fidelity polymerase chain reaction kit (e.g., NEB Phusion polymerase Master Mix kit).

2. Primers for amplification of overlapping NRPS CDS fragments.

3. Gel extraction kit (e.g., Zymoclean™ Gel DNA Recovery Kit).

4. Yeast expression vector pXW55. The pXW55 vector is derived from yeplac195-ADH2 vector. It has the *S. cerevisiae* 2μ origin of replication and *ura3* gene for selection in uracil dropout media. pXW55 also contains an N-terminal FLAG-tag coding sequence and a C-terminal 6×-His-tag coding sequence flanked by an upstream ADH2 promoter and downstream ADH2 terminator, allowing expression of the megasynthase in yeast. It also has an *E. coli* origin of replication and the β-lactamase gene to allow manipulation of plasmid in *E. coli* (*see* Fig. 2).

5. *Spe*I and *Pml*I restriction enzymes.

6. *Saccharomyces cerevisiae* strain BJ5464-*npgA* (MATα *ura3-52 his3-Δ200 leu2-Δ1 trp1 pep4::HIS3 prb1Δ1.6R can1* GAL *A. terreus npgA*).

7. Yeast peptone dextrose (YPD) medium: 1 % Bacto yeast extract, 2 % Bacto peptone, and 2 % dextrose (*see* **Note 1**).

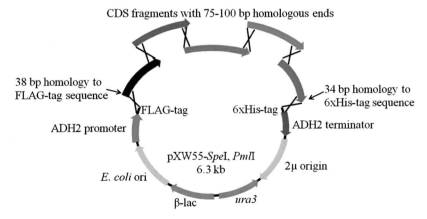

Fig. 2 Scheme for the cloning of the NRPS coding sequence into the yeast expression plasmid pXW55. pXW55, linearized by digestion with *Spe*I and *Pml*I, contains a FLAG-coding sequence downstream of ADH2 promoter that allows recombination with the first CDS fragment containing the designed 5′ sequence. Similarly, the last CDS fragment contains a 3′ sequence allowing recombination with the 6×-His tag coding sequence upstream of the ADH2 terminator

8. Lithium acetate solution: 1.0 M lithium acetate. Sterilize by filtration and keep at room temperature.

9. PEG solution: 50 % PEG3350 (Sigma) solution (*see* **Note 2**).

10. 2.0 mg/mL single-stranded carrier DNA (SS-DNA): Dissolve 2.0 mg/mL of deoxyribonucleic acid from salmon sperm (Sigma) in sterile 10 mM Tris–HCl, 1 mM EDTA, pH 8.0. Keep 1 mL aliquots at –20 °C (*see* **Note 3**).

11. 10× Yeast nitrogen base solution without amino acid with ammonium sulfate (10× YNB): 6.7 % Difco yeast nitrogen base with ammonium sulfate. Filter-sterilize and keep at 4 °C and away from light.

12. Adenine stock solution: 2 % adenine hemisulfate (Sigma). Add HCl dropwise until fully dissolved and filter-sterilize. Keep at 4 °C and away from light.

13. Tryptophan stock solution: 2 % tryptophan. Filter-sterilize and keep at 4 °C and away from light.

14. Uracil dropout media: 0.5 % Bacto casamino acids, 0.67 % yeast nitrogen base with ammonium sulfate, 2 % dextrose, 0.02 % adenine hemisulfate, 0.02 % tryptophan, 2 % Bacto agar (for solid media). Sterilize by autoclaving (*see* **Note 4**).

15. Digestion buffer: 50 mM Tris–HCl, 10 mM EDTA, pH to 8.0. Add 1 μL of β-mercaptoethanol and 10 μL of Zymolyase solution per 1 mL of digestion buffer immediately before use.

16. Lysis buffer: 0.2 M sodium hydroxide, 1 % sodium dodecyl sulfate.

17. Neutralization buffer: 3 M potassium acetate, pH 4.5. Adjust pH using glacial acetic acid and keep at 4 °C.

18. Chemically competent *E. coli* cells.

19. LB Broth.

20. LB plates with 35 μg/mL carbenicillin for selection and screening of *E. coli* clones harboring assembled NRPS expression plasmid.

21. LB Broth with 100 μg/mL ampicillin.

2.2 Expression and Purification of FungalNRPS

1. ANTI-FLAG® M2 Affinity Gel (Sigma-Aldrich, catalog number A2220).

2. Tris-buffered saline buffer (TBS): 50 mM Tris–HCl, 150 mM NaCl, pH 7.4. Adjust pH to 7.4 using HCl and store at 4 °C.

3. 0.1 M Glycine: Adjust pH to 3.5 using HCl and store at 4 °C.

4. Elution buffer: Dissolve 1.5 mg/mL of 3×-FLAG peptide in TBS, pH 7.4. Store at 4 °C. Use the elution buffer within 24 h of preparation of the buffer.

5. FLAG column storage buffer: 50 mM Tris–HCl, 150 mM NaCl, 50 % glycerol, and 0.02 % sodium azide. Store at 4 °C and use within 24 h of preparation of buffer.

6. Amicon ultrafiltration column with 100,000 molecular weight cut-off.

7. Protein storage buffer: 50 mM Tris–HCl, 150 mM NaCl, 20 % glycerol, pH 8.0. Store at 4 °C.

2.3 Characterization of Heterologously Expressed NRPS

1. 500 mM adenosine triphosphate (ATP) stock solution (*see* **Note 5**).

2. Amino acid solution (200 mM) (*see* **Note 6**).

3. Magnesium chloride solution (1 M).

4. Tris–HCl buffer pH 7.5 (1 M).

5. Ethyl acetate–methanol–glacial acetic acid mixture (90:9:1).

6. Reagent grade acetone.

7. LCMS-grade methanol.

8. LCMS-grade acetonitrile with 0.1 % formic acid.

9. LCMS-grade water with 0.1 % formic acid.

10. Phenomenex Luna 5μ 100 × 2 mm C18 reverse-phase column or an equivalent analytical reverse-phase chromatography column.

11. Shimadzu 2010 EV Liquid Chromatography Mass Spectrometer or an equivalent LC-MS.

3 Methods

3.1 Cloning of NRPS Coding Sequence (CDS)

3.1.1 Primer Design

1. Determine the location of putative start codon, stop codon and introns in the fungal NRPS gene using eukaryotic gene prediction software such as Softberry or Augustus (*see* **Note 7**).

2. Design primer sequences for the different NRPS CDS fragment such that each end of the fragment has 75–100 base pair overlap with the adjacent NRPS CDS fragment. The length of each NRPS CDS fragment is typically around 2–5 kb. For regions of the CDS that contains predicted introns, primers for shorter CDS fragment lengths (~2 kb) are designed in order to easily determine if the introns are spliced from amplification from a cDNA sample.

3. Design the forward primer of the first NRPS CDS fragment with the following 39-base overhang (5′-atggctagcgattataaggatgatgatgataagactagt-3′) before the gene specific sequence. This overhang serves as the homologous sequence that allows recombination with the pXW55 such that the CDS will be

in-frame with the first start codon and the FLAG-tag coding sequence (FLAG tag coding sequence is underlined, *see* Fig. 2).

4. Design the reverse primer of the last NRPS CDS primer such that it contains the following 34-base overhang (5′-ttgtcattta-aatta<u>gtgatggtgatggtgatgc</u>-3′) before the gene specific sequence. This overhang allows recombination with the C-terminal 6×-His tag sequence of pXW55 (reverse complement of the 6×-His coding sequence is underlined, *see* Fig. 2).

3.1.2 Preparation of NRPS CDS Fragments

1. Prepare the PCR mix for the fragments by adding 50–100 ng template genomic DNA (or cDNA if the predicted introns are found in the region to be amplified), 0.3 μL of each primer (100 μM) and 25 μL Phusion PCR Master Mix. Add Milli-Q water to obtain a final volume of 50 μL.

2. PCR amplify using cycle given in Table 1.

3. Verify fragment size using agarose gel electrophoresis. If the fragments sizes are correct, then gel-purify PCR products (*see* **Note 8**).

4. Measure concentration of DNA.

3.1.3 Linearization of Yeast Expression Plasmid pXW55

1. Mix 1 μg of yeast expression plasmid pXW55 with 5 μL 10× Cutsmart buffer, 1 μL *Spe*I, 1 μL *Pml*I, and Milli-Q water to a final volume of 50 μL. Incubate at 37 °C for 6 h.

2. Perform agarose gel electrophoresis on the completed digestion mix and gel-purify the 6.3 kb band containing the plasmid backbone of pXW55 (*see* Fig. 2).

3. Measure the concentration of pXW55.

3.1.4 Yeast Transformation

Using a modified method based on established protocol for lithium acetate/PEG-mediated transformation of intact yeast cells [31], the NRPS yeast expression plasmid is assembled via transformation-assisted DNA recombination.

Table 1
PCR temperature cycle for amplification of NRPS CDS fragment

Step	Temperature	Time
1. Initial denaturation	98 °C	2:00
2. Denaturation	98 °C	0:20
3. Annealing	Primer Tm	0:15
4. Extension	72 °C	0:30/kb
Repeat steps 2–4 for 29 times		
5. Final extension	72 °C	8:00

Day 1

1. Inoculate yeast strain BJ5464-*npgA* into 5 mL YPD and incubate overnight on a rotary shaker at 250 rpm and 28 °C.

Day 2

2. Measure the OD_{600} of the yeast culture using a spectrophotometer. Dilute the yeast culture in YPD such that the final OD_{600} is 0.2 and final volume of 5 mL.

3. Incubate the yeast culture in a rotary shaker at 28 °C and 250 rpm to an OD_{600} of ~0.8–1.0 (usually around 4–5 h).

4. Pellet down cells for 5 min at $3000 \times g$ and 4 °C.

5. Decant liquid and resuspend cells in 5 mL sterile Milli-Q water.

6. Repeat **steps 3** and **4**. Spin down cells again and resuspend in 1 mL sterile Milli-Q water.

7. Aliquot 100 µL of cells per assembly reaction in microcentrifuge tubes and spin down cells for 30 s at $15,000 \times g$.

8. Add the components listed in Table 2 to the cell pellet in the order listed.

9. Resuspend cells by vortexing and incubate cells at 42 °C for 45 min.

10. Spin down cells for 30 s at $15,000 \times g$. Remove supernatant by pipetting and resuspend cells in 500 µL Milli-Q water. Plate resuspended cells in uracil dropout media.

11. Incubate plates in 28 °C for at least 3 days or until colonies appear.

3.1.5 Screening of Yeast Transformants

1. Using sterile toothpicks or pipette tips, transfer yeast transformants from original transformation plates to new uracil dropout plates in ~1.0 cm² patches. Incubate the replica plates in 28 °C incubator for 24–48 h.

2. Using a sterile toothpick, pick yeast cells from the replica plate and resuspend into 200 µL digestion buffer.

3. Incubate resuspended cells in 37 °C for 1 h.

Table 2
Setup for LiAc/PEG-mediated yeast transformation

Component	Volume (µL)
PEG 3350 (50 %, w/v)	240
Lithium acetate (1.0 M)	36
Boiled SS-DNA (2 mg/mL)	50
Premixed DNA fragments and water (*see* **Note 9**)	24

4. Add 200 μL lysis buffer to digestion mix and mix by inversion.

5. Add 300 μL neutralization buffer to cells and mix by inversion.

6. Spin down the tubes for 10 min in 20,000×*g*.

7. Transfer supernatant to fresh tubes.

8. Add 800 μL of isopropanol to the supernatant and invert tubes until solution is well-mixed.

9. Incubate tubes in room temperature for 10 min.

10. Spin down tubes at 20,000×*g* for 10 min. After spinning, discard supernatant.

11. Add 400 μL 70 % ethanol to the pellet. Spin down for 2 min at 20,000×*g*.

12. Discard supernatant and dry the pellet. Resuspend pellet in 20 μL TE buffer.

13. Perform PCR screening isolated plasmid DNA or transform plasmid into *E. coli* for restriction digest check and sequencing.

3.2 Expression and Purification of FungalNRPS

After the sequence of the NRPS expression plasmid has been verified, the plasmid is re-transformed into BJ5464-*npgA* strain for heterologous expression. Since the late-phase ADH2 promoter induces a high-expression level when glucose is depleted and once ethanol consumption begins, the yeast cells are typically harvested for protein purification once the cultures reach stationary phase after 72 h of growth in YPD [32]. After lysis of the yeast cells and clarification by centrifugation, the yeast soluble fraction is subjected to FLAG-affinity chromatography. This technique is especially effective for proteins larger than ~300 kDa. The FLAG-tag affinity chromatography method is based on the protocol given by the manufacturers.

3.2.1 Preparation of Yeast Soluble Protein Fraction (See Note 10)

1. Transform 1 μg of NRPS expression plasmid into BJ5464-*npgA* (*see* Subheading 3.1.4 for Procedure).

2. Inoculate a single transformant colony into 4 mL of liquid uracil dropout media and incubate in a rotary shaker for 2 days at 28 °C and 250 rpm. Transfer seed culture to 1 L of YPD and incubate in a rotary shaker for an additional 3 days at 28 °C and 250 rpm.

3. Pellet the yeast cells by centrifugation at 6000×*g* and 4 °C for 20 min.

4. Using a vortex mixer, resuspend the yeast cells in 30 mL of Tris-buffered saline, pH 7.4 per 1 L of yeast culture.

5. Lyse resuspended cells using sonication or three passes in the French press at 7000–10,000 psi (*see* **Note 11**).

6. Clarify yeast lysate by centrifugation at $30,000 \times g$ and 4 °C for 1 h.

7. Filter clarified yeast lysate using a 0.22 μm filter.

8. Run an aliquot of yeast lysate on an SDS-PAGE gel to check if the NRPS is expressed in the large-scale fermentation (*see* **Note 10**).

3.2.2 Preparation of FLAG-Tag Affinity Column

1. Load 5 mL of the M2 FLAG-tag affinity gel into a glass chromatography column (*see* **Note 12**). Wait for the resin to settle in the column.

2. Drain the storage buffer from the column, leaving just a small amount of buffer in the column.

3. Wash the affinity gel column with 3 column volumes of Tris-buffered saline pH 7.4.

4. Wash the affinity gel column with 3 sequential column volumes of 0.1 M glycine, pH 3.5. Let each wash go through the column before adding the next wash. Avoid disturbing the gel bed while loading (*see* **Note 13**).

5. Wash the resin with 5–6 column volumes of Tris-buffered saline, pH 7.4 to equilibrate the resin for use. Test the pH of the effluent with a pH strip to determine if the pH is restored to 7.4. Do not let the column bed run dry. Allow a small amount of buffer to remain on the top of the column.

3.2.3 FLAG-tag Affinity Chromatography

1. Gently load the filtered yeast lysate on the column and let the lysate flow through the column using gravity.

2. Load the filtered yeast lysate again on the column and let it flow through using pressure. Repeat for a total of three times.

3. Gently add 5–10 column volumes of Tris-buffered saline, pH 7.4 to the column bed and collect the wash fraction in a 50 mL Corning tube. Test the eluent using microtiter-plate Bradford assay to assess if additional wash steps are necessary (*see* **Note 14**).

4. If the Bradford assay indicates that protein is still present in the eluent, then repeat **step 3** to eliminate unspecific binding in the column.

5. Change to a new collection tube. Add 2 column volumes of elution buffer and let the buffer flow through the column using only gravity. Using a qualitative Bradford assay, determine if more elution buffer is required (*see* **Note 15**).

6. Analyze the wash and elution fractions using SDS-PAGE to determine the presence of the correct protein size in the elution fraction.

7. If the protein of the correct size as determined by SDS-PAGE (Fig. 3) is detected in the elution fraction, dialyze and concentrate

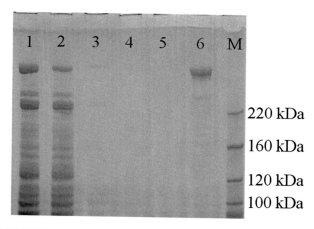

Fig. 3 SDS-PAGE gel from different fractions obtained from FLAG-tag affinity chromatography of a ~375 kDa trimodular NRPS. *1*: Insoluble yeast proteins, *2*: Flowthrough fraction, *3*: Wash fraction 1, *4*: Wash fraction 2, *5*: Wash fraction 3, *6*: Elution fraction

the elution fraction using an Amicon ultracentrifugation column and protein storage buffer as dialysis buffer (*see* **Note 16**).

8. After the volume of the dialyzed and concentrated protein solution is ~200 μL, use the protein directly for assay or flash-freeze protein.

3.2.4 Recycling and Storage of FLAGAffinity Column

1. Change to a new collection tube. Gently add 3–5 column volumes of 0.1 M glycine to the column bed and let the solution flow through until the buffer level is right above the column bed.

2. Gently add 5 column volumes of Tris-buffered saline, pH 7.4 to the column bed. Continue washing the column with Tris-buffered saline, pH 7.4 until the pH of the effluent is restored to pH 7.4 as indicated by a pH strip.

3. Wash column with 5–6 column volumes of storage buffer. Air pressure is necessary to push buffer through column. Let the storage buffer to flow through until the buffer level is right above the column bed.

4. After washing, add 5 mL of storage buffer to top of solution and cap the column. Store column in –20 °C freezer (*see* **Note 17**).

3.3 Characterization of NRPS

Following purification of the NRPS, one can set up an in vitro assay of the megasynthase by incubating it with the amino acid substrates and ATP. One can determine possible amino acid candidates either through ATP:pyrophosphate exchange assay of individual A domains or through results from NRPS gene knockout from the native fungal host [6, 7, 14]. Alternatively, if the heterologously expressed NRPS utilizes the proteinogenic 20 amino acid or nonproteinogenic amino acids that are synthesized by yeast such as anthranilic acid or ornithine, then it may be possible to see the NRPS product via extraction of metabolites from yeast expressing the NRPS and comparison of the metabolites from yeast cultures that are harboring no plasmids.

Following extraction of the reaction or the yeast culture, the extracts are subjected to LC-MS analysis. The use of LC-MS allows characterization of the probable mass and the UV-Vis spectrum of the compound of interest. If the NRPS product is detected in yeast cultures, one can do a large scale fermentation of the yeast culture and purify the compound of interest for further structural characterization.

3.3.1 In Vitro Assay

1. In a microcentrifuge tube, set up the enzymatic reaction using the components given in Table 3 (add in order).

2. After 6–12 h, quench the reaction by addition of equal volume of ethyl acetate–methanol–acetic acid (90:9:1) to the reaction mixture. Mix well by vortexing.

3. Centrifuge the mixture at 20,000×*g* for 5 min to separate the aqueous layer from the organic layer. Remove the organic layer by pipetting. Repeat step for a total of two times.

4. Dry the organic layer in vacuo and redissolve the sample in 10 μL HPLC-grade methanol.

3.3.2 Extraction from Yeast Cultures Expressing the Fungal NRPS

1. Inoculate and grow yeast cultures expressing the fungal NRPS in the same manner as describe for protein purification (*see* **steps 1** and **2** of Subheading 3.2.1).

2. Take 1 mL of yeast culture at different timepoints starting from 24 h after growth in YPD and until 96 h after growth in YPD.

Table 3
Setup for in vitro NRPS assay

Reaction component	Final concentration
Deionized water	To a final volume of 100 μL
Tris–HCl buffer (pH 7.5)	100 mM
MgCl$_2$	10 mM
Amino acid solution	1 mM of each amino acid
NRPS enzyme	1–10 μM
ATP	5 mM

3. Spin down the timepoint samples at $20,000 \times g$ for 1 min. Separate the cell pellet from the supernatant.

4. Resuspend the yeast cell pellet 500 µL of reagent grade acetone by vortexing. Incubate the cell pellet in acetone for 10 min.

5. Centrifuge the cell pellet suspension at $20,000 \times g$ for 1 min. Separate the cell pellet from the acetone and dry the solvent in vacuo.

6. Redissolve the dried extract in 40 µL HPLC-grade methanol.

7. Add an equal volume of ethyl acetate–methanol–acetic acid (90:9:1) to the supernatant from **step 3**. Mix well by vortexing.

8. Centrifuge the mixture at $20,000 \times g$ for 5 min to separate the aqueous layer from the organic layer. Remove the organic layer by pipetting. Repeat step for a total of two times.

9. Dry organic layer in vacuo and redissolve the dried extract in 40 µL HPLC-grade methanol.

10. Analyze samples using liquid-chromatography-mass spectrometry.

3.3.3 LCMS Method

1. Remove solids from LCMS injection samples obtained in Subheading 3.2.1 or 3.2.2 by filtration or centrifugation at $10,000 \times g$ for 10 min.

2. Adjust to the appropriate values the mass range for the MS detector. Since the average weight of an amino acid is ~110 Da, one can predict the mass range of the NRPS product based on the number of modules found in the NRPS.

3. Monitor the 250 nm wavelength using the photodiode array. If the NRPS product contains aromatic amino acids, it is possible to have a chromophore at 280 nm or beyond (Fig. 4).

4. Equilibrate the LCMS column to 5 % acetonitrile with 0.1 % formic acid using a flow rate of 0.10 mL/min for at least 10 min.

5. Use a linear gradient from 5 to 95 % acetonitrile with 0.1 % formic acid for 30 min, followed by isocratic phase of 95 % acetonitrile with 0.1 % formic acid for 15 min (*see* **Note 18**). Use a flow rate of 0.10 mL/min during the run.

4   Notes

1. To prevent caramelization of dextrose, 20 % dextrose solution is autoclaved separately from the dissolved yeast extract and peptone. Typically, 10 g of Bacto-yeast extract and 20 g Bacto-peptone are dissolved in 950 mL deionized water and sterilized by autoclaving. After media has cooled down to room temperature, 50 mL of sterile 20 % dextrose solution is added.

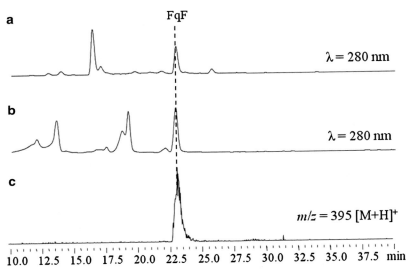

Fig. 4 LCMS chromatograms from characterization of TqaA. (**a**) Chromatogram from in vitro assay of TqaA with anthranilic acid, l-tryptophan, l-alanine and ATP showing production of fumiquinazoline F. (**b**) Chromatogram of metabolites from yeast cells expressing TqaA showing production of fumiquinazoline F. (**c**) Extracted ion chromatogram of fumiquinazoline F ($m/z = 395$, [M + H]$^+$)

2. Dissolve PEG 3350 in Milli-Q water while stirring on a hot plate. Sterilize by autoclaving. To maintain the concentration of PEG, make sure that the container is tightly capped at room temperature after autoclaving to prevent evaporation of water.

3. Before performing yeast transformation, denature the SS-DNA in a boiling water bath for 5 min and chill in ice. SS-DNA can be reboiled for three times without loss of transformation efficiency.

4. In order to prevent degradation of nutrients during autoclaving, uracil dropout media is prepared by autoclaving Bacto casamino acid solution with agar then 20 % dextrose, 10× YNB solution, adenine stock solution and tryptophan solution are added to final concentration after autoclaving.

5. The ATP stock solution has to be adjusted to pH 7. Typically, 60.5 mg of adenosine triphosphate disodium salt trihydrate is dissolved in 120 μL deionized water and 40 μL 5 N KOH is added. Check pH and adjust final volume to 200 μL. Aliquot the ATP stock solution in 20 μL portions and store in −80 °C.

6. While it is recommended to buffer the amino acid solution to pH 7.5 using 50 mM Tris–HCl buffer, a few of the hydrophobic amino acids such as tyrosine are not soluble at neutral pH. Thus one needs to be aware of the pH change in the in vitro assay due to addition of a high concentration of unbuffered amino acid.

7. It is recommended to perform prediction with multiple gene prediction software in order to determine if other introns might be found in the NRPS gene. We typically do gene prediction and re-annotation of deposited sequences in order to verify whether alternative coding sequences might be possible, as was in the case of ApdA from *Aspergillus nidulans* [28].

8. While not required, it is recommended to subclone and sequence the amplified CDS fragments in order to determine correct intron splicing to insure that the CDS fragment will be intron-less prior to assembly by recombination. Moreover, sequencing of the amplified CDS fragments helps in determining if there are discrepancies between the sequence of the amplicon and the deposited sequence or draft genome sequence.

9. Pre-combine ~0.1 pmol of each DNA fragment and vector in a separate tube before addition to yeast cell pellet.

10. In order to determine if the NRPS is expressed in the yeast culture and is soluble, it is highly recommended to perform SDS-PAGE analysis of yeast lysate from a small scale (50 mL) fermentation. In our experience with purification of megasynthases from *S. cerevisiae* BJ5464-*npgA*, the biggest detectable protein band in the SDS-PAGE analysis of *S. cerevisiae* BJ5464-*npgA* harboring no expression plasmid is ~230 kDa. Thus, if the megasynthase size is greater than 300 kDa, then one should see an extra band protein band above ~230 kDa band in the yeast lysate of the BJ5464-*npgA* harboring the expression plasmid after SDS-PAGE.

11. To see if the resuspended yeast cells are properly lysed, dilute yeast lysate 1000× and compare it to the diluted resuspended yeast cells under the microscope. The yeast lysate should have <1 % of the total number of yeast compared to the diluted resuspended yeast cells before lysis.

12. Typical yields for the purification is around 1 mg of protein per liter of yeast culture while the binding capacity of the M2 FLAG-affinity gel is 0.6 mg of protein per mL of gel. Thus, one has to adjust the volume of minimum volume of the affinity gel to account for the fermentation culture size.

13. Changing the pH of the gel to 4.5 removes the remaining protein that is bound to the antibody. However, the antibody cannot stay at pH 4.5 for more than 20 min since it can denature the anti-FLAG antibody (per manufacturer's instruction).

14. Typically, 10 μL of eluent is added to 100 μL of Bradford assay solution.

15. The protein will start eluting once the first column volume of elution buffer has passed through the column.

16. If the elution fraction from the FLAG-tag purification show protein bands of size smaller than what is expected in the SDS-PAGE gel, then one can add fungal protease inhibitor cocktail (Sigma) after resuspension of yeast cells in Tris-buffered saline pH 7.4 to prevent proteolysis.

17. Based on our experience, the FLAG-tag affinity column can be reused for up to a year.

18. Addition of formic acid to the mobile phase decreases the pH of the mobile phase and keeps the carboxylic acid groups in the extracts in the protonated form, thus increasing the retention time of compounds with acidic groups. However, some compounds may not be stable in acidic conditions.

Acknowledgements

We think Dr. Wei Xu for the construction of pXW55 plasmid. Work from our group is supported by the NIH grants 1R01GM085128 and 1DP1GM106413. R.A.C. is supported by the NRSA grant GM008496 and the UCLA Graduate Division.

References

1. Fischbach MA, Walsh CT (2006) Assembly-line enzymology for polyketide and nonribosomal peptide antibiotics: logic, machinery, and mechanisms. Chem Rev 106:3468–3496

2. Felnagle EA, Jackson EE, Chan YA et al (2008) Nonribosomal peptide synthetases involved in the production of medically relevant natural products. Mol Pharm 5:191–211

3. Schwarzer D, Finking R, Marahiel MA (2003) Nonribosomal peptides: from genes to products. Nat Prod Rep 20:275–287

4. Hur GH, Vickery CR, Burkart MD (2012) Explorations of catalytic domains in nonribosomal peptide synthetase enzymology. Nat Prod Rep 29:1074–1098

5. Koglin A, Walsh CT (2009) Structural insights into nonribosomal peptide enzymatic assembly lines. Nat Prod Rep 26:987–1000

6. Balibar CJ, Walsh CT (2006) GliP, a multimodular nonribosomal peptide synthetase in Aspergillus fumigatus, makes the diketopiperazine scaffold of gliotoxin. Biochemistry 45:15029–15038

7. Balibar CJ, Howard-Jones AR, Walsh CT (2007) Terrequinone A biosynthesis through L-tryptophan oxidation, dimerization and bis-prenylation. Nat Chem Biol 3:584–592

8. Fang F, Salmon K, Shen MWY et al (2011) A vector set for systematic metabolic engineering in Saccharomyces cerevisiae. Yeast 28:123–136

9. Kealey JT, Liu L, Santi DV et al (1998) Production of a polyketide natural product in nonpolyketide-producing prokaryotic and eukaryotic hosts. Proc Natl Acad Sci U S A 95:505–509

10. Jones EW (1991) Tackling the protease problem in Saccharomyces cerevisiae. Methods Enzymol 194:428–453

11. Johnson RE, Washington MT, Prakash S, Prakash L (2000) Fidelity of human DNA polymerase eta. J Biol Chem 275:7447–7450

12. Shusta EV, Raines RT, Pluckthun A et al (1998) Increasing the secretory capacity of Saccharomyces cerevisiae for production of single-chain antibody fragments. Nat Biotechnol 16:773–777

13. Mootz HD, Schörgendorfer K, Marahiel MA (2002) Functional characterization of 4'-phosphopantetheinyl transferase genes of bacterial and fungal origin by complementation of Saccharomyces cerevisiae lys5. FEMS Microbiol Lett 213:51–57

14. Gao X, Haynes SW, Ames BD et al (2012) Cyclization of fungal nonribosomal peptides by

a terminal condensation-like domain. Nat Chem Biol 8:823–830

15. Ishiuchi K, Nakazawa T, Ookuma T et al (2012) Establishing a new methodology for genome mining and biosynthesis of polyketides and peptides through yeast molecular genetics. Chembiochem 13:846–854

16. Haynes SW, Gao X, Tang Y et al (2013) Complexity generation in fungal peptidyl alkaloid biosynthesis: a two-enzyme pathway to the hexacyclic MDR export pump inhibitor ardeemin. ACS Chem Biol 8:741–748

17. Gao X, Jiang W, Jiménez-Osés G et al (2013) An iterative, bimodular nonribosomal peptide synthetase that converts anthranilate and tryptophan into tetracyclic asperlicins. Chem Biol 20:870–878

18. Yu D, Xu F, Gage D, Zhan J (2013) Functional dissection and module swapping of fungal cyclooligomer depsipeptide synthetases. Chem Commun 49:6176–6178

19. Zhou H, Qiao K, Gao Z et al (2010) Enzymatic synthesis of resorcylic acid lactones by cooperation of fungal iterative polyketide synthases involved in hypothemycin biosynthesis. J Am Chem Soc 132:4530–4531

20. Zhou H, Qiao K, Gao Z et al (2010) Insights into radicicol biosynthesis via heterologous synthesis of intermediates and analogs. J Biol Chem 285:41412–41421

21. Cacho RA, Chooi YH, Zhou H et al (2013) Complexity generation in fungal polyketide biosynthesis: a spirocycle-forming P450 in the concise pathway to the antifungal drug griseofulvin. ACS Chem Biol 8:2322–2330

22. Xu Y, Zhou T, Zhou Z et al (2013) Rational reprogramming of fungal polyketide first-ring cyclization. Proc Natl Acad Sci U S A 110:5398–5403

23. Xu Y, Zhou T, Espinosa-Artiles P et al (2014) Insights into the biosynthesis of 12-membered resorcylic acid lactones from heterologous production in *Saccharomyces cerevisiae*. ACS Chem Biol 9(5):1119–1127

24. Ma SM, Li JW, Choi JW et al (2009) Complete reconstitution of a highly reducing iterative polyketide synthase. Science 326:589–592

25. Xie X, Meehan MJ, Xu W et al (2009) Acyltransferase mediated polyketide release from a fungal megasynthase. J Am Chem Soc 131:8388–8389

26. Zhou H, Gao Z, Qiao K et al (2012) A fungal ketoreductase domain that displays substrate-dependent stereospecificity. Nat Chem Biol 8:331–333

27. Lin HC, Chooi YH, Dhingra S et al (2013) The fumagillin biosynthetic gene cluster in *Aspergillus fumigatus* encodes a cryptic terpene cyclase involved in the formation of beta-trans-bergamotene. J Am Chem Soc 135:4616–4619

28. Xu W, Cai X, Jung ME et al (2010) Analysis of intact and dissected fungal polyketide synthase-nonribosomal peptide synthetase in vitro and in *Saccharomyces cerevisiae*. J Am Chem Soc 132:13604–13607

29. Gibson DG (2012) Oligonucleotide assembly in yeast to produce synthetic DNA fragments. Methods Mol Biol 852:11–21

30. Shao Z, Zhao H (2012) DNA assembler: a synthetic biology tool for characterizing and engineering natural product gene clusters. Methods Enzymol 517:203–224

31. Gietz RD, Woods RA (2006) Yeast transformation by the LiAc/SS Carrier DNA/PEG method. Methods Mol Biol 313:107–120

32. Lee KM, DaSilva NA (2005) Evaluation of the Saccharomyces cerevisiae ADH2 promoter for protein synthesis. Yeast 22:431–440

Chapter 8

The Continuing Development of *E. coli* as a Heterologous Host for Complex Natural Product Biosynthesis

Haoran Zhang, Lei Fang, Marcia S. Osburne, and Blaine A. Pfeifer

Abstract

Heterologous biosynthesis of natural products is meant to enable access to the vast array of valuable properties associated with these compounds. Often motivated by limitations inherent in native production hosts, the heterologous biosynthetic process begins with a candidate host regarded as technically advanced relative to original producing organisms. Given this requirement, *E. coli* has been a top choice for heterologous biosynthesis attempts as associated recombinant tools emerged and continue to develop. However, success requires overcoming challenges associated with natural product formation, including complex biosynthetic pathways and the need for metabolic support. These two challenges have been heavily featured in cellular engineering efforts completed to position *E. coli* as a viable surrogate host. This chapter outlines steps taken to engineer *E. coli* with an emphasis on genetic manipulations designed to support the heterologous production of polyketide, nonribosomal peptide, and similarly complex natural products.

Key words Heterologous biosynthesis, Cellular design, Natural products, Polyketide, Nonribosomal peptide, *E. coli*

1 Introduction

Heterologous natural product biosynthesis enables access to the beneficial properties of compounds that derive from their nonideal original hosts [1, 2]. The host selected for heterologous biosynthesis should provide a combination of innate and potential advantages relative to the native host in order to motivate the technical steps required for genetic content transfer and pathway reconstitution. The establishment of heterologous biosynthesis is the focal point of this chapter. Specifically, the latest alterations to the genetic background of an *Escherichia coli* host designed to support complex natural product biosynthesis are described.

E. coli holds an advantage over nearly every possible heterologous host because of its rapid growth kinetics. In turn, it is also the most thoroughly characterized bacterial host and features an impressive range of genetic manipulation protocols. These features

Bradley S. Evans (ed.), *Nonribosomal Peptide and Polyketide Biosynthesis: Methods and Protocols*, Methods in Molecular Biology, vol. 1401, DOI 10.1007/978-1-4939-3375-4_8, © Springer Science+Business Media New York 2016

provide both a motivation and an experimental basis for enacting heterologous biosynthesis. However, *E. coli* is not a native producer of complex natural products (such as polyketides and nonribosomal peptides). As such, the advantages offered by this potential surrogate host are offset by concerns regarding heterologous pathway transfer and reconstitution. Key issues include (1) enabling expression of foreign clusters that contain both numerous and individually large genes, (2) ensuring active protein products, and (3) providing the needed metabolic substrates required for complex natural product formation.

The cellular engineering steps outlined below describe the latest alterations to an *E. coli* strain, termed BAP1, capable of supporting both polyketide and nonribosomal peptide biosynthesis [3]. The BAP1 strain contains the following features designed to assist heterologous natural product biosynthesis: (1) a strong and processive T7 RNA polymerase [4–6] to aid in the expression of complex foreign natural product genetic pathways; (2) a promiscuous 4′-phosphopantetheinyl transferase [7, 8] to posttranslationally modify and, hence, activate polyketide synthases and nonribosomal peptide synthetases; and (3) a re-engineered *prp* operon that allows inducible upregulation of a propionyl-CoA synthetase gene (*prpE*) while deleting the remaining catabolic capabilities of the operon, in order to provide a key polyketide precursor, propionyl-CoA, while also eliminating a primary consumption pathway for this same metabolite. To this strain, additional genetic manipulations have been made to further support and advance heterologous biosynthesis by this host. The genotypes of all strains described in this chapter are presented in Table 1. As opposed to describing a single method used during the procedures to generate these strains, we more generally emphasize the goals associated with each (also outlined in Table 1) and the series of experimental steps completed during construction.

Table 1
***E. coli* strains constructed in this study**

Strain	Description	Note
BAP1	F– ompT hsdSB (rB–mB–) gal dcm (DE3) Δ *prpRBCD*::T7prom-*sfp*, T7prom-*prpE*	Parent strain; capable of polyketide substrate support and posttranslational modification
BT2	BAP1 (*araBADC*::*Tn10*)	Disrupted arabinose degradation pathway for sustained induction
BT3	BAP1 (*araBADC*::*Tn10*, *lacZ*::*trfA-Kan*)	Integration of *trfA* gene for plasmid copy-up
BTRA	BAP1 (*araBADC*::*Tn10*, *lacZ*::*trfA-Kan*, Δ*recA*)	Deletion of *recA* to increase plasmid stability
BTRAP	BAP1 (*araBADC*::*Tn10*, *lacZ*::*trfA-Kan*, Δ*recA*, *pccAB*<>*codB*)	Inclusion of PCC to enable enhanced substrate support

2   Materials

2.1 Chromosomal Engineering

1. Ampicillin: 100 mg/L in water.
2. Kanamycin: 50 mg/L in water.
3. Chloramphenicol: 20 mg/L in ethanol.
4. Tetracycline: 10 mg/L in ethanol.
5. L-arabinose: 2 mg/L in water for plasmid copy-up studies.
6. IPTG: isopropyl β-D-1-thiogalactopyranoside, 100 μM in water.
7. Lysogeny broth (LB) medium: 10 g Bacto-tryptone, 5 g yeast extract, and 10 g NaCl per liter.
8. LB agar plates: 10 g Bacto-tryptone, 5 g yeast extract, 10 g NaCl, and 1 % agarose per liter.
9. M9 medium: 12.8 g $Na_2HPO_4 \cdot 7H_2O$, 3 g KH_2PO_4, 0.5 g NaCl, 1.0 g NH_4Cl, 2 mM $MgSO_4$, and 0.1 mM $CaCl_2$ per liter.
10. P1 phage lysate prepared using chloroform and stored at 4 °C.
11. Plasmids pKD46, pCP20, and pKD3 used for chromosomal DNA manipulation as described by Datsenko et al. [9].
12. Phusion DNA polymerase obtained from New England Biolabs (Ipswich, MA).
13. Restriction enzymes.
14. Qiaprep Spin miniprep kit for plasmid preparation, Qiagen (Valencia, CA).
15. QIAquick gel extraction kit for PCR product purification, Qiagen (Valencia, CA).
16. GeneJET gel extraction kit, Thermo Scientific (Pittsburgh, PA).
17. MicroPulser electroporator from Bio-Rad (Hercules, CA).
18. Thermo Scientific Barnstead Micropure Water Purification System used to provide water for reagents and reactions.

2.2 Phenotypic Assays

1. L-arabinose: 1 g/L in water.
2. pCC1FOS™ fosmid vector and the accompanying EPI3000 strain, Epicenter (Madison, WI).
3. UVP GelDoc-It TS™ Imaging System Transilluminator (Upland, CA).
4. API 3000 Triple Quad LC-MS with a Turbo Ion Spray source (PE Sciex) coupled with a Shimadzu Prominence LC system.

3   Methods

3.1   *ara* Operon Deletion

The first step for the desired strain construction is the removal of the *araBADC* operon from the *E. coli* BAP1 chromosome in order to prevent arabinose degradation and to thus enable a constant concentration of arabinose inducer during later gene expression or plasmid copy-up steps in the context of heterologous biosynthesis.

1. For this step, transfer the *araBADC::Tn10* cassette from *E. coli* LMG194 (Life Sciences, Grand Island, NY) to BAP1 by P1 transduction. Grow *E. coli* LMG194 overnight in LB medium containing tetracycline at 37 °C with shaking. Dilute the overnight culture at a volume ratio of 1:100 in LB medium containing 5 mM CaCl$_2$. Incubate the diluted culture at 37 °C for 1 h.

2. Add approximately 10^7 phage from a previous lysate to 1 mL of the bacterial subculture in **step 1** and incubate at 37 °C for 1 h until the culture becomes clear, indicating cell lysis. Add chloroform (50 μL) to the cell lysate (*see* **Note 1**), followed by vortexing for 1 min. Clear cell debris after lysis by centrifugation; save the supernatant in a fresh tube at 4 °C as the P1 phage lysate solution.

3. Collect 1 mL overnight LB culture of recipient BAP1 cells by centrifugation and resuspend in 200 μL fresh LB medium containing 5 mM CaCl$_2$ and 100 mM MgSO$_4$. Mix this culture (100 μL) with 100 μL of the P1 phage lysate solution prepared in **step 2**. After a 30 min incubation at 37 °C, add 300 μL of 0.1 M sodium citrate and 1 mL LB medium to the reaction and incubate the mixture at 37 °C for 1 h to allow sufficient time for expression of the tetracycline resistance marker.

4. Harvest the transduced cells after P1 transduction by centrifugation and resuspend in 0.1 M sodium citrate solution. Plate the resulting suspension on LB agar containing tetracycline for isolation of individual colonies.

5. After overnight incubation, re-streak several colonies from the plate onto a fresh LB agar plate containing 0.1 M sodium citrate and tetracycline. Individual colonies are then picked and tested for the desired gene knockout by PCR and phenotypic analysis.

3.2   *trfA* Chromosomal Integration

1. The second step is to equip *E. coli* with the ability to increase the copy number of specific copy-up plasmids, enabling the production of varying metabolic levels of newly introduced natural product pathways. For this step, integrate the *trfA* gene into the chromosome of the *E. coli* strain produced according to Subheading 3.1 above. The *trfA* gene product

initiates plasmid replication from *oriV* and thus increases copy number by 10–50-fold [10–13]. Based upon a method described previously [14], replace the *lacZ* gene in the chromosome of BAP1 (*araBADC::Tn10*) with a cassette containing the *trfA* gene and a kanamycin resistance gene. Specifically, transfer plasmid pJW410, a derivative of pBRINT-TsKm (ampicillin resistant), containing the *araC-para-trfA-Kan* cassette flanked by *lacZ* homology sequences into BAP1 (*araBADC::Tn10*) by electroporation. Grow the transformants on an LB plate containing ampicillin and kanamycin at 30 °C.

2. Pick single colonies and re-streak onto a fresh LB plate containing kanamycin only. Grow re-streaked single colonies in LB medium containing kanamycin at 30 °C for 3 h. Dilute the culture with fresh LB medium and plate on LB agar plates containing kanamycin, followed by an overnight incubation at 42 °C.

3. Pick individual ampicillin-sensitive colonies (indicating pJW410 plasmid loss) and test for the desired *trfA* gene integration by PCR and the copy-up phenotype assay.

3.3   *recA* Deletion

1. Delete the *recA* gene from the chromosome of BAP1 (*araBADC::Tn10, lacZ::trfA-Kan*) to improve the stability of extrachromosomal plasmid DNA [15]. This step will potentially limit loss or rearrangements of large foreign natural product gene clusters localized to plasmids during reconstitution efforts.

2. PCR amplify a DNA fragment containing a chloramphenicol-resistance gene (*cat*) from plasmid pKD3 [9] using primers recA1 and recA2.

3. Gel-purify the resulting PCR product using a QIAquick gel extraction kit.

4. The PCR product is transformed via electroporation (*see* **Note 2**) into *E. coli* K-12 strain BW25113 (obtained from the *E. coli* Genetic Stock Center, Yale University) (*see* **Note 3**) harboring pKD46.

5. Once recombination is completed and confirmed, prepare genomic DNA from the *recA*– BW25113 strain for a second PCR by centrifuging 50 μL of an overnight 2 mL LB culture, washing the pelleted cells once, and resuspending in 50 μL water prior to incubation at 95 °C for 10 min using a heating block. The resulting supernatant can be used as a PCR template. Amplify the *recA::cat* cassette using primers recA3 and recA4 and insert into BAP1 (*araBADC::Tn10, lacZ::trfA-Kan*) by another round of lambda Red recombination (*see* **Note 4**).

6. Remove the *cat* resistance marker by FRT-site-mediated recombination with pCP20, a FLP synthesis plasmid [9]. Specifically, the pCP20 plasmid is transformed into BAP1 (*araBADC::Tn10, lacZ::trfA-Kan, recA::cat*) to express the desired flippase and remove the *cat* resistance gene. The resulting transformant is incubated at 42 °C to allow the loss of the temperature-sensitive pCP20 plasmid. The resulting strain *E. coli* BAP1 (*araBADC::Tn10, lacZ::trfA-Kan, ΔrecA*) is referred to as BTRA.

7. Confirm the deletion of the *recA* gene by PCR and a UV exposure assay.

3.4 Propionyl-CoA Carboxylase (PCC) Chromosomal Integration

The final and most recent step in augmenting the genetic and metabolic background of *E. coli* in support of complex natural product biosynthesis is the replacement of the nonessential chromosomal gene *codB* with a *pccAB-cat* cassette. The encoded PCC enables the conversion of propionyl-CoA to (2S)-methylmalonyl-CoA, which is an important metabolic substrate for polyketide biosynthesis [16, 17]. The steps in the integration process are outlined below:

1. Amplify the *pccAB* genes by PCR from pBP144 [3] using the primers specified in Table 2. The primers are designed to provide flanking *Nde*I (5′) and *Bam*HI (3′) sites to the resulting PCR product.

2. Isolate the PCR fragment using the GeneJET gel extraction kit, digest with the flanking restriction sites, and ligate into plasmid pET21c.

3. PCR amplify the *cat* gene, flanked by FRT sites, from pKD3 using primers indicated in Table 2. This set of primers introduces *Sac*I (5′) and *Xho*I (3′) flanking restriction sites to the PCR product.

4. Digest the *cat* PCR product using the flanking restriction sites and insert into the plasmid containing the *pccAB* genes to generate the *pccAB-cat* cassette.

5. PCR amplify the *pccAB-cat* cassette (Table 2), to be used for homologous recombination by means of the lambda Red procedure.

6. Complete homologous recombination into strain BW25113 using the protocol for lambda Red recombination described previously (*see* **Note 5**).

7. Confirm integration of the *pccAB-cat* cassette into strain BW25113 by PCR amplification and sequencing of the inserted cassette. Use the genomic DNA of this confirmed strain for a second round of lambda Red recombination.

Table 2
Primers used in the engineering of *E. coli*

Name	Sequence
recA1	Forward: 5′-ATGCGACCCTTGTATCAAACAAGACGATTAAAAATCTTCGTTAGTTTCGTGTAGGCTGGAGCTGCTTC-3′
recA2	Reverse: 5′-CAGAACATATTGACTATCCGGTATTACCCGGCATGACAGGAGTAAAAATGATGGGAATTAGCCATGGTCC-3′
recA3	Forward: 5′-CTTGTCGAGCCCAAGGAACA-3′
recA4	Reverse: 5′-GAACCCGTCGTGGTGGAAAT-3′
pccAB1	Forward: 5′-GATCATATGGCTGCTCCGGCTTCTG-3′
pccAB2	Reverse: 5′-GTAGGATCCTTACAGCGGGATG-3′
cat1	Forward: 5′-GCGGAGCTCGTGTAGGCTGGAGCTGCTTC-3′
cat2	Reverse: 5′-GCGCTCGAGCATATGAATATCCTCCTTGG-3′
pccAB-cat1	Forward: 5′-TAGAATGCGCGCGGATTTTTTGGGTTTCAAACAGCAAAAAGGGGAATTTCAGATCTCGATCCGCGG-3′
pccAB-cat2	Reverse: 5′-ACCGGGCGTTAATAATTGTTTGTAAAGCGTTATTCGACACTGTTAGCCTCCAAAAACCCCTCAAG-3′
pccAB-cat3	Forward: 5′-TCCTGCGTCGTTGGATCAGA-3′
pccAB-cat4	Reverse: 5′-GTGCCGGACTGATTCCAGTT-3′

8. Prepare BW25113 (*pccAB-cat*) genomic DNA for a second round of PCR and lambda Red recombination: wash 500 µL of a 3 mL overnight LB culture (incubated at 37 °C with shaking) once and resuspend in 200 µL water. Boil the sample at 95 °C for 5 min in a heating block; use the resulting supernatant for PCR.

9. Complete PCR using the BW25113 (*pccAB-cat*) genomic DNA template with primers (Table 2) that will yield an amplified product with a minimum of 200 base pairs on either side of the *pccAB-cat* cassette (*see* **Note 6**).

10. Repeat the lambda Red procedure within BTRA using a PCR product which has a 278-nt left homology arm and a 252-nt right homology arm (Fig. 1).

11. Transform plasmid pCP20 into BTRA (*pccAB-cat<>codB*) to eliminate the *cat* chloramphenicol resistance gene. The resulting *E. coli* strain, named BTRAP, is confirmed by diagnostic PCR and tested for (2*S*)-methylmalonyl-CoA metabolism through the phenotypic assay described below.

3.5 Genetic and Phenotypic Analysis of Strain Construction

Assays to confirm the above genetic manipulations can be divided into genetic and phenotypic categories. At the genetic level, PCR amplification and fragment sequencing allow confirmation of gene insertion or deletion (Fig. 2). Complementary assays, to be described below, use a phenotype associated with the desired genetic change. Eventually, the phenotypic assays should support the larger objective of complex natural product heterologous biosynthesis.

The simplest phenotypic assay associated with the various genetic construction steps is the correct antibiotic resistance patterns associated with each step. Though simple, this is the first and essential signature needed for strain verification.

As an example, strain BAP1 (*araBADC::Tn10*) or BT2 is resistant to the antibiotic tetracycline. The deletion of the *araBADC* cassette is also confirmed by a phenotype assay in which the desired inability of the strain to grow in M9 medium with arabinose as the sole carbon source is tested (Fig. 3), indicating that the arabinose utilization pathway is successfully disrupted (*see* **Note 7**).

1. Likewise, strain BAP1 (*araBADC::Tn10, lacZ:: trfA-Kan*) or BT3 can be selected for resistance to tetracycline and kanamycin. In addition, a copy-up assay can be performed to confirm the induced copy-up capability of a fosmid with an *oriV* origin of replication (Fig. 4). In this assay, *E. coli* BAP1 (negative control) and EPI300 (positive control) are transformed with the pCC1FOS™ fosmid and grown in LB medium with and without the inducer arabinose (*see* **Note 8**). Strains can be grown overnight, adjusted to the same optical density, and

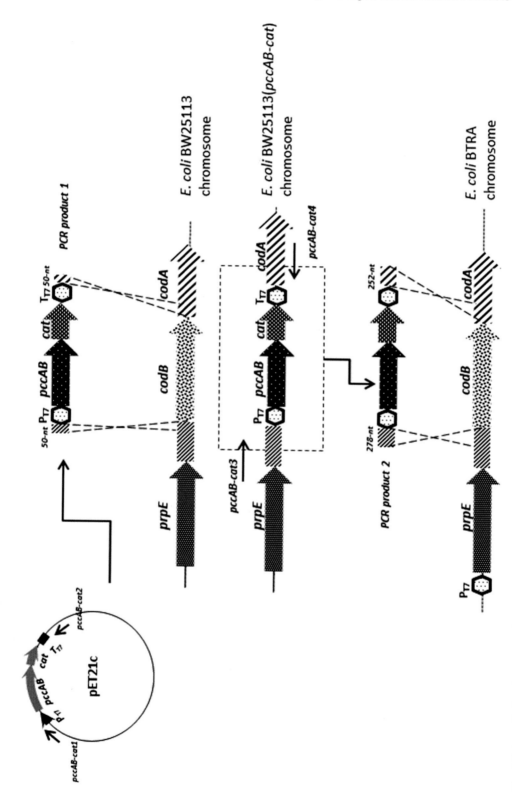

Fig. 1 Schematic for chromosomal engineering using lambda Red recombination and a sequential process of genetic transfer between K and B strains of *E. coli*. Table 2 primers are indicated in association with first and second PCR steps. P$_{T7}$; T7 promoter; T$_{T7}$; T7 terminator

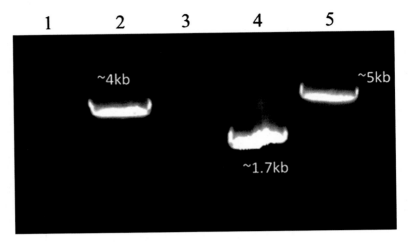

Fig. 2 PCR verification of chromosomal insertion of *pccAB*. *Lanes 1* and *3*: marker (*top band*, 10 kb); *Lane 2*: PCR of final BTRAP genomic DNA containing *pccAB* (post-*cat* removal); *Lane 4*: negative control: PCR of starting BTRA genomic DNA insertion region; *Lane 5*: PCR of BTRAP genomic DNA containing *pccAB-cat*

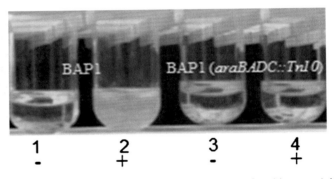

Fig. 3 Engineered *araBADC* disruption and assessment of arabinose catabolism. BAP1 [*1* and *2*] and BAP1 (*araBADC::Tn10*) [*3* and *4*] cultured in M9 minimal medium without [−] and with [+] L-arabinose

fosmid DNA can be isolated from each strain using a Qiaprep Spin miniprep kit (*see* **Note 9**). As indicated, the original BAP1 is unable to increase the copy number of the pCC1FOS™ fosmid vector; whereas, strains EPI300 and BAP1 (*araBADC::Tn10, lacZ::trfA-Kan*) containing the inducible *trfA* gene show copy-up capability (*see* **Note 10**).

2. To confirm the *recA* deletion in BTRA, assay for sensitivity to UV light, as *recA* strains are more sensitive than *recA*+ strains. Grow BAP1 (*araBADC::Tn10*) and BTRA overnight at 37 °C with shaking in LB medium and then streak strains horizontally across an LB agar plate. Expose one half of the plate to UV light (UVLMS-38 8-W lamp (UVP, Upland, CA)) at 254 nm wavelength and at a distance of 22 cm for 15 s; shield

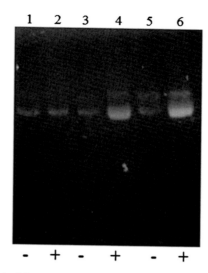

Fig. 4 Copy-up test of the constructed *E. coli* BAP1 (*araBADC::Tn10, lacZ::trfA-Kan*) by agarose gel electrophoresis. *Lanes 1* and *2*: negative control BAP1 containing pCC1FOS™ fosmid vector with (+) and without (−) induction (no DNA yield improvement was observed); *Lanes 3* and *4*: positive control EPI300 containing the pCC1FOS™ fosmid vector with and without induction (DNA yield was improved after induction); *Lanes 5* and *6*: BAP1 (*araBADC::Tn10, lacZ::trfA-Kan*) containing the pCC1FOS™ fosmid vector with and without induction (DNA yield was improved after induction)

UV exposure

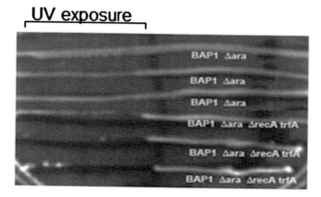

Fig. 5 UV test for *recA* gene deletion. Comparison of BAP1 (*araBADC::Tn10*) and BAP1 (*araBADC::Tn10, lacZ::trfA-Kan, ΔrecA*) after UV exposure. The removal of *recA* results in UV sensitivity

the other half of the plate from UV exposure. Compare the growth of UV-exposed and non-UV-exposed cells after overnight incubation at 37 °C (Fig. 5).

3. In order to test functionality of the *pccAB* gene products in BTRAP, obtain the DEBS1 gene from pBP144 [3] via digestion with *Nde*I and *Eco*RI and introduce into pET21c to generate

a   **b**

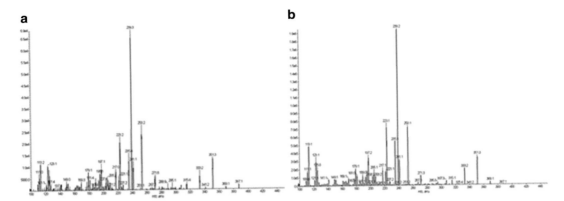

Fig. 6 Mass spectrum of 6dEB produced from BTRAP (pLF04/pBPJW144) (**a**) when compared to 6dEB standard (**b**)

plasmid pLF04. The polyketide product 6-deoxyerythronolide B (6dEB), which requires propionyl- and (2S)-methylmalonyl-CoA, is produced by the DEBS1, 2, and 3 enzymes encoded on plasmids pBPJW130 [18] and pLF04. As such, 6dEB production is tested by culturing BTRAP (pLF04/pBPJW130) in 2 mL of LB medium containing selection antibiotics and 20 mM sodium propionate for 5 days at 22 °C with shaking prior to extraction (1:1 by volume) with ethyl acetate, evaporating the solvent under vacuum, and resuspending the extract in methanol for LC-MS analysis (Fig. 6); successful production then confirms the functionality of the introduced *pccAB* genes (*see* **Note 11**).

4. All MS analyses should be conducted in positive ion mode and chromatography performed on a Waters X Terra C18 column (5 μm, 2.1 mm×250 mm). After an injection of 10 μL of extract or 6dEB standard, a linear gradient of 30 % buffer A (95 % water–5 % acetonitrile–0.1 % formic acid) to 100 % buffer B (5 % water–95 % acetonitrile–0.1 % formic acid) should be used at a flow rate of 0.2 mL/min.

4   Notes

1. The chloroform used in the P1 transduction procedure may lyse recipient cells, reducing the transduction efficiency. As such, it is recommended that chloroform be evaporated from the P1 phage lysate by air drying before adding P1 to recipient cells.

2. It is important to give the lambda Red recombination system ample time to function. Therefore, let electroporated cells incubate at 37 °C for a minimum of 3 h; this extends the recovery period and also provides more time for chromosomal

integrates to produce sufficient amounts of the selection antibiotic resistance protein.

3. The BW25113 strain was chosen based upon its common usage and higher efficiency in the lambda Red recombination protocols outlined by Datsenko et al. [9].

4. Our group has observed a reduced efficiency for the lambda Red procedure when using B strains of *E. coli* (such as BL21(DE3)) when compared to K strains. Thus, we have adopted an approach in which we first complete the procedure in a K strain of *E. coli* and then transfer the chromosomal modification to the desired B strain by a second round of lambda Red recombination or P1 transduction.

5. This is another example of a means of transferring an engineered cassette from a K to a B strain of *E. coli*.

6. In our experience, 50 bp homology arms, though efficient when using K *E. coli* strains, are considered too short for chromosomal engineering in a B strain. We therefore use PCR fragments with a minimum of 200 bp homology for both flanking regions. As a second example, the deletion of *recA* outlined in Subheading 3.3 featured flanking PCR fragment homology regions of 340 and 850 bp.

7. For the phenotypic assay to test growth on arabinose, it is important that the inoculum prepared from the LB starter culture is thoroughly washed before inoculating into M9 medium containing arabinose. Residual amounts of nutrients in the LB medium may cause false negative results if not completely removed in the wash step.

8. Although a high concentration of arabinose may improve the signal associated with the copy-up assay, the deletion of the arabinose operon ensures that even a low concentration of arabinose (2 mg/L) is sufficient for copy number amplification.

9. For the copy-up comparison, cultures capable of arabinose induction displayed a lower cell density, as the increased copy number of the fosmid presumably imposed intracellular metabolic burden. As such, it is essential that the cell density of the cultures being compared be normalized before isolation and assessment of the fosmid.

10. In addition to the copy-up test, the replacement of the *lacZ* gene with the *trfA-Kan* cassette could also be confirmed by a blue-white screening test. The parent strain with an intact *lacZ* gene would generate blue colonies on an X-gal plate; whereas, the BT3 strain would only generate white colonies.

11. Intracellular generation of the (2S)-methylmalonyl-CoA metabolite could also be tested directly by LC-MS analysis, as described previously [19].

Acknowledgement

This work was supported by grants from the NIH (AI074224) and NSF (0712019 and 0924699) in collaboration with EarthGenes Pharmaceuticals. Guidance in strain design and assessment was provided by Drs. Asuncion Martinez, David Rothstein, and Lance Davidow of EarthGenes.

References

1. Zhang H, Boghigian BA, Armando J et al (2011) Methods and options for the heterologous production of complex natural products. Nat Prod Rep 28:125–151

2. Ongley SE, Bian X, Neilan BA et al (2013) Recent advances in the heterologous expression of microbial natural product biosynthetic pathways. Nat Prod Rep 30:1121–1138

3. Pfeifer BA, Admiraal SJ, Gramajo H et al (2001) Biosynthesis of complex polyketides in a metabolically engineered strain of E. coli. Science 291:1790–1792

4. Golomb M, Chamberlin M (1974) Characterization of T7-specific ribonucleic acid polymerase. IV. Resolution of the major in vitro transcripts by gel electrophoresis. J Biol Chem 249:2858–2863

5. McAllister WT, Morris C, Rosenberg AH et al (1981) Utilization of bacteriophage T7 late promoters in recombinant plasmids during infection. J Mol Biol 153:527–544

6. Studier FW, Moffatt BA (1986) Use of bacteriophage T7 RNA polymerase to direct selective high-level expression of cloned genes. J Mol Biol 189:113–130

7. Lambalot RH, Gehring AM, Flugel RS et al (1996) A new enzyme superfamily—the phosphopantetheinyl transferases. Chem Biol 3:923–936

8. Quadri LE, Weinreb PH, Lei M et al (1998) Characterization of Sfp, a Bacillus subtilis phosphopantetheinyl transferase for peptidyl carrier protein domains in peptide synthetases. Biochemistry 37:1585–1595

9. Datsenko KA, Wanner BL (2000) One-step inactivation of chromosomal genes in Escherichia coli K-12 using PCR products. Proc Natl Acad Sci U S A 97:6640–6645

10. Hradecna Z, Wild J, Szybalski W (1998) Conditionally amplifiable inserts in pBAC vectors. Microb Comp Genomics 3:58

11. Wild J, Hradecna Z, Szybalski W (2002) Conditionally amplifiable BACs: switching from single-copy to high-copy vectors and genomic clones. Genome Res 12:1434–1444

12. Wild J, Szybalski W (2004) Copy-control pBAC/oriV vectors for genomic cloning. Methods Mol Biol 267:145–154

13. Wild J, Hradecna Z, Szybalski W (2001) Single-copy/high-copy (SC/HC) pBAC/oriV novel vectors for genomics and gene expression. Plasmid 45:142

14. Le Borgne S, Palmeros B, Valle F et al (1998) pBRINT-Ts: a plasmid family with a temperature-sensitive replicon, designed for chromosomal integration into the lacZ gene of Escherichia coli. Gene 223:213–219

15. Phue JN, Lee SJ, Trinh L et al (2008) Modified Escherichia coli B (BL21), a superior producer of plasmid DNA compared with Escherichia coli K (DH5alpha). Biotechnol Bioeng 101:831–836

16. Rodriguez E, Gramajo H (1999) Genetic and biochemical characterization of the alpha and beta components of a propionyl-CoA carboxylase complex of Streptomyces coelicolor A3(2). Microbiology 145:3109–3119

17. Chan YA, Podevels AM, Kevany BM et al (2009) Biosynthesis of polyketide synthase extender units. Nat Prod Rep 26:90–114

18. Zhang H, Wang Y, Wu J, Skalina K et al (2010) Complete biosynthesis of erythromycin A and designed analogs using E. coli as a heterologous host. Chem Biol 17:1232–1240

19. Armando JW, Boghigian BA, Pfeifer BA (2012) LC-MS/MS quantification of short-chain acyl-CoA's in Escherichia coli demonstrates versatile propionyl-CoA synthetase substrate specificity. Lett Appl Microbiol 54:140–148

Chapter 9

Screening for Expressed Nonribosomal Peptide Synthetases and Polyketide Synthases Using LC-MS/MS-Based Proteomics

Yunqiu Chen, Ryan A. McClure, and Neil L. Kelleher

Abstract

Liquid chromatography–mass spectrometry (LC-MS)-based proteomics is a powerful technique for the profiling of protein expression in cells in a high-throughput fashion. Herein we report a protocol using LC-MS/MS-based proteomics for the screening of enzymes involved in natural product biosynthesis, such as nonribosomal peptide synthetases (NRPSs) and polyketide synthases (PKSs) from bacterial strains. Taking advantage of the large size of modular NRPSs and PKSs (often >200 kDa), size-based separation (SDS-PAGE) is employed prior to LC-MS/MS analysis. Based upon the protein identifications obtained through software search, we can accurately pinpoint the expressed NRPS and/or PKS gene clusters from a given strain and growth condition. The proteomics screening result can be used to guide the discovery of potentially new nonribosomal peptide and polyketide natural products.

Key words Proteomics, Liquid chromatography, Mass spectrometry, Nonribosomal peptide, Polyketide, Natural product

1 Introduction

Many natural products with antibiotic, anticancer, and antifungal activities are synthesized by nonribosomal peptide synthetases (NRPSs) and polyketide synthases (PKSs) [1]. Multimodular NRPS and PKS enzymes are often very large proteins (>200 kDa) and contain many functional domains for the assembly of simple building blocks into natural products [2]. Traditionally, natural product discovery has relied on a bioactivity-guided screening approach, which includes repeated fractionation to isolate the compounds of interest [3]. This approach often suffers from a high frequency of rediscovering known natural products [4]. With the advent of the genomic revolution and whole genome sequencing, researchers have realized that microorganisms possess a far greater genetic potential for natural product biosynthesis than what had been observed [5]. Through rational prediction followed by tar-

Bradley S. Evans (ed.), *Nonribosomal Peptide and Polyketide Biosynthesis: Methods and Protocols*, Methods in Molecular Biology, vol. 1401, DOI 10.1007/978-1-4939-3375-4_9, © Springer Science+Business Media New York 2016

geted detection of the natural products based on the genetic information, an approach known as "genome mining" has led to the discovery of a number of new natural products [6–8]. On the other hand, genome mining approaches also have faced roadblocks in discovery as some biosynthetic pathways are not actively expressed or expressed at extremely low levels in laboratory conditions, also known as "cryptic" pathways [9, 10].

To complement the genome mining and bioassay-based natural product discovery approaches, a proteomics approach was developed by the Kelleher group, termed Proteomic Investigation of Secondary Metabolism (PrISM) [11]. By initially identifying the actively expressed biosynthetic proteins, PrISM avoids pursuing "cryptic" pathways. Natural products are predicted according to the genetic information and discovered in a targeted way. The PrISM approach has led to the discovery of several new natural products including koranimine, flavopeptins, gobichelin A and B, as well as the identification of biosynthetic pathways for many known natural products [11–15]. Most recently, the "depth" at which PrISM could be applied (i.e., detecting expression of gene clusters at low levels) was evaluated using a single Actinomycete strain with its genome sequence; this was called Genome-enabled PrISM [16].

In this chapter, we present the detailed protocol for the PrISM screening of Actinobacteria strains for the expressed NRPS and PKS proteins. In this protocol, Actinobacteria strains are grown under a variety of culture conditions and harvested at several time points. Cells are lysed by mechanical disruption using bead beating, and the proteomes are fractionated by size using one dimensional sodium dodecyl sulfate polyacrylamide gel electrophoresis (SDS-PAGE). The high molecular weight proteins (>150 kDa), often representing NRPSs or PKSs, are subjected to in-gel trypsin digestion followed by liquid chromatography coupled with mass spectrometric (LC-MS) analysis. The LC-MS data are automatically analyzed using proteomics software and the protein identifications reveal which NRPS and/or PKS pathways are actively expressed under the growth conditions—guiding the discovery of the corresponding natural products.

2   Materials

2.1   Culture Media

1. *Malt Yeast Glucose*: 10 g/L malt extract, 4 g/L yeast extract, 4 g/L glucose, pH to 7.

2. *ATCC 172*: 10 g/L glucose, 20 g/L soluble starch, 5 g/L yeast extract, 5 g/L N-Z amine type A, 1 g/L $CaCO_3$.

3. *ISP4*: 10 g/L soluble starch, 1 g/L K_2HPO_4, 1 g/L $MgSO_4$, 1 g/L NaCl, 2 g/L $(NH_4)_2SO_4$, 2 g/L $CaCO_3$, 0.001 g/L

$FeSO_4 \cdot 7H_2O$, 0.001 g/L $ZnSO_4 \cdot 7H_2O$, 0.001 g/L $MnCl_2 \cdot 4H_2O$, pH to 7.0.

4. *Arginine Glycerol Salts*: 0.85 g/L l-arginine, 12.5 g/L glycerol, 1 g/L K_2HPO_4, 1 g/L NaCl, 0.5 g/L $MgSO_4 \cdot 7H_2O$, 0.01 g/L $FeSO_4 \cdot 7H_2O$, 0.001 g/L $CuSO_4 \cdot 5H_2O$, 0.001 g/L $ZnSO_4 \cdot 7H_2O$, 0.001 g/L $MnSO_4 \cdot H_2O$, pH to 7.0.

5. *Soy Flour Mannitol*: 20 g/L d-mannitol, 20 g/L soy flour, pH to 7.0 (This medium needs to be autoclaved twice).

6. *4× R2A*: 2 g/L peptone, 2 g/L starch, 2 g/L glucose, 2 g/L yeast extract, 2 g/L casein hydrolysate, 1.2 g/L K_2HPO4, 1.2 g/L sodium pyruvate, 0.1 g/L $MgSO_4$.

7. *MSB medium*: To make 1 L MSB medium, mix separately sterilized A (40 mL), B (10 mL), C (5 mL), and ddH_2O. Supplement with 10 mM (final concentration) of sodium succinate and 0.05 % casamino acids.

   (a) 1 M $Na_2HPO_4 + KH_2PO_4$, pH 7.3. Mix ~38.25 mL 1 M Na_2HPO_4 stock solution and ~11.5 mL KH_2PO_4 stock solutions and adjust pH to 7.3.

   (b) Dissolve 20 g nitrilotriacetic acid and 14 g KOH in 700 mL H_2O. Then add the following chemicals individually in order: 28 g $MgSO_4$, 6.67 g $CaCl_2 \cdot 2H_2O$ (then add KOH to raise pH till dissolved), 0.0185 g $(NH_4)_6Mo_7O_{24} \cdot 4H_2O$, 0.198 g $FeSO_4 \cdot 7H_2O$, 100 mL Hunter's metals 44*. Adjust pH to 6.8 with KOH. Adjust volume to 1 L.

   *Hunter's metals 44: 2.5 g EDTA (free acid) dissolved in 800 mL dH_2O, add 10.95 g $ZnSO_4 \cdot 7H_2O$, 5 g $FeSO_4 \cdot 7H_2O$, 1.54 g $MnSO_4 \cdot H_2O$, 0.392 g $CuSO_4 \cdot 5H_2O$, 0.250 g $Co(NO_3)_2 \cdot 6H_2O$, 0.177 g $Na_2B_4O_7 \cdot 10H_2O$, add five drops of concentrated sulfuric acid, adjust volume to 1 L and store at 4 °C.

   (c) 20 % (wt/vol) $(NH_4)_2SO_4$.

2.2 Reagents

All solutions are prepared using ultrapure water (prepared by purifying deionized water to attain a resistivity of 18 MΩ • cm at 25 °C) and analytical grade reagents.

1. SDS lysis buffer (4×): 125 mM Tris (pH 6.8), 4 % (wt/vol) SDS, 40 % (vol/vol) glycerol, 10 % (vol/vol) β-mercaptoethanol, 0.01 % (wt/vol) bromophenol blue.

2. SDS-PAGE running buffer: 25 mM Tris, pH 8.3, 192 mM glycine, 0.1 % SDS.

3. Ammonium bicarbonate (NH_4HCO_3) solution: 100 mM in water, pH 7.8 and 50 mM in water, pH 7.8.

4. Tris-(2-carboxyethyl) phosphine (TCEP): 10 mM in 50 mM NH_4HCO_3, pH 7.

5. Iodoacetamide (IAA): 50 mM solution in water.

6. Trypsin: dissolve one vial of lyophilized trypsin in 40 μL trypsin resuspension buffer provided by manufacturer, and heat at 30 °C for 15 min, then dilute 40-fold using 50 mM NH_4HCO_3 to a final concentration of 12.5 μg/mL. This solution should be freshly prepared.

7. 1:2 5 % formic acid–acetonitrile: 1 part 5 % v/v formic acid solution in water to 2 parts LC-MS grade acetonitrile.

8. LC-MS buffer A: LC-MS grade water with 0.1 % formic acid.

9. LC-MS buffer B: LC-MS grade acetonitrile with 0.1 % formic acid.

2.3 Equipment

1. LTQ-Orbitrap mass spectrometer (Thermo-Fisher Scientific, MA, USA) with a nanoelectrospray source.

2. Nano LC system with autosampler.

3. Nano LC trap column (2 cm length × 100 μm ID, 5 μm C18 particle).

4. Nano LC analytical column (10 cm length × 75 μm ID, 5 μm C18 particle).

5. Vortexer.

6. Vortexer adapter for 1.5 mL tubes, horizontal orientation (MOBIO, CA, USA).

7. Sonic dismembrator with 1/8 in. microtip.

8. Electrophoresis system and power supply.

9. SpeedVac concentrator.

10. Ultrasonic cleanser.

11. Shaker incubator.

12. 18 × 150 mm glass tubes with caps (autoclave).

13. Centrifuge for 1.5 mL tube.

14. Protein LoBind Tubes (Eppendorf, Germany).

15. 15 mL conical centrifuge tubes.

16. 14 mm culture dish.

17. Razor blades.

18. Carbide beads (0.25 mm, MOBIO, CA, USA).

3 Methods

3.1 Bacteria Strain Growth

This protocol uses Actinobacteria strains as an example, but the PrISM methodology can be applied to other species of bacteria as well. We suggest to use at least three types of culture media and to collect samples at 4 time points (*see* **Note 1**).

1. Inoculate 5 mL of culture media in an 18×150 mm sterile glass tube with a few single colonies of the strain of interest to generate a starting culture.

2. The tubes are placed at 45° angle in an incubator shaker and are allowed to grow at 30 °C with rotation at 250 rpm.

3. After ~2–3 days, when the strain has reached log growth phase, transfer an aliquot (500 µL) of the starting culture to the screening media (5 mL each) which are also contained in 18×150 mm glass tubes.

4. For each screening medium, four tubes are prepared to be harvested at 4 time points (see **Note 2**). For a typical Actinobacteria strain, we recommend harvesting at 24 h, 48 h, 72 h, and 96 h after transferring to screening media.

3.2 Cell Lysis

1. At each time point for harvesting, the entire culture is transferred to a 15 mL conical centrifuge tube and subjected to centrifugation at $>4000 \times g$ for 10 min (see **Note 3**). The culture supernatants are removed and the cell pellets can be stored at –80 °C until ready for lysis.

2. Resuspend the cell pellet in an equal volume of 4× SDS lysis buffer, and transfer it to a 1.5 mL centrifuge tube. Place the sample on a hotplate set at 95 °C for 10 min (see **Note 4**).

3. Add a small amount (the same volume as the cell pellet) of carbide beads (0.25 mm) to the sample, and place the tubes on a vortexer with a horizontal orientation. Vortex for 30 min at maximum speed (see **Note 5**).

4. Sonicate the sample for 1 min using a sonic dismembrator with a 1/8 in. microtip (see **Note 6**).

5. Heat the samples at 95 °C for 10 min.

6. Centrifuge the samples at $20,000 \times g$ for 10 min. Extract the supernatant to a new 1.5 mL centrifuge tube for SDS-PAGE analysis. This is the proteomic sample for a given strain under a certain growth condition. The samples can be stored at –80 °C until ready for SDS-PAGE separation.

3.3 Separation of the Proteomic Samples Using SDS-PAGE

1. Assemble the 10 % T precast gels (15 well) onto the electrophoresis system following the manufacturer's instructions. Fill the inner and outside chambers of the cassette with the SDS running buffer.

2. Load 5 µL of the prestained protein standard into the first lane. Load 15 µL of the proteomic samples from Subheading 3.3 into each other lane. To avoid cross-contamination, try to load samples from the same strain on the same gel. Perform the electrophoresis at a constant voltage of 180 V, until the bromophenol blue dye reaches the bottom of the gel.

3. After SDS-PAGE, pry the gel plates open with a spatula. Place the gel in a clean 14 mm culture dish. Wash the gel with 200 mL of water three times for 5 min each. Remove the water and add ~50 mL of Bio-Safe™ Coomassie stain to the dish and let shake for 1 h. After staining, remove the stain and rinse with water for 3×10 min. Take a picture of the Coomassie stained gel and check the efficiency of the cell lysis (*see* **Note 7**) (Fig. 1).

3.4 In-Gel Tryptic Digestion

1. Remove most of the water from the dish—leaving a small amount to prevent the gel from drying. For each lane, a razor blade is used to excise the region above 250 kDa as one slice and the 150–250 kDa region as another slice (*see* **Note 8**). Cut each gel slice into 1×1 mm gel pieces and transfer using the razor blade into a 1.5 ml LoBind Eppendorf tube (*see* **Note 9**). Always use a new razor blade and a new tube for each lane from a gel sample.

2. Add 300 μL H₂O to each tube and vortex for 15 min.

3. Add 300 μL acetonitrile and wash for another 15 min.

4. Remove the supernatant carefully without picking up the gel pieces. Wash the gel pieces with 300 μL of 100 mM NH₄HCO₃ for 15 min. Discard the supernatant. Wash the gel pieces with 300 μL of 100 mM NH₄HCO₃–acetonitrile (50:50 vol/vol) for 15 min (*see* **Note 10**).

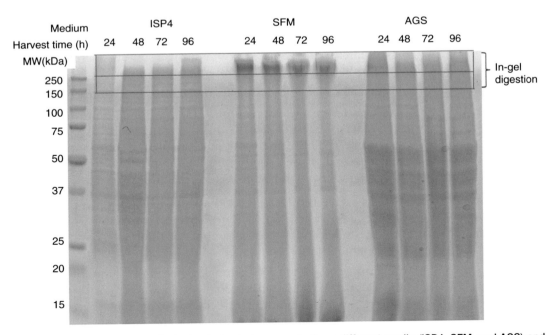

Fig. 1 PrISM screening for an Actinobacteria strain grown in three different media (ISP4, SFM, and AGS) and harvested at 4 time points (24, 48, 72, and 96 h). The proteome samples are separated on an SDS-PAGE gel and stained by Coomassie. The high molecular weight regions (>150 kDa, shown within *red* box) are subjected to in-gel trypsin digestion

5. Add 100 μL acetonitrile to dehydrate the gel pieces for 5 min.

6. Discard the supernatant. Dry the gel pieces in a SpeedVac for 5 min. The samples can be stored at –80 °C until ready to move onto the next step.

7. Reduce the disulfide bonds of proteins by adding 50 μL of 10 mM TCEP in 50 mM NH_4HCO_3, incubate at room temperature for 30 min. Discard the supernatant.

8. Add 50 μL of 50 mM iodoacetamide in 50 mM NH_4HCO_3 (freshly prepared). Incubate at room temperature in the dark for 30 min. Discard the supernatant.

9. Wash the gel pieces with 300 μL of 100 mM NH_4HCO_3 for 15 min. Discard the supernatant.

10. Wash the gel pieces with 300 μL of 50 mM NH_4HCO_3–acetonitrile (50:50 vol/vol) for 15 min. Discard the supernatant.

11. Add 100 μL of acetonitrile to dehydrate the gel pieces for 5 min. Discard the supernatant.

12. Dry the gel pieces in a SpeedVac for 5 min. The samples can be stored at –80 °C until ready to move onto the next step.

13. Add 20 μL of the trypsin solution (12.5 μg/mL) to each tube. Allow the gel bands to rehydrate in trypsin solution for 30 min on ice.

14. If the digestion solution is fully absorbed after 30 min, add an additional 10 μL trypsin solution to cover the gel pieces. Let the gel sit on ice for an additional 1 h.

15. Remove the excessive trypsin solution and add 50 mM NH_4HCO_3 (sufficient volume to cover the gel pieces). Keep a record of the quantity of liquid added to the gel pieces.

16. Incubate at 30 °C overnight.

17. Cool the sample to room temperature and add 2 μL of 100 % formic acid to quench the digestion.

18. Add acetonitrile (an equal volume as added from 15) to the tube and shake vigorously on a vortexer for 30 min.

19. Transfer the supernatant to a new clean Eppendorf LoBind tube. This contains the extracted peptides.

20. Add 100 μL of 1:2 5 % formic acid–acetonitrile to the gel pieces and let shake for 20 min. Transfer the supernatant to the tube in **step (19)**.

21. Repeat **step (20)**.

22. Add 100 μL of acetonitrile to the gel pieces. Shake for 10 min. Transfer the supernatant to the tube in **step (19)**.

23. Repeat **step(22)**.

24. Dry the peptides in the tube at **step (19)** completely using the SpeedVac. Store the sample in –80 °C freezer until ready for LC-MS/MS analysis.

3.5 LC-MS/MS

1. Add 20 µL of LC-MS buffer A to each tube. To solubilize the peptides, place the tubes in an ultrasonic cleanser containing ice water and sonicate for 10 min.

2. Centrifuge the samples at $20,000 \times g$ for 10 min to precipitate any insoluble materials. Transfer the supernatant to an LC sample vial.

3. The nanoLC method is configured as follows: the samples are initially loaded using 100 % LC-MS buffer A to the nanoLC trap column with a flow rate of 3 µL/min for 10 min. The eluent is diverted to the waste at this step.

4. Switch the valve so that the eluent is diverted to the nanoLC analytical column. The LC gradient is set as follows, with a flow rate at 300 nL/min:

Time (min)	% B
0	0
55	45
63	80
67	0
90	0

5. The eluted peptides are subjected to positive nanoelectrospray ionization (nanoESI) for mass spectrometric analysis using a Thermo LTQ-Orbitrap instrument (*see* **Note 11**). The instrument settings are configured as follows:

FTMS full scan Automatic Gaining Control (AGC): 1e6.

FTMS full scan maximum ion time: 2000 ms.

Ion trap MS^n AGC: 1e4.

Ion trap MS^n maximum ion time: 300 ms.

Mass spectrometry method setup:

FTMSfull scan from m/z 400–2000 using a 30,000 resolving power. For each full scan, the top five most intense ions are selected for data dependent fragmentation using collision induced dissociation (CID) at 35 eV. The fragments are detected in the ion trap. Minimum signal required for MS^2 is 500. Dynamic exclusion is set to 60 s, repeat count = 2. 1+ and unassigned charge states are rejected for MS^2 events.

3.6 Data Analysis

1. To process the LC-MS data, extract the peak file lists from the .raw data files by using the DTA generator module from COMPASS software suite (http://www.chem.wisc.edu/~coon/software.php). Assume the precursor charge state to be 2+ through 4+.

2. Search the data using a proteomics software. In this protocol we use the Mascot software (Matrix Science, UK) for demon-

stration but other software can be used. If the target strain has a sequenced genome or draft genome, the target protein database should be used for the search. Otherwise, the publicly available NCBI nonredundant protein database (NCBInr) can be used, selecting bacteria as the taxonomy.

3. The following searching parameters are recommended: 10 ppm precursor mass error tolerance; 0.8 Da MS² error tolerance; carbamidomethylation of cysteines as fixed modification; oxidation of methionines as variable modifications; maximal number of missed cleavages is 2; allowed peptide charge states are 2+, 3+, and 4+; data format is Mascot generic; and instrument type is ESI-TRAP.

4. After the Mascot search has finished, open the search results and perform filtering. We recommend the "ion score or expect cut-off" to be set to 30.

5. Browse the "protein family summary" and look for proteins that are associated with NRPS or PKS biosynthesis. For strains with sequenced genomes, the protein identification list will inform which high molecular weight NRPS and/or PKS proteins are expressed. The number of peptides identified from each proteomic sample is generally correlated with the expression level of the NRPS/PKS proteins under each growth conditions (Table 1).

Table 1

An example of protein identification list (partial) from Mascot search for an Actinobacteria strain using its own sequenced genome as the database

Protein description	Protein mass (kDa)	#Matches (sig)	#Sequence (sig)	%Coverage
NAD-glutamate dehydrogenase	185	262	68	48.60
GAF sensor hybrid histidine kinase	194	35	15	10.33
Protein of unknown function DUF3686	175	30	16	12.54
Hypothetical protein Sfla_5835	185	27	14	9.96
Hypothetical protein Sfla_5465	169	18	15	11.81
DNA-directed RNA polymerase, beta subunit	144	18	12	11.39
Amino acid adenylation domain protein	265	15	8	3.80
Bacterioferritin	18	8	6	39.62
Amino acid adenylation domain protein	340	8	5	1.93
Amino acid adenylation domain protein	511	8	5	1.60

Three "amino acid adenylation proteins" (i.e., NRPSs, bolded), 511 kDa, 340 kDa, and 265 kDa each, are identified, indicating these three NRPS proteins are expressed under this growth condition

6. If the genome sequence of the target strain is not available and the Mascot search is performed using NCBInr as the database, the protein identification list usually contains NRPS/PKS proteins from different organisms (Table 2). Depending on the sequence homology between the actual protein sequence in your sample and the sequence in the database, you may get one or two protein identifications that contain many peptide hits, or you may get many protein identifications each containing only one or two peptide hits. In the latter case and when the LC-MS/MS data are of high quality (as shown in Table 2), it is likely that the NRPS/PKS in your sample is a novel gene cluster that is not represented in the database.

4   Notes

1. For a general PrISM screening of Actinobacteria strains, the typical medium for starting culture is Malt Yeast Glucose (MYG) medium or ATCC 172 medium, in which most strains grow vigorously. For the screening media, the researchers can use whichever media they prefer. In our practice, we have used ISP4, Arginine Glycerol Salts (AGS), Soy Flour Mannitol (SFM), 4× R2A, and MSB media. All media are prepared using tap deionized water (do not use ultrapure water) and autoclaved.

2. This growth condition (30 °C, 3 day, 250 rpm) is suitable for most Actinobacteria strains. However, if the strain of interest has special growth needs (e.g., higher/lower temperature, special growth factors, light/dark environment, or faster/slower growth rate), the specific growth conditions should be adjusted accordingly.

3. For some strains, the cell pellet can be very loose. Be careful when pouring out the culture supernatant. Use a pipette when necessary. If the cells do not pellet down completely, repeat the centrifugation step at a higher speed then remove the medium completely.

4. Estimate the volume of the cell pellet. Then add an equal volume of the lysis buffer, e.g., 25–100 μL.

5. Bead beating has been demonstrated as the most powerful and universal method for lysis of cells and tissue. We have tested that carbide bead with 0.25 mm diameter is best for the lysis of bacteria samples. Placing the tubes in the horizontal orientation increases lysis efficiency. If the cells are not lysed using bead beating, other methods should be considered, sonication, freeze–thaw, French press, manual grinding, etc.

Table 2
An example of Mascot search result (partial) using NCBInr as the database for an Actinobacteria strain which has no sequenced genome available

Peptide sequence	Protein description	Accession #
DAEALVAYcDR	Nonribosomal peptide synthetase [Streptomyces kitasatoensis	BAH68474
FVADPFGEPGER	Nonribosomal peptide synthetase [Streptomyces griseolus]	BAH68409
ADGAVEYIGR	Pyoverdine sidechain peptide synthetase I, epsilon-Lys module [Pseudomonas syringae pv. tabaci ATCC 11528]	ZP_05637427
GVGPESVVGVAVPR	Nonribosomal peptide synthetase [Rhodococcus opacus B4]	YP_002779301
ADGNVDFLGR	Amino acid adenylation domain-containing protein [Opitutus terrae PB90-1]	YP_001818855
FVADPYGAPGSR	Nonribosomal peptide synthetase [Streptomyces gibsonii]	BAH68663
LGAGDDIPIGTPVAGR	Peptide synthetase 2 [Streptomyces roseosporus NRRL 11379]	AAX31558
DVVFGTTVSGR	Amino acid adenylation domain protein [Acetivibrio cellulolyticus CD2]	ZP_07326042
FTADPYGPAGSR	Putative NRPS [Streptomyces griseus subsp. griseus NBRC 13350]	YP_001824777
GAGPETLVAVALPR	Mannopeptimycin peptide synthetase MppB [Streptomyces hygroscopicus]	AAU34203
TVAALAALAR	Putative nonribosomal peptide synthetase [Nocardia farcinica IFM 10152]	YP_119006
LGAGDDIPIGSPVAGR	Amino acid adenylation domain-containing protein [Frankia sp. EAN1pec]	YP_001510190
VVAIALPR	Peptide synthetase ScpsA [Saccharothrix mutabilis subsp. capreolus]	AAM47272
VAGPALTDFLADQR	Nonribosomal peptide synthetase [Streptomyces griseolus]	BAH68411
LVLITDEAR	Nonribosomal peptide synthetase [Streptomyces griseolus]	BAH68411
DIATAYAAR	Amino acid adenylation domain-containing protein [Streptomyces roseosporus NRRL 15998]	ZP_04696845
IPLSYAQR	Nonribosomal peptide synthetase [Streptomyces fungicidicus]	ABD65957

All the peptides related to NRPS/PKS are extracted. These peptides match to NRPS proteins from different organisms, indicating the detected NRPS share a low sequence homology with any known NRPS gene clusters in the database and likely represent a new NRPS biosynthetic pathway

6. This short sonication step is to disrupt DNA in the cells. The undisrupted DNA molecules will cause streaking bands on the SDS-PAGE.

7. The Coomassie staining step is primarily to check the efficiency of cell lysis. A successfully lysed sample should show an abundant amount of proteins on the SDS-PAGE gel. If the cells are not lysed properly, little or no protein bands will show on the gel. In this case, there is no need to proceed to the following steps. Other cell lysis protocols should be attempted to extract the proteome. If the proteome sample is much diluted (>100 μL volume), and higher sensitivity of detection is desired, the sample can be separated on a preparative SDS-PAGE gel, which contains a single lane with 250 μL loading capacity.

8. We recommend performing in-gel trypsin digestion to all samples, whether they show visible protein bands at the high molecular weight region by Coomassie stain. Coomassie stain has a much lower sensitivity of detection than nanoLC-MS. We also suggest cutting the high molecular weight region into two separate gel slices, one contains proteins >250 kDa and one contains 150–250 kDa proteins. Additional size based fractionation will result in more protein identifications.

9. In-gel digestion procedure is very sensitive to the contamination of keratin. To minimize contamination, always wear gloves when handling gels and performing in-gel digestion. Always use LoBind tubes and use clean spatula when preparing reagents.

10. If the band pieces are still blue after this step, repeat the washing steps using 300 μL of 100 mM NH_4HCO_3 for 15 min and then 300 μL of 100 mM NH_4HCO_3–acetonitrile (50:50 vol/vol) for 15 min.

11. Other types of mass spectrometers that are capable of proteomic analysis can also be used, Q-Exactive, QqTOF, Ion trap, FT-ICR, etc. Users need to adjust the mass spectrometric method according to the type of instrument used.

Acknowledgement

This work was supported by National Institute of General Medical Sciences of the National Institutes of Health under award number R01 GM067725. We thank the Agricultural Research Service, US Department of Agriculture for providing the bacterial strains.

References

1. Walsh CT (2004) Polyketide and nonribosomal peptide antibiotics: modularity and versatility. Science 303:1805–1810

2. Fischbach MA, Walsh CT (2006) Assembly-line enzymology for polyketide and nonribosomal peptide antibiotics: logic, machinery, and mechanisms. Chem Rev 106:3468–3496

3. Koehn FE (2008) High impact technologies for natural products screening. Prog Drug Res 65:177–210

4. Baltz RH (2006) Marcel Faber Roundtable: is our antibiotic pipeline unproductive because of starvation, constipation or lack of inspiration? J Ind Microbiol Biotechnol 33:507–513

5. Nett M, Ikeda H, Moore BS (2009) Genomic basis for natural product biosynthetic diversity in the actinomycetes. Nat Prod Rep 26: 1362–1384

6. Lautru S, Deeth RJ, Bailey LM et al (2005) Discovery of a new peptide natural product by Streptomyces coelicolor genome mining. Nat Chem Biol 1:265–269

7. Corre C, Challis GL (2009) New natural product biosynthetic chemistry discovered by genome mining. Nat Prod Rep 26:977–986

8. Gross H, Stockwell VO, Henkels MD et al (2007) The genomisotopic approach: a systematic method to isolate products of orphan biosynthetic gene clusters. Chem Biol 14:53–63

9. Scherlach K, Hertweck C (2009) Triggering cryptic natural product biosynthesis in microorganisms. Org Biomol Chem 7:1753–1760

10. van Wezel GP, McDowall KJ (2011) The regulation of the secondary metabolism of Streptomyces: new links and experimental advances. Nat Prod Rep 28:1311–1333

11. Bumpus SB, Evans BS, Thomas PM et al (2009) A proteomics approach to discovering natural products and their biosynthetic pathways. Nat Biotechnol 27:951–956

12. Evans BS, Ntai I, Chen Y et al (2011) Proteomics-based discovery of koranimine, a cyclic imine natural product. J Am Chem Soc 133:7316–7319

13. Chen Y, McClure RA, Zheng Y et al (2013) Proteomics guided discovery of flavopeptins: anti-proliferative aldehydes synthesized by a reductase domain containing non-ribosomal peptide synthetase. J Am Chem Soc 135: 10449–10456

14. Chen Y, Ntai I, Ju KS et al (2012) A proteomic survey of nonribosomal peptide and polyketide biosynthesis in Actinobacteria. J Proteome Res 11:85–94

15. Chen Y, Unger M, Ntai I et al (2013) Gobichelin A and B: mixed-ligand siderophores discovered using proteomics. Medchemcomm 4:233–238

16. Albright JC, Goering AW, Doroghazi JR et al (2014) Strain-specific proteogenomics accelerates the discovery of natural products via their biosynthetic pathways. J Ind Microbiol Biotechnol 41:451–459

Chapter 10

Enhancing Nonribosomal Peptide Biosynthesis in Filamentous Fungi

Alexandra A. Soukup, Nancy P. Keller, and Philipp Wiemann

Abstract

Filamentous fungi are historically known as rich sources for production of biologically active natural products, so-called secondary metabolites. One particularly pharmaceutically relevant chemical group of secondary metabolites is the nonribosomal peptides synthesized by nonribosomal peptide synthetases (NRPSs). As most of the fungal NRPS gene clusters leading to production of the desired molecules are not expressed under laboratory conditions, efforts to overcome this impediment are crucial to unlock the full chemical potential of each fungal species. One way to activate these silent clusters is by overexpressing and deleting global regulators of secondary metabolism. The conserved fungal-specific regulator of secondary metabolism, LaeA, was shown to be a valuable target for sleuthing of novel gene clusters and metabolites. Additionally, modulation of chromatin structures by either chemical or genetic manipulation has been shown to activate cryptic metabolites. Furthermore, NRPS-derived molecules seem to be affected by cross talk between the specific gene clusters and some of these metabolites have a tissue- or developmental-specific regulation. This chapter summarizes how this knowledge of different tiers of regulation can be combined to increase production of NRPS-derived metabolites in fungal species.

Key words LaeA regulation, Cross talk, Chromatin regulation, Transcription factor

1 Introduction

In filamentous fungi, the genes required for biosynthesis of a given secondary metabolite are predominantly contiguously aligned in the genome [1]. These gene clusters typically contain one or more backbone gene(s), including nonribosomal peptide synthetases (NRPSs), polyketide synthases (PKSs), dimethylallyl tryptophan synthases (DMATs), and terpene cyclases (TCs). Adjacent to backbone genes are those genes encoding additional modifying enzymes (e.g., oxidoreductases, monooxygenases, etc.), as well as regulatory and resistance elements. The multiple levels of regulation involved in coexpression of these cluster genes can facilitate demarcation of cluster boundaries and can be used to facilitate activation of otherwise silent gene clusters, to be discussed in further detail

Bradley S. Evans (ed.), *Nonribosomal Peptide and Polyketide Biosynthesis: Methods and Protocols*, Methods in Molecular Biology, vol. 1401, DOI 10.1007/978-1-4939-3375-4_10, © Springer Science+Business Media New York 2016

below. The reader is encouraged to peruse recent reviews [2–6] for additional information.

NRPS-derived metabolites display a vast chemical diversity as they can incorporate proteinogenic and non-proteinogenic amino acids in their D- and L-configurations [7]. The incorporated amino acids can be connected in a linear or cyclic fashion that can undergo additional modifications by tailoring enzymes encoded in the respective gene cluster. This structural diversity of nonribosomal peptides is also reflected in their broad spectrum of biological activities utilized in medicinal and pharmaceutical research [8, 9], with penicillin, cephalosporin, and cyclosporin as the most prominent examples [10–13]. Despite the chemical variety produced by NRPSs, the standard NRPS enzyme itself is composed of three discrete canonical domains (adenylation (A), thiolation (T) or peptidyl carrier protein (PCP), and condensation (C) domains), each of which is referred to as one NRPS module. Each module is responsible for the recognition (via the A domain) and incorporation of a single amino acid into the growing peptide product. In general, an NRPS consists of one or more modules and can terminate in a condensation-like (C_T) domain that releases the peptide. Occasionally, epimerase (E) and N-methyltransferase (M) domains that convert L- to D-amino acids and N-methylate peptide bonds, respectively, are present within the NRPS [14, 15]. Deviations of the classical NRPS composition can be found in hybrid PKS/NRPS enzymes [16] (e.g., Fus1 responsible for fusarin C biosynthesis) and stand-alone monomodular NRPS-like enzymes [17] (e.g., TdiA responsible for terrequinone A biosynthesis) where not all canonical domains are present (Fig. 1).

The presence of at least one A domain in each NRPS enzyme facilitates identification in sequenced fungal genomes. However, in contrast to bacterial NRPS, knowledge about amino acid specificity of fungal A domains is limited up to date with only the anthranilate-activating A domain identified with some confidence [21]. Despite the variety of NRPS, NRPS-like, and PKS/NRPS hybrid enzymes predicted to be encoded by many fungal genomes (e.g., 20–30 predicted in *Aspergillus* and *Fusarium* spp. [22–26]), relatively little is known about the metabolites produced by these enzymes.

The first characterized NRPS-derived compounds identified from fungal species are produced under standard laboratory conditions. Pioneering work using protein purification of NRPS enzymes demonstrated their involvement in cyclosporine, enniatin, and beauvericin biosynthesis [13, 27, 28]. With genetic manipulation of fungal genomes becoming available, genetic and chemical pathways leading to NRPS-derived metabolites could be identified through relatively straightforward gene deletions of NRPS-encoding genes coupled with analytical screenings for loss of compound production. Examples of compounds and pathways that have been identified in this traditional fashion include those of cyclosporine

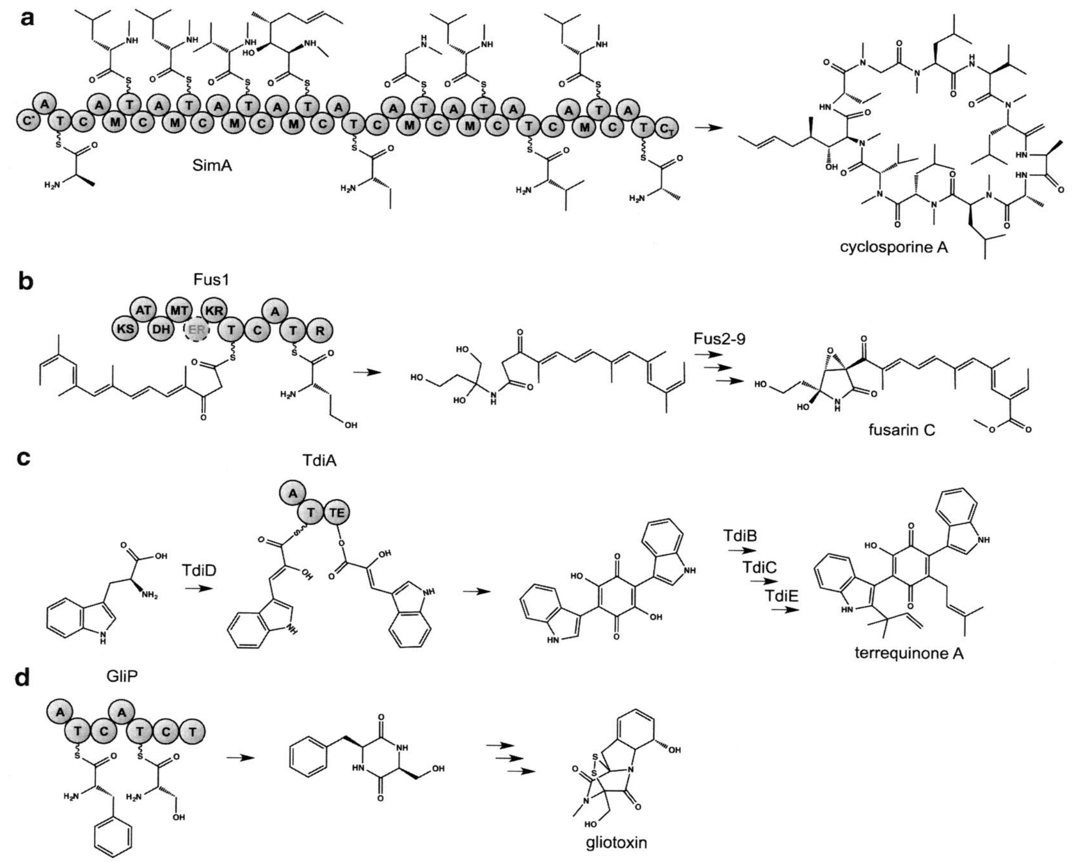

Fig. 1 Domain architecture of select fungal NRPS and their chemical products. (**a**) Domain organization of the SimA from *Tolypocladium inflatum* responsible for cyclosporine A production modified from Hoffmann et al. [18]. C* represents a truncated and presumably inactive C domain. (**b**) Domain architecture of the hybrid PKS/NRPS Fus1 from *Fusarium fujikuroi* responsible for fusarin C production modified from Niehaus et al. [19]. *KS* β-ketoacyl synthase domain; *AT* malonyl coenzyme A-acyl carrier protein transacylase domain; *DH* dehydratase domain; *ER* enoyl reductase domain, most likely nonfunctional in Fus1; *KR* keto reductase domain; *R* reductase domain; Fus2-9, tailoring enzymes encoded in the fusarin C gene cluster of *F. fujikuroi*. (**c**) Domain organization of the NRPS-like TdiA involved in terrequinone A production in *Aspergillus nidulans* modified from Balibar et al. [17]. *TE* thioesterase domain, TdiB-D, tailoring enzymes encoded in the terrequinone A gene cluster of *A. nidulans*. (**d**) Domain organization of the NRPS GliP responsible for the first committed step in gliotoxin production in *Aspergillus fumigatus* modified from Scharf et al. [20]

[27], HC-toxin [29], AM-toxin [30], peptaibols [31], ergotpeptine [32], fusarin C [33], equisetin [34], peramine [35], sirodesmin [36], gliotoxin [37], fumitremorgin [38], tenellin [39], pseurotin A [40], cytochalasin [41], cyclopiazonic acid [42], aureobasidin A [43], fumiquinazolines [44], apicidin [45], tryptopquialanine [46], ochratoxin A [47], fumigaclavines [48], ardeemin [49], nidulanin A [50], and pneumocandin [51]. Although many of the aforementioned NRPS-derived fungal secondary

metabolites display biological activity, the need for discovery of new antibiotics is becoming more urgent [52] and calls for innovative avenues of uncovering cryptic metabolites.

In order to overcome the hurdle of activating silent gene clusters in fungal genomes, information about cluster boundaries and regulatory mechanisms is crucial. The discovery of a novel global regulator of secondary metabolism came in 2004 with the identification of the putative methyltransferase LaeA [53]. Since then, manipulation of LaeA has been used to identify and demarcate a number of NRPS-derived metabolites in a multitude of species [54]. Identified metabolites and corresponding gene clusters include penicillin [53, 55, 56], gliotoxin [53], terrequinone A [57, 58], NRPS9- and NRPS11-derived metabolites [59], fusarin C [60, 61], pseurotin [62], ochratoxin A [63], tyrosine-derived alkaloids [64], beauvericin [65], as well as lolitrem and ergot alkaloids [66].

Additionally, approaches of directly activating clusters through overexpressing the cluster backbone gene itself and/or any transcription factors associated within the cluster have been undertaken. Although not always successful, this approach has led to the identification of aspyridone A [67], apicidins [45, 68], microperfuranone [69], pyranonigrin E [70], and hexadehydroastechrome [71] and has confirmed the enzymes responsible for fumitremorgin [38], apicidin [45], fusarin C [19], and cytochalasin [72, 73] production.

In the protocol below, we will describe methods for identifying and characterizing fungal NRPS enzymes and methods for activating these NRPS and corresponding clusters through both cluster specific and global regulators.

2 Materials

1. YG medium: Prepare 20 g/L D-glucose, 5 g/L yeast extract, 1 mL/L trace elements, supplements as needed; other rich media may be substituted, sterilized by autoclaving.

2. KCl citric acid solution: Prepare 82 g/L KCl, 21 g/L citric acid monohydrate, pH to 5.8 with 1 M KOH.

3. Protoplasting solution: Prepare by mixing 8 mL KCl citric acid solution, 8 mL YG medium, and 1.3 g VinoTaste Pro.

4. KCl calcium chloride solution: 0.6 M KCl, 50 mM CaCl$_2$.

5. PEG solution: Prepare by making a solution of 44.7 g/L KCl, 5.5 g/L CaCl$_2$, 10 mM Tris pH 7.5, 250 g/L PEG (MW 3350) in water.

6. KCl Minimal Medium (KMM): Prepare 10 g/L D-glucose, 50 mL 20× nitrate salts, 1.0 mL/L trace elements, 44.7 g/L KCl, 15 g/L agar, supplements as needed, pH 6.5 sterilized by autoclaving.

3   Methods

3.1   Identification and Characterization of NRPS

1. Identification of NRPS-encoding genes in the sequenced fungal genome of choice can be achieved using BLAST-based algorithms [74]. Previously analyzed NRPS, PKS/NRPS, and NRPS-like sequences should function as archetypical input sequences to ensure identification of all predicted gene calls in the sequenced genome.

2. In order to parse and assign specific domains in the predicted protein sequences, a variety of Web-based tools are available and listed in Table 1. These tools use more precise and refined statistical models for functional assignment of domains and substrate predictions (*see* **Note 1**).

3.2   Creation of Transformation Cassettes

1. Amplify flanking regions, marker gene, and promoter region (if used) using expand long template polymerase. Refer to Fig. 2 for reaction setup (*see* **Note 2**).

2. Use agarose gel electrophoresis to confirm PCR products and purify using gel extraction (such as Qiagen QIAquick Gel Extraction Kit).

3. Quantify the concentration of each PCR fragment. This may be done either by spectrophotometry or by visual comparison to a known quantity of DNA during electrophoresis.

4. Set up second PCR reaction using a fragment copy number ratio of 1:2:1 (for deletion cassettes, 5′ flank/marker gene/3′ flank) or 1:2:2:1 (for overexpression cassettes, 5′ flank/marker/promoter/3′ flank).

5. Amplify the final product using nested primers (*see* **Note 3**).

6. Confirm desired PCR product through agarose gel electrophoresis of a small amount (~2 μL) of the reaction (*see* **Note 4**).

7. Remove buffer components and salts from the final PCR product by using a gel filtration column such as Sephadex G-50 superfine resin (GE Healthcare) (*see* **Note 5**).

8. Purified product should be kept on ice and may be stored at −20 °C for several months.

3.3   Transformation Procedure (*See* Note 6)

1. In a sterile 125 mL Erlenmeyer flask, inoculate 1×10^8 spores into 10 mL rich medium plus appropriate supplements (e.g., YG medium, supplement if needed).

2. Incubate overnight at room temperature shaking at 150 rpm depending on strain background and type of auxotrophy (if present). Extended, actively growing hyphae should be present in the sample.

3. Collect mycelia by filtration through sterile mirocloth (Calbiochem).

Table 1
Web-based tools for NRPS domain analysis

Tool	Website	Reference
Conserved Domain Database (CDD)	www.ncbi.nlm.nih.gov/Structure/cdd/wrpsb.cgi	[76]
Universal Protein Resource (UniProt)	http://www.uniprot.org	[77]
Protein sequence analysis and classification (InterPro)	http://www.ebi.ac.uk/interpro/	[78]
PKS/NRPS analysis website	http://nrps.igs.umaryland.edu/nrps/	[79]
NRPSpredictor	http://ab.inf.uni-tuebingen.de/software/NRPSpredictor/	[80]
NRPSpredictor2	http://nrps.informatik.uni-tuebingen.de/Controller?cmd=SubmitJob	[81]
Norine	http://bioinfo.lifl.fr/norine/index.jsp	[82]
Nonribosomal peptide synthetase substrate predictor (NRPSsp)	http://www.nrpssp.com/	[83]
Antibiotics and secondary metabolite analysis shell (antiSMASH)	http://antismash.secondarymetabolites.org/	[84]
Secondary metabolite unique regions finder (SMURF)	www.jcvi.org/smurf	[85]
Natural product domain seeker (NaPDoS)	http://napdos.ucsd.edu/napdos_home.html	[86]

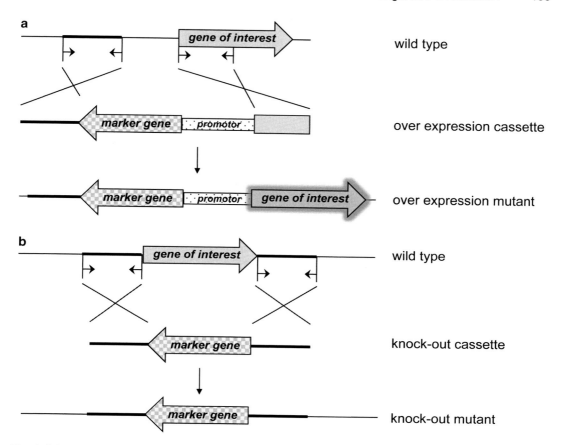

Fig. 2 Schematic overview for gene manipulation strategies. (**a**) Schematic depiction of construction of an overexpression construct for targeted promoter replacement at the gene locus. (**b**) Schematic depiction of construction of a knockout construct for targeted gene replacement

4. Rinse with additional medium.

5. Transfer mycelia into sterile 125 mL flask.

6. Add 16 mL fresh protoplasting solution.

7. Incubate 2–4 h at 30 °C at 100 rpm (*see* **Note 7**).

8. Add 10 mL 1.2 M sucrose solution to a sterile 50 mL centrifuge tube.

9. Gently overlay protoplast solution, taking care not to disrupt layer formation.

10. Centrifuge at $1800 \times g$ for 10 min. Protoplasts will appear as a cloudy layer at the interphase.

11. Gently collect the protoplasts with a pipette and transfer to a fresh sterile tube.

12. Add at least 1 volume of 0.6 M KCl. Mix gently.

13. Centrifuge at $1800 \times g$ for 10 min to pellet the protoplasts.

14. Resuspend protoplasts in 1 mL of 0.6 M KCl.

15. Centrifuge at $2300 \times g$ for 3 min to pellet the protoplasts. Carefully remove supernatant.

16. Resuspend protoplasts in an appropriate volume of 0.6 M KCl, 50 mM citric acid solution. 100 µL per transformation (including negative control) should be used. Typically up to 10 transformations may be performed per batch of protoplasts.

17. Add DNA (10 µL of PCR product; 1–2 µg) to 100 µL protoplasts.

18. Vortex four to five times (1 s each time) at maximum speed.

19. Add 50 µL of PEG solution (*see* **Note 8**).

20. Vortex four to five times (1 s each time) at maximum speed.

21. Incubate on ice for 25 min.

22. Add 1 mL PEG solution. Mix by gently pipetting up and down.

23. Incubate at room temperature for 25 min.

24. Plate transformation mixture on selection KCl Minimal Medium (KMM) with appropriate selection supplements by spreading transformation sample over selection plates with a glass spreader (*see* **Note 9**).

25. Incubate at 37 °C. Transformants will be visible after 2–3 days. Proceed with your desired method of transformant confirmation and growth for metabolite production.

4   Notes

1. None of the bacterial substrate codes for A domains matches the fungal A domains. (An exception is the penicillin NRPS, which is likely derived by horizontal gene transfer from bacteria.) Fungal species most likely use a degenerated specificity code since different A domain codes have been reported for binding the same amino acid.

2. Alternative high-fidelity polymerases may be used for all reactions, but note that SDS in the PCR buffer is not easily removed by gel filtration and may affect transformation efficiency.

3. Using nested primers decreases amplification of nonspecific products, although in most cases the outermost primers may also be used successfully for amplification.

4. The final product should be easily visualized. Additional bands may be present, but if over 50 % of final concentration, the product may be purified using gel extraction. Restriction digests may be used to further confirm the correct assembly of the final PCR product.

5. Although gel extraction may be used to purify products, in our hands efficiency is decreased relative to gel filtration purification.

6. This procedure has been optimized for *Aspergillus nidulans* adapted from [75]; adjust as appropriate for your fungus.

7. Protoplasts are normally visible after 1 h of incubation. Protoplast clumping can be inhibited by gentle pipetting several times during incubation. Do not overdigest.

8. PEG solution should be freshly filtered prior to transformation to remove precipitates which may rupture protoplasts.

9. We typically plate ~100 µL per medium size (15 cm) plate. This allows for sufficient separation of transformants.

Acknowledgments

This research was funded by NIH PO1 GM084077 to NPK, by R01 Al065728-01 to NPK, and by NIH NRSA AI55397 to AAS.

References

1. Keller NP, Hohn TM (1997) Metabolic pathway gene clusters in filamentous fungi. Fungal Genet Biol 21:17–29

2. Cichewicz RH (2010) Epigenome manipulation as a pathway to new natural product scaffolds and their congeners. Nat Prod Rep 27:11–22

3. Strauss J, Reyes-Dominguez Y (2011) Regulation of secondary metabolism by chromatin structure and epigenetic codes. Fungal Genet Biol 48:62–69

4. Yin W, Keller NP (2011) Transcriptional regulatory elements in fungal secondary metabolism. J Microbiol 49:329–339

5. Wiemann P, Keller NP (2014) Strategies for mining fungal natural products. J Ind Microbiol Biotechnol 4:301–313

6. Brakhage AA, Schroeckh V (2011) Fungal secondary metabolites - strategies to activate silent gene clusters. Fungal Genet Biol 48: 15–22

7. Schwarzer D, Finking R, Marahiel MA (2003) Nonribosomal peptides: from genes to products. Nat Prod Rep 20(3):275–287

8. Schwarzer D, Marahiel MA (2001) Multimodular biocatalysts for natural product assembly. Naturwissenschaften 88:93–101

9. von Döhren H, Keller U, Vater J, Zocher R (1997) Multifunctional peptide synthetases. Chem Rev 97:2675–2706

10. Díez B, Gutiérrez S, Barredo JL, van Solingen P, van der Voort LH, Martín JF (1990) The cluster of penicillin biosynthetic genes. Identification and characterization of the pcbAB gene encoding the alpha-aminoadipyl-cysteinyl-valine synthetase and linkage to the pcbC and penDE genes. J Biol Chem 265:16358–16365

11. MacCabe AP, Riach MB, Unkles SE et al (1990) The Aspergillus nidulans npeA locus consists of three contiguous genes required for penicillin biosynthesis. EMBO J 9(1):279–287

12. Smith DJ, Burnham MK, Edwards J et al (1990) Cloning and heterologous expression of the penicillin biosynthetic gene cluster from *Penicillium chrysogenum*. Biotechnology (N Y) 8:39–41

13. Zocher R, Nihira T, Paul E et al (1986) Biosynthesis of cyclosporin A: partial purification and properties of a multifunctional enzyme from *Tolypocladium inflatum*. Biochemistry 25:550–553

14. Keating TA, Ehmann DE, Kohli RM et al (2001) Chain termination steps in nonribosomal peptide synthetase assembly lines: directed acyl-S-enzyme breakdown in antibiotic and siderophore biosynthesis. Chembiochem 2:99–107

15. Grünewald J, Marahiel MA (2006) Chemoenzymatic and template-directed syn-

thesis of bioactive macrocyclic peptides. Microbiol Mol Biol Rev 70(1):121–146. doi:10.1128/MMBR.70.1.121-146.2006

16. Boettger D, Bergmann H, Kuehn B et al (2012) Evolutionary imprint of catalytic domains in fungal PKS-NRPS hybrids. Chembiochem 13:2363–2373

17. Balibar CJ, Howard-Jones AR, Walsh CT (2007) Terrequinone A biosynthesis through L-tryptophan oxidation, dimerization and bis-prenylation. Nat Chem Biol 3:584–592

18. Hoffmann K, Schneider-Scherzer E, Kleinkauf H et al (1994) Purification and characterization of eucaryotic alanine racemase acting as key enzyme in cyclosporin biosynthesis. J Biol Chem 269:12710–12714

19. Niehaus EM, Kleigrewe K, Wiemann P et al (2013) Genetic manipulation of the *Fusarium fujikuroi* fusarin gene cluster yields insight into the complex regulation and fusarin biosynthetic pathway. Chem Biol 20:1055–1066

20. Scharf DH, Heinekamp T, Remme N et al (2012) Biosynthesis and function of gliotoxin in Aspergillus fumigatus. Appl Microbiol Biotechnol 93:467–472

21. Ames BD, Walsh CT (2010) Anthranilate-activating modules from fungal nonribosomal peptide assembly lines. Biochemistry 49:3351–3365

22. Galagan JE, Calvo SE, Cuomo C et al (2005) Sequencing of *Aspergillus nidulans* and comparative analysis with *A. fumigatus* and *A. oryzae*. Nature 438:1105–1115

23. Pel HJ, de Winde JH, Archer DB et al (2007) Genome sequencing and analysis of the versatile cell factory *Aspergillus niger* CBS 513.88. Nat Biotechnol 25:221–231

24. Cleveland TE, Yu J, Fedorova N et al (2009) Potential of *Aspergillus flavus* genomics for applications in biotechnology. Trends Biotechnol 27:151–157

25. von Döhren H (2009) A survey of nonribosomal peptide synthetase (NRPS) genes in *Aspergillus nidulans*. Fungal Genet Biol 46(Suppl 1):45–52

26. Wiemann P, Albermann S, Niehaus EM et al (2012) The Sfp-type 4′-phosphopantetheinyl transferase Ppt1 of *Fusarium fujikuroi* controls development, secondary metabolism and pathogenicity. PLoS One 7:37519

27. Zocher R, Keller U, Kleinkauf H (1982) Enniatin synthetase, a novel type of multifunctional enzyme catalyzing depsipeptide synthesis in *Fusarium oxysporum*. Biochemistry 21:43–48

28. Peeters H, Zocher R, Kleinkauf H (1988) Synthesis of beauvericin by a multifunctional enzyme. J Antibiot (Tokyo) 41:352–359

29. Panaccione DG, Scott-Craig JS, Pocard JA et al (1992) A cyclic peptide synthetase gene required for pathogenicity of the fungus *Cochliobolus carbonum* on maize. Proc Natl Acad Sci U S A 89:6590–6594

30. Johnson RD, Johnson L, Itoh Y, Kodama M, Otani H, Kohmoto K (2000) Cloning and characterization of a cyclic peptide synthetase gene from *Alternaria alternata* apple pathotype whose product is involved in AM-toxin synthesis and pathogenicity. Mol Plant Microbe Interact 13:742–753

31. Wiest A, Grzegorski D, Xu BW et al (2002) Identification of peptaibols from *Trichoderma virens* and cloning of a peptaibol synthetase. J Biol Chem 277:20862–20868

32. Correia T, Grammel N, Ortel I et al (2003) Molecular cloning and analysis of the ergopeptine assembly system in the ergot fungus *Claviceps purpurea*. Chem Biol 10:1281–1292

33. Song Z, Cox RJ, Lazarus CM et al (2004) Fusarin C biosynthesis in *Fusarium moniliforme* and *Fusarium venenatum*. Chembiochem 5:1196–1203

34. Sims JW, Fillmore JP, Warner DD et al (2005) Equisetin biosynthesis in *Fusarium heterosporum*. Chem Commun (Camb) 2:186–188

35. Tanaka A, Tapper BA, Popay A et al (2005) A symbiosis expressed non-ribosomal peptide synthetase from a mutualistic fungal endophyte of perennial ryegrass confers protection to the symbiotum from insect herbivory. Mol Microbiol 57:1036–1050

36. Gardiner DM, Cozijnsen AJ, Wilson LM et al (2004) The sirodesmin biosynthetic gene cluster of the plant pathogenic fungus *Leptosphaeria maculans*. Mol Microbiol 53:1307–1318

37. Cramer RA, Gamcsik MP, Brooking RM et al (2006) Disruption of a nonribosomal peptide synthetase in *Aspergillus fumigatus* eliminates gliotoxin production. Eukaryot Cell 5:972–980

38. Maiya S, Grundmann A, Li SM et al (2006) The fumitremorgin gene cluster of *Aspergillus fumigatus*: identification of a gene encoding brevianamide F synthetase. Chembiochem 7:1062–1069

39. Eley KL, Halo LM, Song Z et al (2007) Biosynthesis of the 2-pyridone tenellin in the insect pathogenic fungus *Beauveria bassiana*. Chembiochem 8:289–297

40. Maiya S, Grundmann A, Li X, Li SM, Turner G (2007) Identification of a hybrid PKS/NRPS required for pseurotin A biosynthesis in the human pathogen *Aspergillus fumigatus*. Chembiochem 8:1736–1743

41. Schümann J, Hertweck C (2007) Molecular basis of cytochalasin biosynthesis in fungi: gene

cluster analysis and evidence for the involvement of a PKS-NRPS hybrid synthase by RNA silencing. J Am Chem Soc 129: 9564–9565

42. Tokuoka M, Seshime Y, Fujii I et al (2008) Identification of a novel polyketide synthase-nonribosomal peptide synthetase (PKS-NRPS) gene required for the biosynthesis of cyclopiazonic acid in *Aspergillus oryzae*. Fungal Genet Biol 45:1608–1615

43. Slightom JL, Metzger BP, Luu HT et al (2009) Cloning and molecular characterization of the gene encoding the Aureobasidin A biosynthesis complex in *Aureobasidium pullulans* BP-1938. Gene 431:67–79

44. Ames BD, Liu X, Walsh CT (2010) Enzymatic processing of fumiquinazoline F: a tandem oxidative-acylation strategy for the generation of multicyclic scaffolds in fungal indole alkaloid biosynthesis. Biochemistry 49:8564–8576

45. Jin JM, Lee S, Lee J (2010) Functional characterization and manipulation of the apicidin biosynthetic pathway in *Fusarium semitectum*. Mol Microbiol 76:456–466

46. Gao X, Chooi YH, Ames BD et al (2011) Fungal indole alkaloid biosynthesis: genetic and biochemical investigation of the tryptoquialanine pathway in *Penicillium aethiopicum*. J Am Chem Soc 133:2729–2741

47. Gallo A, Bruno KS, Solfrizzo M et al (2012) New insight into the ochratoxin A biosynthetic pathway through deletion of a nonribosomal peptide synthetase gene in *Aspergillus carbonarius*. Appl Environ Microbiol 78:8208–8218

48. O'Hanlon KA, Gallagher L, Schrettl M (2012) Nonribosomal peptide synthetase genes *pesL* and *pes1* are essential for Fumigaclavine C production in *Aspergillus fumigatus*. Appl Environ Microbiol 78:3166–3176

49. Haynes SW, Gao X, Tang Y et al (2013) Complexity generation in fungal peptidyl alkaloid biosynthesis: a two-enzyme pathway to the hexacyclic MDR export pump inhibitor ardeemin. ACS Chem Biol 8:741–748

50. Andersen MR, Nielsen JB, Klitgaard A et al (2013) Accurate prediction of secondary metabolite gene clusters in filamentous fungi. Proc Natl Acad Sci U S A 110:E99–E107

51. Chen L, Yue Q, Zhang X et al (2013) Genomics-driven discovery of the pneumocandin biosynthetic gene cluster in the fungus Glarea lozoyensis. BMC Genomics 14:339

52. WHO (2014) Antimicrobial resistance: global report on surveillance 2014

53. Bok JW, Keller NP (2004) LaeA, a regulator of secondary metabolism in *Aspergillus spp*. Eukaryot Cell 3:527–535

54. Jain S, Keller N (2013) Insights to fungal biology through LaeA sleuthing. Fungal Biol Rev 27:51–59

55. Kosalkova K, Garcia-Estrada C, Ullan RV et al (2009) The global regulator LaeA controls penicillin biosynthesis, pigmentation and sporulation, but not roquefortine C synthesis in *Penicillium chrysogenum*. Biochimie 91: 214–225

56. Kawauchi M, Nishiura M, Iwashita K (2013) Fungus-specific sirtuin HstD coordinates secondary metabolism and development through control of LaeA. Eukaryot Cell 12:1087–1096

57. Bouhired S, Weber M, Kempf-Sontag A et al (2007) Accurate prediction of the *Aspergillus nidulans* terrequinone gene cluster boundaries using the transcriptional regulator LaeA. Fungal Genet Biol 44:1134–1145

58. Bok JW, Hoffmeister D, Maggio-Hall LA, Murillo R, Glasner JD, Keller NP (2006) Genomic mining for Aspergillus natural products. Chem Biol 13:31–37

59. Lee I, Oh JH, Shwab EK et al (2009) HdaA, a class 2 histone deacetylase of *Aspergillus fumigatus*, affects germination and secondary metabolite production. Fungal Genet Biol 46:782–790

60. Wiemann P, Brown DW, Kleigrewe K et al (2010) FfVel1 and FfLae1, components of a velvet-like complex in *Fusarium fujikuroi*, affect differentiation, secondary metabolism and virulence. Mol Microbiol 9:1236–1250

61. Butchko RA, Brown DW, Busman M et al (2012) Lae1 regulates expression of multiple secondary metabolite gene clusters in *Fusarium verticillioides*. Fungal Genet Biol 49:602–612

62. Wiemann P, Guo CJ, Palmer JM et al (2013) Prototype of an intertwined secondary-metabolite supercluster. Proc Natl Acad Sci U S A 110:17065–17070

63. Crespo-Sempere A, Marín S, Sanchis V, Ramos AJ (2013) VeA and LaeA transcriptional factors regulate ochratoxin A biosynthesis in *Aspergillus carbonarius*. Int J Food Microbiol 166:479–486

64. Forseth RR, Amaike S, Schwenk D et al (2013) Homologous NRPS-like gene clusters mediate redundant small-molecule biosynthesis in *Aspergillus flavus*. Angew Chem Int Ed Engl 52:1590–1594

65. López-Berges MS, Schäfer K, Hera C et al (2014) Combinatorial function of velvet and AreA in transcriptional regulation of nitrate utilization and secondary metabolism. Fungal Genet Biol 62:78–84

66. Chujo T, Scott B (2014) Histone H3K9 and H3K27 methylation regulates fungal alkaloid biosynthesis in a fungal endophyte-plant symbiosis. Mol Microbiol 92:413–434

67. Bergmann S, Schümann J, Scherlach K et al (2007) Genomics-driven discovery of PKS-NRPS hybrid metabolites from *Aspergillus nidulans*. Nat Chem Biol 3:213–217

68. Wiemann P, Sieber CM, von Bargen KW et al (2013) Deciphering the cryptic genome: genome-wide analyses of the rice pathogen *Fusarium fujikuroi* reveal complex regulation of secondary metabolism and novel metabolites. PLoS Pathog 9, e1003475

69. Yeh HH, Chiang YM, Entwistle R et al (2012) Molecular genetic analysis reveals that a nonribosomal peptide synthetase-like (NRPS-like) gene in *Aspergillus nidulans* is responsible for microperfuranone biosynthesis. Appl Microbiol Biotechnol 96:739–748

70. Awakawa T, Yang XL, Wakimoto T et al (2013) Pyranonigrin E: a PKS-NRPS hybrid metabolite from *Aspergillus niger* identified by genome mining. Chembiochem 14:2095–2099

71. Yin WB, Baccile JA, Bok JW et al (2013) A nonribosomal peptide synthetase-derived iron(iii) complex from the pathogenic fungus *Aspergillus fumigatus*. J Am Chem Soc 135(6):2064–2067. doi:10.1021/ja311145n

72. Qiao K, Chooi YH, Tang Y (2011) Identification and engineering of the cytochalasin gene cluster from Aspergillus clavatus NRRL 1. Metab Eng 13:723–732

73. Zhang D, Ge H, Xie D et al (2013) Periconiasins A-C, new cytotoxic cytochalasins with an unprecedented 9/6/5 tricyclic ring system from endophytic fungus *Periconia sp.* Org Lett 15:1674–1677

74. Altschul SF, Gish W, Miller W, Myers EW, Lipman DJ (1990) Basic local alignment search tool. J Mol Biol 215:403–410

75. Szewczyk E, Nayak T, Oakley CE et al (2006) Fusion PCR and gene targeting in Aspergillus nidulans. Nat Protoc 1:3111–3120

76. Marchler-Bauer A, Zheng C, Chitsaz F, Derbyshire MK et al (2013) CDD: conserved domains and protein three-dimensional structure. Nucleic Acids Res 41(Database issue): D348–352

77. UniProt Consortium U (2012) Reorganizing the protein space at the Universal Protein Resource (UniProt). Nucleic Acids Res 40(Database issue):D71–75

78. Hunter S, Jones P, Mitchell A et al (2012) InterPro in 2011: new developments in the family and domain prediction database. Nucleic Acids Res 40(Database issue):D306–312

79. Bachmann BO, Ravel J (2009) Chapter 8. Methods for in silico prediction of microbial polyketide and nonribosomal peptide biosynthetic pathways from DNA sequence data. Methods Enzymol 458:181–217

80. Rausch C, Weber T, Kohlbacher O et al (2005) Specificity prediction of adenylation domains in nonribosomal peptide synthetases (NRPS) using transductive support vector machines (TSVMs). Nucleic Acids Res 33:5799–5808

81. Röttig M, Medema MH, Blin K et al (2011) NRPSpredictor2–a web server for predicting NRPS adenylation domain specificity. Nucleic Acids Res 39(Web Server issue): W362–367

82. Caboche S, Pupin M, Leclère V et al (2008) NORINE: a database of nonribosomal peptides. Nucleic Acids Res 36(Database issue):D326–331

83. Prieto C, García-Estrada C, Lorenzana D et al (2012) NRPSsp: non-ribosomal peptide synthase substrate predictor. Bioinformatics 28:426–427

84. Blin K, Medema MH, Kazempour D et al (2013) antiSMASH 2.0–a versatile platform for genome mining of secondary metabolite producers. Nucleic Acids Res 41:W204–212

85. Khaldi N, Seifuddin FT, Turner G (2010) SMURF: Genomic mapping of fungal secondary metabolite clusters. Fungal Genet Biol 47:736–741

86. Ziemert N, Podell S, Penn K et al (2012) The natural product domain seeker NaPDoS: a phylogeny based bioinformatic tool to classify secondary metabolite gene diversity. PLoS One 7:e34064

Chapter 11

In Situ Analysis of Bacterial Lipopeptide Antibiotics by Matrix-Assisted Laser Desorption/Ionization Mass Spectrometry Imaging

Delphine Debois, Marc Ongena, Hélène Cawoy, and Edwin De Pauw

Abstract

Matrix-assisted laser desorption/ionization mass spectrometry imaging (MALDI MSI) is a technique developed in the late 1990s enabling the two-dimensional mapping of a broad variety of biomolecules present at the surface of a sample. In many applications including pharmaceutical studies or biomarker discovery, the distribution of proteins, lipids or drugs, and metabolites may be visualized within tissue sections. More recently, MALDI MSI has become increasingly applied in microbiology where the versatility of the technique is perfectly suited to monitor the metabolic dynamics of bacterial colonies. The work described here is focused on the application of MALDI MSI to map secondary metabolites produced by *Bacilli*, especially lipopeptides, produced by bacterial cells during their interaction with their environment (bacteria, fungi, plant roots, etc.). This chapter addresses the advantages and challenges that the implementation of MALDI MSI to microbiological samples entails, including detailed protocols on sample preparation (from both microbiologist and mass spectrometrist points of view), matrix deposition, and data acquisition and interpretation. Lipopeptide images recorded from confrontation plates are also presented.

Key words Matrix-assisted laser desorption/ionization mass spectrometry imaging, Lipopeptides, Bacteria

1 Introduction

Antimicrobial compounds represent a broad range of molecules, probably not completely discovered nor explored. Antibiotics may be produced by different kinds of microorganisms, which display very different lifestyles and evolve in various environments. Some of these microorganisms live in the rhizosphere, the zone of soil that surrounds and is influenced by the roots of plants [1]. This complex ecosystem gathers diverse types of microbes, such as fungi, nematodes, and others, including bacteria [2]. The interactions existing between the plants and these microorganisms may be positive or negative (or even neutral); bacteria having a beneficial

Bradley S. Evans (ed.), *Nonribosomal Peptide and Polyketide Biosynthesis: Methods and Protocols*, Methods in Molecular Biology, vol. 1401, DOI 10.1007/978-1-4939-3375-4_11, © Springer Science+Business Media New York 2016

effect on plants are called plant growth-promoting rhizobacteria (PGPR) [3]. Some isolates of the bacterial genera *Bacillus* and *Paenibacillus* may be found in soil and are considered powerful biocontrol agents, owing to their beneficial behavior toward plants [4]. How do they do it to protect their host? Multiple mechanisms are involved in that process but the best characterized and understood is the direct antagonism against phytopathogens. To that end, soil *Bacilli* and *Paenibacilli* produce antibiotics, mobilizing up to 8 % of their genome to antibiotic synthesis [5–7]. Among these antimicrobial compounds are found cyclic lipopeptides (cLP) which are synthesized by nonribosomal peptide synthetases (NRPS) or hybrid polyketide synthases/nonribosomal peptide synthetases (PKS/NRPS) [8]. Their particular chemical structure, made of a cyclic peptide linked to a fatty acid chain, leads to a wide structural variety. Especially, modifications to the amino acids' structure and/or sequence explain the occurrence of different variants within a family (fusaricidins A and C; fengycins A, B, C, and S; or iturins A and C). Also, the diversity in length and isomery (linear, iso, or anteiso) of the acyl chain generates, for each variant, different homologues (Fig. 1) [9–11].

Actually, these compounds are well characterized and have been studied for decades [12]; however, the producing strains are cultivated in "lab" conditions which are not representative of what bacteria face in their natural habitat where nutrients are limited and the growth rates are slowed down. Besides, bacteria evolve in biofilm-related structures on the surface of plant roots [13]. The role of cLPs in the biocontrol activity of their producers has already been investigated, and surfactins were proven to be elicitors of the immune-related responses in the host plant (ISR, induced systemic resistance) [14–16]. Nevertheless, adequate models and innovative analytical methods are still needed to determine if these compounds are actually produced and accumulate in the microenvironment in a biologically relevant way.

One available analytical technique for the analysis of secondary metabolites originating from microbes is matrix-assisted laser desorption/ionization mass spectrometry (MALDI MS). MALDI MS relies on the use of a laser (most of the time, a UV laser), to desorb and ionize the molecules of interest present in the sample. The energy of the laser beam is transferred to the analyte through a MALDI matrix. Usually, the MALDI matrix is a small organic molecule, exhibiting an acidic (to work in positive ion mode) or a basic (to work in negative ion mode) character and absorbing at the wavelength of the laser. Different chemical reactions take place in the ionization source, such as proton transfer from the ionized matrix to the analyte, leading to protonated ions, noted $[M+H]^+$. In the case of biological samples analyzed without purification steps, $[M+Na]^+$ or $[M+K]^+$ ions are usually detected. MALDI MS

a Surfactins

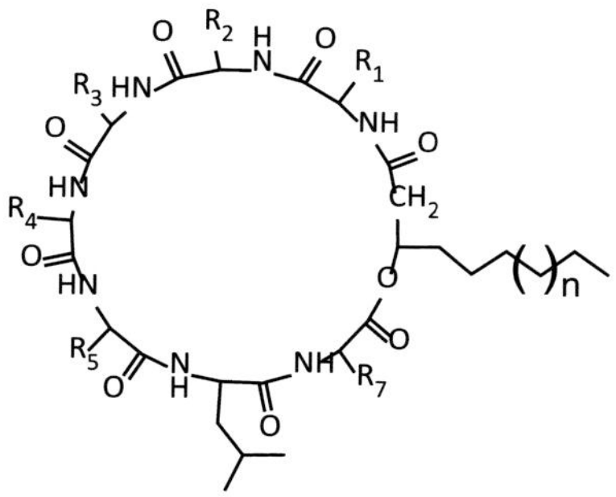

R_1 = Glu, Gln; R_2 = Val, Leu, Ile; R_3 = Met, Leu; R_4 = Val, Ala, Leu, Ile; R_5 = Asp, Pro; R_7 = Val, Leu, Ile

b Iturins

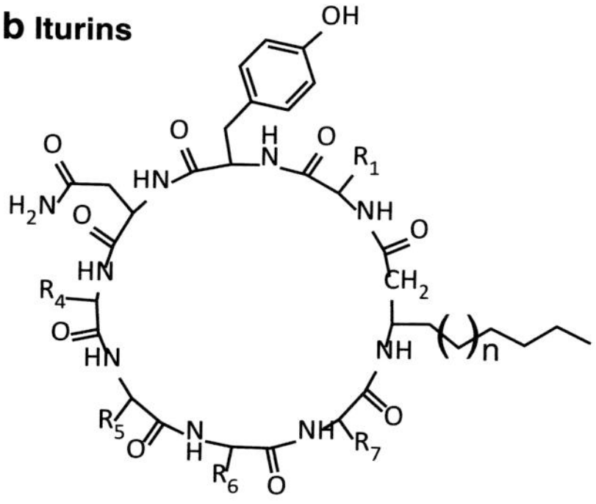

R_1 = Asn, Asp; R_4 = Pro, Gln, Ser; R_5 = Glu, Pro; R_6 = Ser, Asn; R_7 = Thr, Ser, Asn

c Fengycins

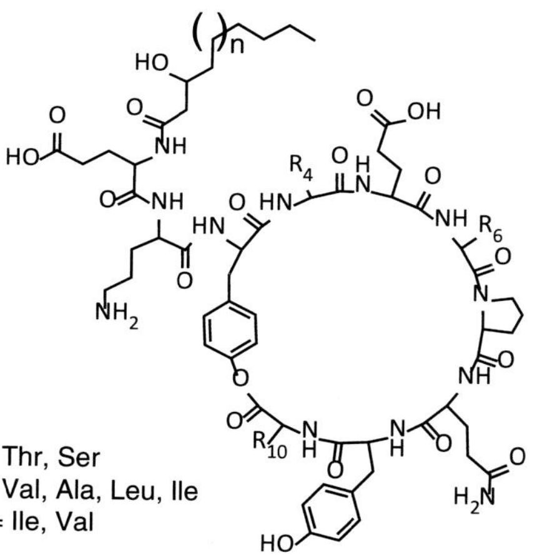

R_4 = Thr, Ser
R_6 = Val, Ala, Leu, Ile
R_{10} = Ile, Val

d Fusaricidins and closely-related LI-F-type LP

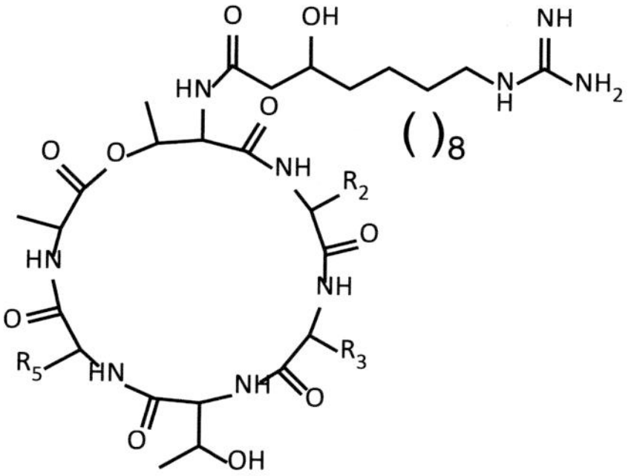

R_2 = Val, Ile
R_3 = Val, Tyr, Ile, Phe
R_5 = Asn, Gln

	Surfactins	Iturins	Fengycins	Fusaricidins
Number of amino acids	7	7	10	6
Number of variants	4	7	4	12
Chain lengths	C_{12} to C_{15}	C_{14} to C_{17}	C_{14} to C_{18}	C_{15}
Isomeries	Linear, iso, anteiso	Linear, iso, anteiso	Linear, iso, anteiso	-

Fig. 1 Examples of structures of lipopeptides from *Bacillus* and *Paenibacillus*. General chemical structures are represented for (**a**) surfactins, (**b**) iturins, (**c**) fengycins, and (**d**) fusaricidins. The table summarizes the structural heterogeneity of each family

is nowadays routinely used in microbiology labs as a tool for bacterial identification, based on the detection of a species-characteristic pattern of ribosomal proteins [17]. Applications of this method include the detection of environmental toxins in water by analyzing peptides and polyketides from cyanobacteria [18] or the establishment of protein profiles from intact fungal spores harvested from the surface of fruits, enabling the identification of unknown fungi [19]. More recently, MALDI MS applied to microorganisms has known new developments in the implementation of MALDI MS imaging (MALDI MSI) to microbial samples. Over the last 15 years, MALDI MSI has become a very powerful tool allowing the localization of potentially hundreds to thousands of compounds (peptide, proteins, lipids, drugs, metabolites, etc.) simultaneously. The principle of MALDI MS imaging relies on the acquisition of an array of coordinate-specific mass spectra, the information of which is integrated to get the spatial distribution of any detected compound. Usually dedicated to the analysis of tissues in proteomic, metabolomic, or lipidomic studies [20], MALDI MSI has gained interest in the microbiological field. Recently, articles have been published, exploiting the unique features of MALDI MSI, to map metabolites, including notably the surfactin-type cLP, secreted during bacterial competition [21, 22], even in 3D [23]. Even more recently, high-resolution MALDI MSI was used to map the secondary metabolites (triterpenoids) of two basidiomycetes *Cyathus striatus* and *Hericium erinaceus* [24]. For our part, we developed a method in which MALDI MSI allows localizing and identifying cLPs produced by bacteria when confronted by a phytopathogen [25] or during the interaction with plant roots [26]. This last methodology uses an *in planta* culture model in which the different partners are grown together on semisolid agar-based medium in a Petri dish.

In this chapter, we describe the methodology and the step-by-step protocols used from the preparation of confrontation plates to MALDI images acquisition.

2   Materials

2.1   Confrontation Plate Preparation

1. Experiments described here were conducted with the *Paenibacillus polymyxa* strain Pp56, isolated from field soil in Ohio and kindly provided by Dr. B. McSpadden-Gardener from Ohio State University, USA. The bacterial strain was confronted by a phytopathogen, *Fusarium oxysporum* f.sp. *radicis-lycopersici* (Forl).

2. Agar.

3. $Ca(NO_3)_2$; KNO_3; $MgSO_4$; KH_2PO_4.

4. H_3BO_3, $MnCl_2 \cdot 4H_2O$; $ZnSO_4 \cdot 7H_2O$; $CuSO_4 \cdot 5H_2O$; $NaMoO_4 \cdot 2H_2O$; $EDTA \cdot 2Na$, $FeSO_4 \cdot 7H_2O$.

5. Petri dishes (10 cm diameter).

6. 1 L boxes.

7. Incubator.

8. Autoclave.

9. ITO (indium tin oxide)-coated glass slide, beforehand sterilized.

2.2 Sample Preparation for MALDI MS Imaging

1. Scalpel.

2. Vacuum desiccator.

3. Pump.

2.3 MALDI Matrix Deposition

1. HPLC grade acetonitrile (ACN).

2. Trifluoroacetic acid (TFA).

3. HPLC grade methanol (MeOH).

4. α-Cyano-4-hydroxycinnamic acid (α-CHCA).

5. Milli-Q water.

6. Flatbed scanner.

7. ImagePrep (Bruker Daltonics, Bremen, Germany).

2.4 Data Acquisition and Processing

1. MALDI Slide Adapter II (target plate).

2. MALDI mass spectrometer (MALDI TOF/TOF Ultraflex II or FTMS Solarix 9.4T, both from Bruker Daltonics, Bremen, Germany).

3. For image acquisition and visualization: FlexImaging (Bruker Daltonics, Bremen, Germany).

4. For data processing: Flex Analysis (for data recorded with Ultraflex II mass spectrometer).

5. For data processing: Data Analysis (for data recorded with Solarix mass spectrometer).

3 Methods

In this example, the antibiotics secreted by *Paenibacillus polymyxa* Pp56 were studied in the context of the interaction of the bacteria with a phytopathogen. The aim was thus to study the secretome in real conditions, which are very different from lab conditions, notably according to nutrients scarcity. Consequently, the first step consisted in producing plant exudates in order to supplement the culture medium with these naturally occurring nutriments.

Table 1
Concentrations of salts in the agar medium for the preparation of culture medium used in confrontation plates

$Ca(NO_3)_2$	5 mM
KNO_3	5 mM
$MgSO_4$	2 mM
KH_2PO_4	1 mM
H_3BO_3	1.4 mg/L
$MnCl_2 \cdot H_2O$	0.9 mg/L
$ZnSO_4 \cdot 7H_2O$	0.1 mg/L
$CuSO_4 \cdot 5H_2O$	0.05 mg/L
$NaMoO_4 \cdot 2H_2O$	0.02 mg/L
EDTA, 2Na	5.2 mg/L
$FeSO_4 \cdot 7H_2O$	3.9 mg/L

3.1 Production of Plant Exudates

1. Prepare an agar solution at a concentration of 15 g/L.
2. Dissolve the salts in the agar solution, in order to get the concentrations specified in Table 1.
3. Pour this nutrient medium into Petri dish.
4. Deposit tomato seeds on this gelified medium and let it germinate for 5 days at room temperature in the dark.
5. Transfer germinated seeds to 1 L boxes filled with 500 mL of nutrient solution (four plants per box).
6. Grow plants for 4 weeks with a 16 h photoperiod alternating sunlight and fluorescent light.
7. Collect the hydroponic liquid and centrifuge to remove debris.

3.2 Preparation of Confrontation Plates

1. Dilute the exudate solution to fivefold volume and supplement with agar to reach a concentration of 15 g/L.
2. Sterilize in an autoclave.
3. Insert a sterilized ITO-coated glass slide into a 10 cm diameter Petri dish. Add gently the nutrient medium in order to cover it with 2–3 mm of gelified medium.
4. Prepare a fresh bacterial suspension from overnight pre-culture in order to obtain a final concentration of 5×10^7 cells/mL.
5. Streak the bacterial suspension over the width of the glass slide.
6. Inoculate the fungus at one extremity of the plate.
7. Leave the Petri dish to incubate the required duration at 25 °C.

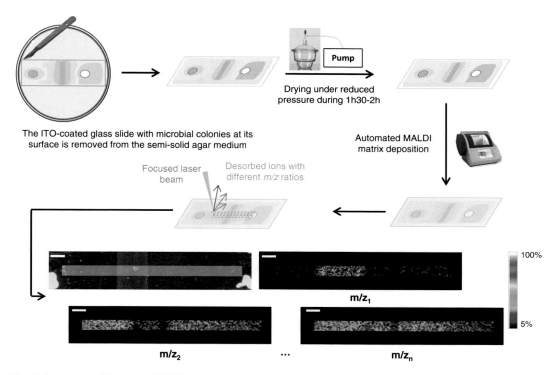

The ITO-coated glass slide with microbial colonies at its surface is removed from the semi-solid agar medium

Pump

Drying under reduced pressure during 1h30-2h

Automated MALDI matrix deposition

Focused laser beam

Desorbed ions with different *m/z* ratios

100%

5%

m/z$_1$

m/z$_2$ ...   m/z$_n$

Fig. 2 General workflow of a MALDI mass spectrometry imaging experiment

3.3 Sample Preparation for MALDI MS Imaging (Figs. 2 and 3)

The general principle of sample preparation for MALDI MS imaging is given in Fig. 2. Figure 3 shows pictures of the sample, at different steps of the preparation.

1. With a scalpel, cut the agar around the glass slide in order to remove it from the Petri dish (Fig. 3b) (*see* **Note 1**).

2. Discard the useless agar (Fig. 3c).

3. Place the glass slide into the vacuum desiccator.

4. Dry the agar under reduced pressure until complete dryness (*see* **Notes 2** and **3**). The time needed for drying completion depends on the quantity of agar present on the slide; less agar, faster drying. In the case of Fig. 3d, the drying took 2 h at a reduced pressure of 300 mbar.

5. Use a white pen to mark off the area to be analyzed from underneath.

6. Take a picture of this area with the scanner.

3.4 MALDI Matrix Deposition

1. Prepare a fresh CHCA solution at 5 mg/mL in ACN/0.2 % TFA 70:30 vol/vol. Sonicate the solution for 10 min to ensure a perfect dissolution (*see* **Notes 4** and **5**).

2. If working with a MALDI TOF/TOF instrument, spot 1 μL of a calibration mixture right to the area of interest (on top of the white marks) (*see* **Note 6**).

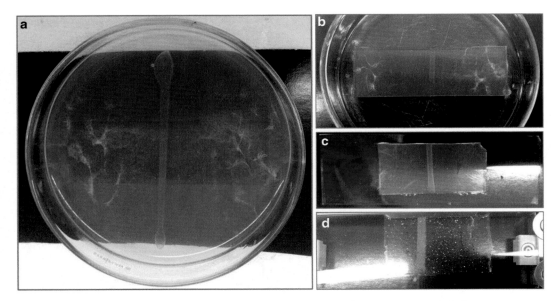

Fig. 3 The different steps of the MALDI MSI sample preparation workflow: (**a**) the sample before any processing, (**b**) the sample after cutting the agar around the ITO-coated glass slide, (**c**) the sample after discarding the useless agar, and (**d**) the sample after drying

3. Coat the slide with the ImagePrep device (*see* **Note 7**). The method used was optimized for these particular samples and was adapted from the manufacturer's standard protocol. The complete parameters of the method are given in Table 2.

4. At the end of preparation, check for the quality of the coating with a microscope. If not homogeneous or if the layer of matrix is too weak, run the last phase of the method again (*see* **Note 8**).

3.5 MALDI MS Imaging Acquisition

1. Introduce the ITO slide with the matrix-coated agar film into the MALDI adapter plate.

2. Insert the adapter plate into the ion source of the MALDI mass spectrometer.

3. Test the quality of signal. (Is the signal-to-noise ratio high? Do you detect matrix adduct peaks?)

4. Calibrate the instrument. Either externally with the calibration mixture (for TOF/TOF instrument) or internally with matrix adduct peaks (for FTMS instrument) (*see* **Note 9**).

5. Design the sequence file in FlexImaging (import the optical image of the sample, teach the sample, define the area of analysis and the raster width) (*see* **Note 10**).

6. Start automated image acquisition (*see* **Note 11**).

7. At the end of acquisition, protect the slide from dust and store it at room temperature.

Table 2
Detailed parameters of the ImagePrep method for the coating with the MALDI matrix

Phase	1	2	3	4
Final ddp	–	0.3 V	0.4 V	0.5 V
Number of cycles	12	6–18	12–40	24–60
Spray power	25 ± 25 %	27 ± 27 %	28 ± 28 %	30 ± 30 %
Spray duration	2.1 s			
Voltage drop per spray	–	0.16 V	0.3 V	0.35 V
Incubation duration	35 s	30 ± 30 s	30 ± 30 s	30 ± 30 s
Drying duration	20 s			
Residual humidity	–	20 ± 60 %	20 ± 60 %	20 ± 40 %
Complete drying every…	1 cycle	2 cycles	4 cycles	6 cycles
Safe dry duration	–	20 s	30 s	30 s

3.6 MALDI Images Visualization

1. In case of TOF data, perform data processing using batch mode in FlexAnalysis (*see* **Note 12**).
2. Load the sequence file in FlexImaging.
3. If wanted, perform a normalization of data.
4. Select the compounds of interest and display ion image.

Images of Fig. 4 display the distribution of some of the detected compounds in the average mass spectrum (Fig. 4b). All ion images show a lower signal in the bacterial colony, meaning that antibiotics diffuse into the medium right after their production by the bacterial cells. However, ion images also show that these compounds don't diffuse the same way. The ion at m/z 911.5954 (mixture of different cLPs sharing the same mass) has an intense signal on the left side of the colony (Fig. 4d), where the mycelium of the pathogen was the closest, demonstrating a likely involvement of this compound in the growth inhibition. A similar behavior is observed for the ion at m/z 925.6113 (LIF 08b peptide, Fig. 4e), albeit the compound stays closer to the bacterial cells than the previous one, suggesting a weaker involvement in the antagonism against the fungus. At last, the ion at m/z 905.5495 (unidentified, Fig. 4f) has the most intense signal far from the colony (on the right of the image), meaning it gets a higher power of diffusion into the medium, compared to the other compounds. The meaningfulness of this observation may be that this compound is probably not involved in the defense mechanisms of the bacteria but may be useful to the colonization process, as a wetting agent, for example.

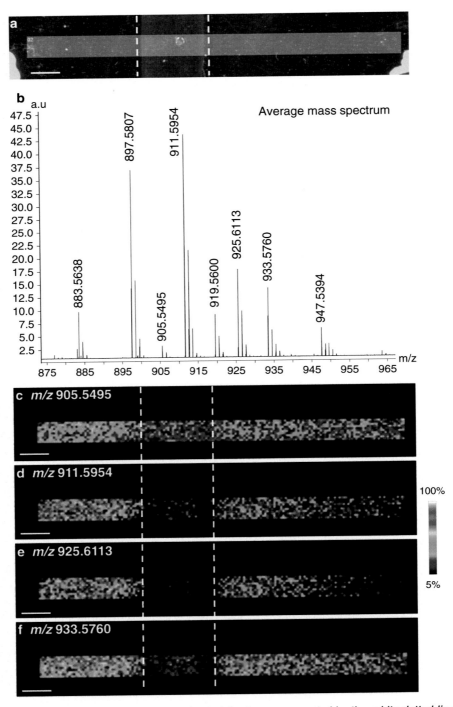

Fig. 4 (a) Picture of the sample, showing the bacterial colony, represented by the *white dotted lines* and the *blue square* represents the analyzed area, (b) MALDI average mass spectrum recorded on the area (zoom on the 875–965 *m/z* range), (c) MALDI image of the ion at *m/z* 905. 5495, (d) MALDI image of the ion at *m/z* 911.5954, (e) MALDI image of the ion at *m/z* 925.6113, and (f) MALDI image of the ion at *m/z* 933.5760. Scale bar: 2 mm. Color scale represents relative intensity for each signal: 5–100 %

3.7 Compound Identification

The identification of a compound is achieved thanks to exact mass measurements, requiring an instrument with a high mass accuracy, and on tandem mass spectrometry experiments. Exact mass measurements allow for elemental composition searches and MS/MS spectra give information on the chemical structure of the molecule. In our case, both kinds of experiments were performed using the FTMS instrument.

1. In Data Analysis, open a mass spectrum corresponding to a pixel in which the compound of interest is highly abundant.

2. Internally recalibrate the mass spectrum using matrix adduct peaks in order to reach an average error less than 0.2 ppm.

3. Use the SmartFormula tool included in Data Analysis to perform elemental composition searches.

4. To perform MS/MS experiments, reinsert the plate into the mass spectrometer.

5. Record a MS/MS spectrum in a region where the unknown compound is highly abundant.

4 Notes

1. During the cutting of agar around the glass slide, the movement of the scalpel blade may displace the bacterial colony or the fungal mycelium. Rather than a sliding motion, prefer a lever movement.

2. The pressure has to be carefully set. Beyond 5–10 mbar, the water contained in the agar gel freezes. Once dried, the resulting agar film is very fragile and does not support the vacuum of the MALDI ionization source. It cracks and comes off the glass slide.

3. This drying procedure is suited for poor culture medium without sugars or peptone-like compounds. The presence of these molecules in the agar-based medium leads to "caramel," which will never completely dry. Introducing such a sample into the MALDI ionization source prevents establishing a good vacuum and the MS analysis cannot be performed.

4. The choice of CHCA was done according to the results of an optimization process concerning the nature of the MALDI matrix, the concentration of the solution, and the solvent mixture. For other kinds of molecules, a novel optimization process should be undertaken.

5. A perfect dissolution of CHCA crystals in solvent is necessary to avoid clogging of the aluminum foil of the ImagePrep.

6. The calibration mixture may ideally be a mixture of commercial lipopeptides. Otherwise, a mixture of compounds whose molecular masses cover the targeted mass range may also be used.

7. The placement of the slide covered with the agar must not prevent the light-scattering sensor to work. Consequently, it is preferable to put on top of the sensor a part of the slide without agar and to add a coverslip, according to the manufacturer's recommendations.

8. If needed, the slide may be kept overnight at room temperature and atmospheric pressure.

9. Acceptable average error values are 5 ppm on the TOF/TOF instrument and 0.5 ppm for the FTMS instrument.

10. The smaller the pixel size, the longer the acquisition and the larger the dataset (especially true for FTMS data).

11. If multiple areas are to be analyzed on the sample carrier, a batch acquisition should be considered if working with Ultraflex mass spectrometer. This option is not yet available on the Solarix mass spectrometer.

12. For TOF data, it may be necessary to submit the whole dataset to a processing method including smoothing, baseline correction, and external calibration, using if possible a lock mass.

References

1. Bais HP, Fall R, Vivanco JM (2004) Biocontrol of *Bacillus subtilis* against infection of *Arabidopsis* Roots by *Pseudomonas syringae* is facilitated by biofilm formation and surfactin production. Plant Physiol 134:307–319

2. Mendes R, Garbeva P, Raaijmakers JM (2013) The rhizosphere microbiome: significance of plant beneficial, plant pathogenic, and human pathogenic microorganisms. FEMS Microbiol Rev 37:634–663

3. Lugtenberg B, Kamilova F (2009) Plant-growth-promoting rhizobacteria. Annu Rev Microbiol 63:541–556

4. McSpadden Gardener BB (2004) Ecology of *Bacillus* and *Paenibacillus* spp. in agricultural systems. Phytopathology 94:1252–1258

5. Chen XH, Koumoutsi A, Scholz R et al (2009) Genome analysis of *Bacillus amyloliquefaciens* FZB42 reveals its potential for biocontrol of plant pathogens. J Biotechnol 140:27–37

6. Rückert C, Blom J, Chen X et al (2011) Genome sequence of *B. amyloliquefaciens* type strain DSM7T reveals differences to plant-associated *B. amyloliquefaciens* FZB42. J Biotechnol 155:78–85

7. Stein T (2005) *Bacillus subtilis* antibiotics: structures, syntheses and specific functions. Mol Microbiol 56:845–857

8. Finking R, Marahiel MA (2004) Biosynthesis of nonribosomal peptides. Annu Rev Microbiol 58:453–488

9. Ongena M, Jacques P (2008) *Bacillus* lipopeptides: versatile weapons for plant disease biocontrol. Trends Microbiol 16:115–125

10. Raaijmakers J, De Bruin I, Nybroe O et al (2010) Natural functions of cyclic lipopeptides from *Bacillus* and *Pseudomonas*: more than surfactants and antibiotics. FEMS Microbiol Rev 34:1037–1062

11. Hashizume H, Nishimura Y (2008) Cyclic lipopeptide antibiotics. Stud Nat Prod Chem 35:693–751

12. Bionda N, Cudic P (2011) Cyclic lipodepsipeptides in novel antimicrobial drug discovery. Croat Chem Acta 84:315–329

13. Ramey BE, Koutsoudis M, Bodman S et al (2004) Biofilm formation in plant-microbe associations. Curr Opin Microbiol 7:602–609

14. Henry G, Deleu M, Jourdan E et al (2011) The bacterial lipopeptide surfactin targets the lipid fraction of the plant plasma membrane to trigger immune-related defence responses. Cell Microbiol 13:1824–1837

15. Jourdan E, Henry G, Duby F et al (2009) Insights into the defense-related events occurring in plant cells following perception of surfactin-type lipopeptide from *Bacillus subtilis*. Mol Plant-Microbe Interact 22:456–468

16. Ongena M, Jourdan E, Adam A et al (2007) Surfactin and fengycin lipopeptides of *Bacillus subtilis* as elicitors of induced systemic resistance in plants. Environ Microbiol 9:1084–1090

17. Braga PAC, Tata A, Goncalves dos Santos V et al (2013) Bacterial identification: from the agar plate to the mass spectrometer. RSC Advances 3:994–1008

18. Erhard M, von Döhren H, Jungblut PR (1999) Rapid identification of the new anabaenopeptin G from *Planktothrix agardhii* HUB 011 using matrix-assisted laser desorption/ionization time-of-flight mass spectrometry. Rapid Commun Mass Spectrom 13:337–343

19. Chen H-Y, Chen Y-C (2005) Characterization of intact *Penicillium* spores by matrix-assisted laser desorption/ionization mass spectrometry. Rapid Commun Mass Spectrom 19: 3564–3568

20. Norris JL, Caprioli RM (2013) Analysis of tissue specimens by matrix-assisted laser desorption/ionization imaging mass spectrometry in biological and clinical research. Chem Rev 113:2309–2342

21. Barger SR, Hoefler BC, Cubillos-Ruiz A et al (2012) Imaging secondary metabolism of *Streptomyces sp.* Mg1 during cellular lysis and colony degradation of competing *Bacillus subtilis*. Antonie van Leeuwenhoek 102:435–445

22. Hoefler BC, Gorzelnik KV, Yang JY et al (2012) Enzymatic resistance to the lipopeptide surfactin as identified through imaging mass spectrometry of bacterial competition. Proc Natl Acad Sci U S A 109:13082–13087

23. Watrous JD, Phelan VV, Hsu CC et al (2013) Microbial metabolic exchange in 3D. ISME J 7:770–780

24. Bhandari DR, Shen T, Römpp A et al (2014) Analysis of cyathane-type diterpenoids from *Cyathus striatus* and *Hericium erinaceus* by high-resolution MALDI MS imaging. Anal Bioanal Chem 406:695–704

25. Debois D, Ongena M, Cawoy H et al (2013) MALDI-FTICR MS imaging as a powerful tool to identify Paenibacillus antibiotics involved in the inhibition of plant pathogens. J Am Soc Mass Spectrom 24: 1202–1213

26. Debois D, Jourdan E, Smargiasso N et al (2014) Spatiotemporal monitoring of the antibiome secreted by *Bacillus* biofilms on plant roots using MALDI mass spectrometry imaging. Anal Chem 86(9):4431–4438. doi:10.1021/ac500290s

Chapter 12

Secondary Metabolic Pathway-Targeted Metabolomics

Maria I. Vizcaino and Jason M. Crawford

Abstract

This chapter provides step-by-step methods for building secondary metabolic pathway-targeted molecular networks to assess microbial natural product biosynthesis at a systems level and to aid in downstream natural product discovery efforts. Methods described include high-resolution mass spectrometry (HRMS)-based comparative metabolomics, pathway-targeted tandem MS (MS/MS) molecular networking, and isotopic labeling for the elucidation of natural products encoded by orphan biosynthetic pathways. The metabolomics network workflow covers the following six points: (1) method development, (2) bacterial culture growth and organic extraction, (3) HRMS data acquisition and analysis, (4) pathway-targeted MS/MS data acquisition, (5) mass spectral network building, and (6) network enhancement. This chapter opens with a discussion on the practical considerations of natural product extraction, chromatographic processing, and enhanced detection of the analytes of interest within complex organic mixtures using liquid chromatography (LC)-HRMS. Next, we discuss the utilization of a chemometric platform, focusing on Agilent Mass Profiler Professional software, to run MS-based differential analysis between sample groups and controls to acquire a unique set of molecular features that are dependent on the presence of a secondary metabolic pathway. Using this unique list of molecular features, the chapter then details targeted MS/MS acquisition for subsequent pathway-dependent network clustering through the online Global Natural Products Social Molecular Networking (GnPS) platform. Genetic information, ionization intensities, isotopic labeling, and additional experimental data can be mapped onto the pathway-dependent network, facilitating systems biosynthesis analyses. The finished product will provide a working molecular network to assess experimental perturbations and guide novel natural product discoveries.

Key words Chemical signaling, Secondary metabolism, Natural product discovery, Comparative metabolomics, Molecular networking, Nonribosomal peptide biosynthesis, Isotopic labeling, High-resolution mass spectrometry

1 Introduction

Microbial genome sequencing efforts are illuminating a growing number of "orphan" biosynthetic gene clusters suspected of biosynthesizing novel natural products with pharmacological, agricultural, and biotechnological values [1]. These unknown small molecules possess the potential to serve as new molecular probes, dietary supplements, flavors, fragrances, signaling agents, commercial products, drugs, and drug leads [2]. Unfortunately, the

Bradley S. Evans (ed.), *Nonribosomal Peptide and Polyketide Biosynthesis: Methods and Protocols*, Methods in Molecular Biology, vol. 1401, DOI 10.1007/978-1-4939-3375-4_12, © Springer Science+Business Media New York 2016

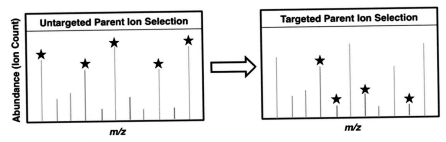

Fig. 1 Untargeted versus pathway-targeted mass spectral data collection. Under untargeted MS/MS fragmentation modes, abundant ions are preferentially selected for fragmentation (highlighted with *stars*), and depending on inclusion and exclusion parameters, many of the less abundant ions go undetected in MS/MS data. Under pathway-targeted fragmentation modes, only pathway-dependent masses are selected (highlighted in *red* and by *stars*), providing higher fragmentation coverage of the secondary metabolic pathway ions of interest for subsequent molecular networking

valuable products of most orphan pathways remain "cryptic," "silent," or simply undetectable in the laboratory environment, undercutting discovery efforts. This chapter describes a "pathway-targeted" structural networking approach, which can be used to aid in the detection and elucidation of these important small molecules known as secondary metabolites (Fig. 1).

Secondary metabolites from microorganisms are most often encoded by biosynthetic gene clusters, including nonribosomal peptide synthetase (NRPS) and polyketide synthase (PKS) systems, among others. Mass spectrometry (MS) and tandem MS (MS/MS) techniques are important tools in secondary metabolite detection, discovery, and characterization. Secondary metabolic pathways often produce a series of structurally related intermediates and products, and consequently, the MS/MS fragmentation data can be used to organize and delineate related molecules into network clusters or "molecular families" [3, 4]. Employing these foundational molecular networking tools developed by Dorrestein and coworkers [3, 4], we recently developed a "pathway-targeted" structural networking approach (Fig. 2) to finely map a bacterial secondary metabolic pathway found in the human gut and linked to colorectal cancer initiation [5, 6]. A defining feature separating microbial secondary metabolism, which is typically involved in host virulence, pathogenicity, mutualism, antibiotic production, and diverse chemical signaling events, from primary metabolism is the general ability to cleanly delete a metabolic pathway from a cell while retaining cell growth and avoiding many of the complications due to primary metabolic rerouting. Consequently, comparative metabolomics analysis among controls, wild-type organisms, and pathway mutant organisms (or heterologous expression systems) can be used to map molecular features dependent on functional pathways (Fig. 3). Here, we use the raw data from our recent studies on the bacterial colibactin pathway to illustrate pathway-targeted structural

a Comparative metabolomics workflow

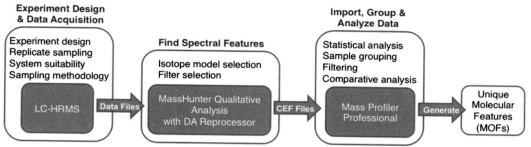

b Pathway-targeted mass spectral networking workflow

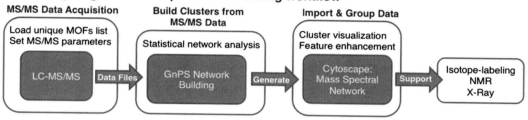

Fig. 2 Workflow overview for pathway-targeted molecular networking

networking (Methodology Overview, Fig. 2) [5, 6], which can be applied to the majority of secondary metabolic pathways of interest in genetically tractable production hosts.

2 Materials

Prepare all culture media and solutions using Milli-Q ultrapure water. High-performance liquid chromatography (HPLC) grade solvents should be used for organic extractions, and LC-MS grade water and solvents should be used for all Q-TOF MS analyses.

2.1 Culture Growth and Extraction

1. General lysogeny broth (LB) or LB agar, Miller (*see* **Note 1**).

2. Difco M9 minimal medium supplemented with casamino acids (5 g/L), 0.4 % (w/v) glucose (filter-sterilized, store at 4 °C until use), 2.0 mM magnesium sulfate ($MgSO_4$, sterile), and 0.1 mM calcium chloride ($CaCl_2$, sterile) (*see* **Note 2**). Add antibiotic(s) to sterile medium if needed.

3. Isopropyl β-D-1-thiogalactopyranoside (IPTG), aqueous stock solution: 1.0 M filter-sterilized and stored in aliquots at –20 °C.

4. Petri dish (100 × 15 mm).

5. Polypropylene culture tubes (14 mL) (*see* **Note 3**).

6. Glass serological pipettes.

a   Untargeted mass spectral network

b   Pathway-targeted mass spectral network

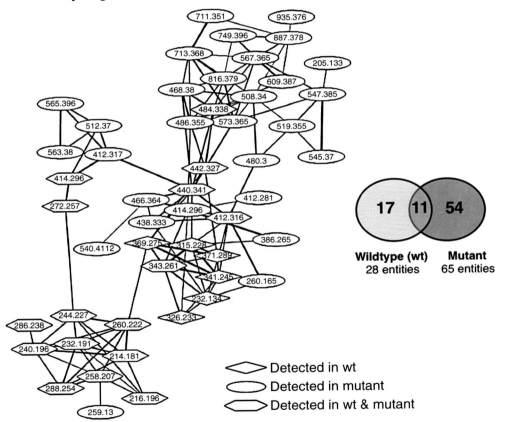

Fig. 3 Pathway-targeted mass spectral networking. (**a**) An untargeted auto MS/MS approach leads to a network with a much larger number of total ion masses (nodes) with often more limited coverage of less abundant pathway-dependent ions. As highlighted in the Venn Diagram, there are a large number of entities associated with controls (media components, primary metabolites, and contaminants). (**b**) By removing all entities detected in any control sample, targeted MS/MS can be performed to access higher MS/MS fragmentation coverage for "pathway-dependent" molecular features (MOFs) associated with the presence of the gene cluster of interest (highlighted in Venn Diagram)

7. 16×100 mm disposable test tubes, borosilicate glass.

8. Ethyl acetate, HPLC grade (*see* **Note 4**).

2.2 Amino Acid Isotope-Labeled Culture Growths

1. Isotopically labeled L-amino acids (e.g., [U-^{13}C$_3$]-L-Cys) (*see* **Note 5**).

2.3 LC-MS Analysis

1. Polypropylene LC-MS vials, 250 μL (*see* **Note 6**).

2. Methanol, LC-MS grade.

3. Mobile phase A: LC-MS grade water with 0.1 % formic acid.

4. Mobile phase B: LC-MS grade acetonitrile with 0.1 % formic acid.

5. Phenomenex Kinetex™ 1.7 μm C18 100 Å LC column (100 × 2.10 mm) or comparable analytical ultrahigh-performance LC column (*see* **Note 7**).

6. HPLC system: Agilent 1290 Infinity Binary HPLC system equipped with Diode Array Detector, autosampler, or comparable system.

7. Mass spectrometer: Agilent 6550 iFunnel quadrupole time-of-flight (Q-TOF) MS instrument equipped with a Dual Agilent Jet Stream (AJS) electrospray ionization (ESI) source, running in positive mode (or negative mode) scanning from 25 to 1700 *m/z* mass range at 1.00 spectra/s or comparable instrument and conditions.

8. Agilent MS standard reference mass solution: *m/z* 121.05087300 (purine) and *m/z* 922.00979800 (hexakis (1H, 1H, 3H-tetrafluoropropoxy)phosphazene).

9. Software: MassHunter Workstation Data Acquisition (Agilent Technologies), MassHunter Qualitative Analysis (Agilent Technologies), and Mass Profiler Professional (Agilent Technologies).

3 Methods

Comparative metabolomics analysis should be conducted at least between a biological control group (described below) and a wild-type sample group expressing the secondary metabolite pathway of interest; mutated derivatives of the pathway can be included. The conditions described below are for *Escherichia coli* strains heterologously expressing a representative hybrid NRPS-PKS biosynthetic gene cluster under an IPTG-inducible system. Correspondingly, the protocols can be modified if the gene cluster is expressed natively or under other inducible systems and should transfer well to other genetically tractable production hosts. Methods can be modified based on growth requirements for different bacterial systems. Once the conditions have been selected for a given pathway, use the same methods when performing growths and organic extractions within a metabolomics experiment to minimize variables and disparities that can influence data analysis. Selection of biological controls is important when designing your experiment. As a biological control, we recommend using the bacterial strain with the full gene cluster deleted to maximize characterization of pathway-dependent metabolites during analysis. Alternatively, deletion of a selected core biosynthetic gene would suffice. Following a systematic approach, method development should involve three main aspects: (1) extraction method, (2) LC method, and (3) MS detection method. As an example, extraction of whole culture metabolomes (cells plus supernatant) by ethyl acetate is described below (*see* **Note 8**).

3.1 Bacterial Growth and Organic Extraction

1. Streak bacteria from frozen stock (stored at –80 °C) onto LB agar plate (with necessary antibiotic(s)). Incubate overnight at 37 °C or until single colonies are formed (*see* **Note 9**).

2. Inoculate 5 mL LB (with appropriate antibiotic(s)) in a polypropylene culture tube with one single colony as a seed culture. We recommend a total of five biological replicates. Incubate while shaking (250 rpm) at 37 °C for 12–14 h (*see* **Note 10**).

3. Next day, inoculate 5 mL M9 minimal broth (with appropriate antibiotic(s)) in a polypropylene culture tube with overnight seed culture (1:100). Grow at 37 °C (250 rpm) until optical density (OD_{600}) is 0.5 for inducible systems (*see* **Note 11**). Place culture at 4 °C for 10 min before induction (if necessary). Then add IPTG to the desired concentration and allow cultures to grow for 48 h at 25 °C, 250 rpm (*see* **Note 12**).

4. After 48 h (or designated time point), remove cultures from shaker. Add 6 mL of ethyl acetate directly onto the culture in the polypropylene tube. Cap tube tightly and vigorously shake for 30 s. Centrifuge cultures at 4 °C for at least 15 min ($2800 \times g$). Carefully remove tubes, and transfer 4 mL of the organic layer into glass vials. Evaporate to dryness under reduced pressure. Once dried, store extracts at –20 °C until use.

3.2 LC-MS Analysis

1. Prepare samples for LC-MS analysis based on recommended concentration for instrument sensitivity. For the Q-TOF MS system described (Materials Section), dissolve extracted samples in 2.5 mL methanol as a starting point (*see* **Note 13**). Filter samples (*see* **Note 14**). Place 50 µL of sample in HPLC vials for analysis, dry the remaining sample volume, and store at –20 °C for future use.

2. Create an LC and MS method for LC-MS analysis.

(a) General LC method. Using mobile phase A and B, set up method as follows: 95 %/5 % A/B, 2 min; run gradient from 95 % A to 98 % B until 26 min (*see* **Note 15**). Hold for 5 min at 98 % B and then bring back to starting conditions. Recommended flow rate for UHPLC column described in Materials is 0.3 mL/min. Equilibrate column for recommended time (*see* **Note 16**).

(b) General MS method. Set source parameters as follows: gas temperature at 225 °C and flow at 12 mL/min, nebulizer at 50 psig, sheath gas temperature at 275 °C, and flow at 12 L/min. Set scan source parameters as follows: capillary 3500 V, fragmentor 125 V, skimmer 65 V, and OCT RF Peak 750 V.

3. Calibrate and tune your high-resolution mass spectrometer according to the specific settings for metabolomics analysis (*see* **Note 17**).

4. Prepare LC column by performing multiple column washes, and run LC method without an injection. Then set up workflow as follows: blank control (using same solvent used to dissolve samples), control extracts, and sample extracts. Run a blank wash between each sample set.

5. Place samples in autosampler, kept at 4 °C. Inject 3–5 μL based on sample concentration. Create a worklist and run samples (*see* **Note 18**).

3.3 Acquiring Molecular Features

1. First, determine quality of data using the MassHunter Qualitative Analysis program. Overlay chromatograms for each sample set to confirm system repeatability (*see* **Note 19**).

2. Then, process centroid MS data in MassHunter Qualitative Analysis. To do this, go to "Method" and create a method using the Extract Molecular Features protocol (Fig. 4). Recommended settings are outlined in Table 1 (*see* **Note 20**).

3. Open DA Reprocessor (offline program), and set up a sample worklist by clicking "Insert multiple samples." Select samples to be analyzed, and load. Select "Method" created in **step 2**. Click the "Start" icon to process data. This feature will generate compound exchange format (CEF) files for all samples, which are used for differential and statistical analysis (*see* **Note 21**).

3.4 Statistical Analysis Using Mass Profiler Professional (MPP) (See Note 22)

1. Open MPP and create a new Project.

2. Create new experiment (you can have multiple experiments per Project), select "combined (identified + unidentified)" experiment type, and "Data Import Wizard" as the workflow type. This is a good starting point for differential analysis by providing full guidance.

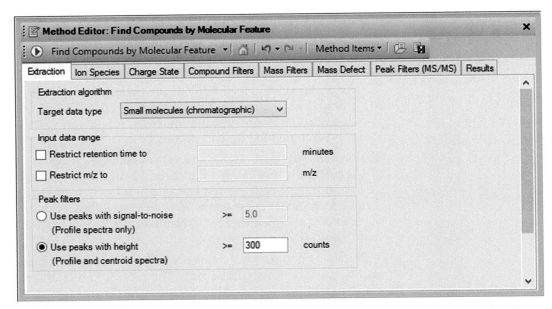

Fig. 4 Find compounds by molecular feature. As described in Table 1 and detailed in the chapter, the different settings (tabs) can be modified to optimize for the in silico extraction of compounds of interests

3. Select program used to acquire data (e.g., MassHunter Qual). Choose organism if applicable.

4. Upload CEF files by selecting "Select Data Files" tab. Then, create grouping for each sample set (i.e., control, wild type, and mutants) by selecting the "Add Parameter" tab.

5. Adjust filters as deemed necessary. Recommended initial parameter for small molecule metabolomics are "Minimum Absolute abundance" set at 5000 counts and "Minimum number of ions" set at two; Select "Multiple charge states forbidden." At this point, it is recommended that you use all available data.

6. Next, set any alignment parameters necessary to align compounds in different samples if their accurate mass and retention times (RTs) fall within the specified tolerance window. Unless you are using an internal standard, recommended parameters are RT window = 0.1 % + 0.15 min and Mass window = 5.0 ppm + 2.0 mDa.

7. The next window shows a summary of data and provides the number of aligned compounds. Continue through the windows until analysis is finished. Once data is uploaded, perform grouping and analysis of data using the Workflow panel (right side of screen). Familiarize yourself with the different tabs within this panel.

8. To group samples.

   (a) First, create a "Control" (absent/nonfunctional gene cluster) interpretation. Under "Experiment Setup," select

Table 1
Recommended settings for *Extract Molecular Features* protocol

Tabs	Selection	Notes
Extraction		
Target data type	Small molecules (chromatographic)	Can also select for proteins or infused samples
Input data range	None	Can modify as seen fit for data
Peak filters	Use peaks with height ≥300 counts	
Ion species	Check positive (+) or negative (−) ions	
Check "sodium species are commonly observed, select for salt dominated ions"		
Charge state		
Isotope model	Common organic molecules	Useful if C, H, O, N, S molecules are present
	Unbiased	Useful if Fe, Br, or other heavy molecules are present
	Can also select for peptides, glycans	Assigns isotopes based on common molecules present
Limit assigned charge states to a maximum of	2	
Compound filters		
Quality score	≥80.0	
Restrict charge states to	1–2 z	
Peak filters (MS/MS)		
Absolute height	≥100 counts	
Results		
Delete previous compounds	Selected	
Highlight all compounds	Selected	
Extract MFE spectrum	Selected	MFE stands for *m*olecular *f*eature *e*xtractor
Extract ECC	Selected	ECC stands for *e*xtracted *c*ompound *c*hromatogram
Extract EIC	Selected	EIC stands for *e*xtracted *i*on *c*hromatogram

Help menu in Agilent is helpful in describing each setting or algorithm in more detail

"Create Interpretation." Deselect any conditions that are not a control; select "Non-Averaged," and deselect "Absent." Name your new parameter (e.g., Control) and click "Finish."

(b) Next, create a "Sample" (with gene cluster) interpretation for wild type and mutant(s), individually, as specified in **step 8a**. Once you are done, the newly created interpretations can be seen on the left panel under your experiment. At this point, there should be one interpretation from each sample set (i.e., control, wild type, mutant, etc.).

9. To perform analysis.

   (a) First, do analysis on control samples. Go to the "Quality Control" tab in the Worklist Panel, and select "Filter by Abundance." In the "Entity List" tab, select "All Entities." Under the "Interpretation" tab, select the control interpretation. Under "Retain entities in which" section, select at least "**1**" out of the **5** samples that have values within range (*see* **Note 23**). Name your analysis and click "Finish."

   (b) Perform a stricter analysis on samples that contain the pathways. Do the same as was done for the control using the "Filter by Abundance" analysis. But, under "Retain entities in which" section, select at least "**5**" out of the **5** samples that have values within range (*see* **Note 24**).

10. Next, create a Venn Diagram of samples to be compared. Go to "Tools," and under "Venn Diagram," select "Entity List." Conversely, click on the "Venn Diagram" icon. Select control, wild type, and mutant samples and click "OK" (Fig. 3a, Venn Diagram).

11. Create a new entity list of molecular features only found in samples with pathway by highlighting the respective section of the Venn Diagram (e.g., the section that corresponds only to wild type). Select the "Create Entity List" icon (icon of Page with pencil). Confirm that entities found only in desired sample are selected, then save new entity list.

12. Repeat **step 11** as many times as necessary (for each desired sample group, e.g., wild type, mutant 1, mutant 2, etc.) to acquire a conservative unique entity list consisting of pathway-dependent molecular features. Molecular features found in all sample strains (with pathway) can be compared (Fig. 3b, Venn Diagram).

13. Examine each unique list, and confirm that ions are not found in control samples (do this in MassHunter Qualitative Analysis software) (*see* **Note 25**). To export molecular feature list into Excel files do as follows:

   (a) Export inclusion list using the tab under "Workflow/Results Interpretation" panel. Save the output file and select "next." On the subsequent page, select the following parameters: (1) limit number of precursor per compound to **1** ion(s); (2) export highest abundance m/z, (3) positive ions [H] and [Na]. The resulting list contains the ion masses to be fragmented and will be used as the precursor list for MS/MS fragmentation.

(b) To download the HRMS, retention time, and abundance (raw and normalized) information, right-click the data file of interest. Select "Export list" option, select the information you want to download (e.g., raw abundance, normalized abundance, mass, retention time), and create file.

3.5 MS/MS Fragmentation

1. Prepare samples for LC-MS/MS analysis as was done in Subheading 3.2, and prepare system using same calibration MS conditions and LC method.

2. Create a MS/MS method by selecting the "Auto MS/MS" setting. At this point, an untargeted metabolomics data set can be collected by performing an "Auto MS/MS" run (*see* **Note 26**).

3. To run a pathway-targeted metabolomics experiment, specifically target masses from the unique ion list. In the Agilent system, in the "Auto MS/MS" setting under the "Preferred/Exclude" tab of the Q-TOF Acquisition section, the preferred ion list can be inserted. Right-click anywhere on the table and upload the exclusion list acquired in MPP. Change the "Delta *m/z* (ppm)" tab to 0.5 (depending on accuracy of system). Check the "Use Preferred ion list" box.

4. Create different MS/MS methods to enhance overall MS/MS data collection (*see* **Note 27**). Use different isotope models (under "Spectral Parameters" tab) and analyze different collision energies (e.g., 10, 20, 40) (*see* **Note 28**).

5. Prepare LC column by performing multiple washes. Then set up workflow for sample extracts to be analyzed. Place samples in autosampler, kept at 4 °C. Inject 3–5 μL based on sample concentration. Run samples.

3.6 Pathway-Targeted Mass Spectral Molecular Networking (*See* Note 29)

1. Download the following programs: MSConvert (ProteoWizard tool) (*see* **Note 30**), Insilicos (*see* **Note 31**), and the open-source platform Cytoscape (*see* **Note 32**).

2. Convert raw MS/MS data in MSConvert to mzXML files. Use same settings as shown in Fig. 5. Open converted files and confirm correct conversion in the Insilicos program.

3. Go to GnPS (Global Natural Products Social Molecular Networking) (www.gnps.ucsd.edu), create a free account, and click on "Data Analysis" (*see* **Note 29**).

4. In the "Workflow Selection," create a descriptive title, and upload files via a FTP connection (*see* **Note 33**). The host name is ccms-ftp01.ucsd.edu, and the username and password are the same as your created account.

5. Select spectrum file(s) to be analyzed and upload. Set parameters for the network including parent mass tolerance (0.1–2.0 Da) and min pairs cos (0.1–1.0) (*see* **Note 34**). For "Minimum

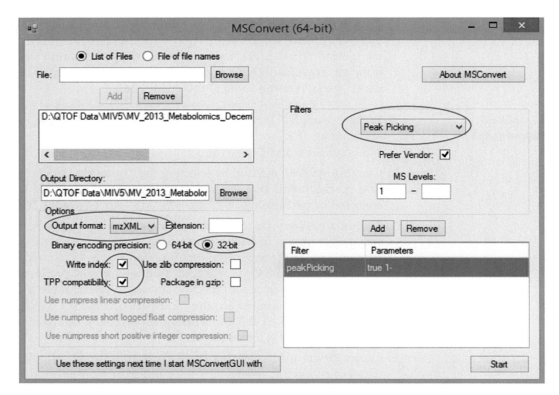

Fig. 5 MS convert settings. MS convert program is used to convert mass spectral data files to CEF files for analyses in Mass Profiler Professional. Selection settings are highlighted (*red ovals*). Once the filter is selected, click "Add," then select "Start"

Cluster Size," type "1." Perform different networks and analyze how each setting affects the clusters.

6. Once the analyses are finished, click "View All Clusters with IDs" and download the files.

7. In Cytoscape, import the "networkedges" file. Select "Column 1" under "Select Source Node Column" and "Column 2" under "Select Target Node Column." Click "Column 3" and "Column 5" to activate (turns blue). Rename columns 3 and 5 as "deltaMZ" and "cosine," respectively. Click OK to load the network. Select "Apply Preferred Layout" to view your initial network template (*see* **Note 35**).

8. Next, load the information parameters by importing table file "clusterinfosummarygroup_attributes_withIDs." Deselect any columns not needed and click "OK."

9. Under the Control Panel, select the "Style" tab. Here you can set different parameters to enhance the appropriate data features for the "node" or "edge" properties.

    (a) To highlight the connectivity strength between ion masses and load cosine settings. To do this, select the "Edge" tab.

For the "width" setting, select *cosine* for the "Column" and *Continuous Mapping* for the "Mapping Type." Double-click the box, and set the min. and max. values to assess the display. The bolder (thicker) edges display stronger connectivity between ion masses.

(b) Load the respective masses onto each node (Fig. 3b). In the "Node" tab, under the "Label" setting, select *parent-mass* for the "Column" and *Passthrough Mapping* for the "Mapping Type."

(c) To distinguish between genetic variants, "node" properties can be modified. Set a distinct node shape (under "Shape") for a specific strain source (e.g., wild type, mutant(s), all) as shown in Fig. 3b.

(d) To highlight differences in parent ion ionization intensities, display each strain metabolite abundances based on a heat-map coloration (Fig. 6a). To accomplish this, first go to the "Table Panel," and manually type in the average raw abundance of each parent ions obtained for each specific strain from the MPP program (*see* Subheading 3.4, **step 13b**). For example, for the wild-type metabolite intensities, manually type in under the "sum (precursor intensity)" column the raw abundance for each unique molecular features. If a feature was not detected in the wild type, type "0." Then select "Fill Color," and perform a "Discrete Mapping" coloration based on the "sum (precursor intensity)" column. Save the network under a different name. Then repeat for each mutant strain, each time saving as a distinct network (Fig. 6a).

3.7 Structural Characterization Support by Isotope-Labeled Feeding Experiments

1. Growth of cultures, organic extraction, and HRMS analysis should be performed as per previous sections (Subheadings 3.1 and 3.2) with few modifications.

2. First, make M9 minimal medium to account for labeled compound, which in the example given, are amino acids. In place of casamino acids, individually add the L-amino acid composition at 5 g/L total concentration as follows: 3.6 % Arg, 21.1 % Glu, 2.7 % His, 5.6 % Ile, 8.4 % Leu, 7.5 % Lys, 4.6 % Phe, 9.9 % Pro, 4.2 % Thr, 1.1 % Trp, 6.1 % Tyr, 5 % Val, 4 % Asn, 4 % Ala, 4 % Met, 4 % Gly, 4 % Cys, and 4 % Ser. Use this media for control cultures.

3. To incorporate isotopically labeled (^{13}C/^{15}N/^{2}H) amino acids into products from NRPS-PKS pathways, modify above media by adding the isotope-labeled amino acid at a specific ratio for a given strain.

(a) For *E. coli* heterologous expression, auxotrophic strain variants can be used by transforming the expression plasmid or

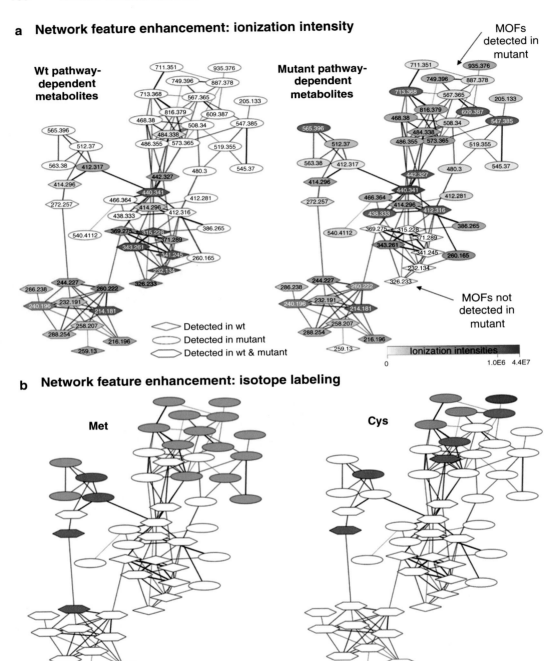

Fig. 6 Molecular network feature enhancement. (**a**) Features, such as average ionization intensities and other experimental data, can be used to enhance the network in Cytoscape to help guide pathway and secondary metabolite characterization. These graphical illustrations allow for the quick identification of metabolic "bottle-necks" in secondary metabolism, such as for a pathway mutant. (**b**) Ion masses (or molecular features, nodes) incorporating labeled amino acid (e.g., methionine (Met) or cysteine (Cys)), as determined by HRMS, are color coded on the network map. Figure was adapted from Vizcaino and Crawford [6]

bacterial artificial chromosome containing the gene cluster of interest into the appropriate mutant strain (*see* **Note 36**). Growth media should then be supplemented with a 1:1 (^{12}C:^{13}C) ratio of nonlabeled/labeled amino acid. For example, add 2 % Cys and 2 % L-[U-^{13}C]-Cys.

(b) If an auxotrophic variant is not available, supplement growth media with a 100 % labeled amino acid (e.g., 4 % L-[U-^{13}C]-Cys).

4. Grow three 5 mL biological replicates per strain using the same growth conditions defined in Subheading 3.1.

5. After 48 h (or designated time point), remove cultures and extract with ethyl acetate. Dry organic layer as described above, then resuspend in 2.5 mL methanol (*see* **Note 37**), and analyze 3–5 μL by LC-Q-TOF-HRMS (as detailed in Subheading 3.2) scanning from *m/z* 25 to 1700.

6. Determine incorporation of labeled amino acid (or other compound) by examining the unique molecular features acquired from MPP.

7. Repeat-labeled feeding experiment in triplicate with only the labeled amino acid (e.g., 4 % L-[U-^{13}C]-Cys) to confirm dose-dependent incorporation.

8. Label mass spectral molecular network to represent incorporation of labeled compound (e.g., L-[U-^{13}C]-Cys incorporation) (Fig. 6b).

4   Notes

1. Miller LB broth and LB agar can be purchased as a premade mix and prepared as per the manufacturer's instructions. We recommend that once the media source is selected, the same brands should be maintained throughout experiment sets.

2. These are typical components for growth of heterotrophic bacteria. Sometimes it is necessary to add other components, such as metals (e.g., Fe^{2+}, Cu^{2+}) or vitamins, depending on requirements for production of secondary metabolites. It is recommended that you test various cultivation conditions to optimize production of relevant metabolites (e.g., *see* ref. [7]).

3. Polypropylene culture tubes are fairly resistant to various extraction solvents including ethyl acetate, butanol, and methanol. Do not use polystyrene culture tubes.

4. For our representative NRPS-PKS pathway, ethyl acetate is routinely used for the enrichment of its relatively nonpolar metabolites. More polar solvents such as butanol or methanol can be used instead to enrich for more polar compounds (*see* **Note 8**).

5. For our representative NRPS-PKS pathway, amino acid incorporation provided additional structural support for all detectable pathway-dependent small molecules. It might be necessary to utilize other labeled substrates, such as uncommon amino acids. Modify experiment as appropriate in a case-by-case basis.

6. Based on your instrument system, make sure to get the correct vials that fit the autosampler. For the Agilent system, we use Agilent vials and caps. Always tightly cap your vials to minimize solvent evaporation during data acquisition.

7. The selection of stationary phase for HPLC analysis will vary depending on the type of molecules being analyzed. As a starting point, reverse-phase C18 columns work well for a median combination of polar and nonpolar compounds. For less polar compounds, a reverse-phase C8 column might be useful. For more polar compounds, Hypercarb porous graphitic carbon (PGC) columns retain polar compounds well.

8. Enrichment of desired secondary metabolites can be optimized through the organic extraction of different culture components (i.e., supernatant, cell, or whole culture) and/or through the use of different extraction solvents (Fig. 7). Most secondary metabolites are produced inside the cells and can be subsequently secreted into the supernatant. If the desired metabolites are found at higher concentrations in the supernatant or cell fraction, it is best to extract the desired culture component as an initial enrichment step. In that case, whole culture should be centrifuged for 15 min at $2800 \times g$. Filter the supernatant into a new polypropylene culture tube and perform extraction. Conversely, if cells are the desired starting material, carefully remove any leftover supernatant from the culture tube containing the cells, and then perform an organic extraction on the cell pellet. Dichloromethane or ethyl acetate solvents are routinely used in natural product isolation for the enrichment of nonpolar compounds and can be used in a solvent-solvent extraction with supernatant/whole culture. Similarly, butanol will enrich for medium polarity compounds through a solvent-solvent extraction of the whole culture or supernatant. Be

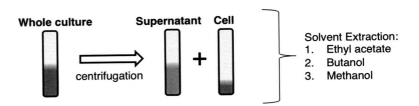

Fig. 7 Schematic for bacterial culture extraction

aware that butanol is slightly miscible with water, so there will be a volume reduction in the top organic layer. Some small molecules are very polar and not extractable with ethyl acetate or butanol. At this point, a methanol extraction might be needed. Since water and methanol are miscible, the whole culture (or supernatant) should be lyophilized before methanol extraction. Once completely dry, add 1 mL methanol and sonicate for 5 min. Centrifuge extract for at least 15 min at 4 °C ($2800 \times g$). Very carefully, remove culture tubes from centrifuge, take a fixed volume aliquot of the methanol extract, making sure not to disturb the pellet, and filter. Dry the filtered extract and store at −80 °C until analysis. Conversely, bacterial cells can be extracted with methanol following same procedure.

9. Select single colonies as biological replicates for analyses. Adjust cultivation time for selected organism as needed.

10. Secondary metabolites are often, but not always, upregulated as the cells transition to stationary phase.

11. OD_{600} range can be between 0.45 and 0.55. We sometimes observe discrepancies between control and sample growth rates, which require cultivation time correction.

12. The method describes the use of an IPTG-inducible system, where IPTG serves to activate transcription of the pathway of interest. If your system is natively expressed, or under the control of another inducible system, this step is not necessary. IPTG induction and growth conditions can be optimized to increase production of small molecules of interest. We have observed that depending on gene cluster, IPTG concentrations can profoundly affect small molecule production. As a good starting range, test 0.05, 0.1, 0.25, 0.5, and 1.0 mM final concentrations to determine optimal metabolite production conditions. Other growth conditions that can be optimized include oxygenation (e.g., shaker revolutions per minute), temperature (typical range from 16 to 30 °C), or growth time, among others.

13. Samples prepared for HRMS instruments, such as the Agilent 6550 Q-TOF HRMS, can detect concentrations in the pg/mL range. Start at lower concentrations initially, and determine what concentrations work best for your extracts. High concentrations of a complex sample can saturate the mass detector and result in lower sensitivity or ion suppression [8, 9]. Low injection volumes of polar solvents will not greatly affect the chromatography, but high injection volumes can affect compound elution times.

Table 2
Some typical column dimensions and column volumes

Column dimension (L(mm) × I.D.)	Column volume (mL)
100 × 2.1	0.24
150 × 2.1	0.37
150 × 4.6	1.75
250 × 4.6	2.91

14. To filter small quantities of sample (50 μL), pass sample through 10 μL filter tips, and collect the flow through.

15. To shorten run, reduce gradient time of LC method. Some ethyl acetate extracts contain nonpolar compounds; therefore, we frequently use a longer 98 % B wash.

16. LC column equilibration times are dependent on the column dimensions. Generally, a column should be flushed with 20 column volumes to ensure equilibration. Table 2 provides column volumes for common analytical column dimensions assuming average pore volume of 0.70. This information can be found online if dimensions are different from the ones described (e.g., see http://www.chiralizer.com/colvol.htm).

17. Calibration should always be performed when changing the instrument state, and we recommend doing it before any metabolomics experiment. General calibration/tuning for the 6500 Series Q-TOF (Agilent) are as follows: (a) Set instrument to 2 GHz, EDR, low mass range for metabolomics (25–1700 m/z). (b) Calibrate in Hi Res mode, in positive or negative mode. Residual error should be less than 1 ppm. Save calibration. Reference ions spectrum should have intensity around two to three million counts. (c) Perform a standard tune (Hi Res mode), and save tune file. (d) Next, recalibrate in positive and negative mode for low masses. Re-save file as a newtunefile.tun.

18. Always make sure that the LC column is intact (no leaking), and check quality of data to confirm that the correct settings are being used.

19. Make sure that TIC chromatography is reproducible among biological replicates. If gross differences can be detected in TIC traces, start over and make sure that column equilibration times are sufficient.

20. The molecular feature extraction algorithm automatically retrieves all spectral information for each component, includ-

ing but not limited to mass, retention time, abundance, and composite spectrum, in a sample mixture, including those in overlapping and co-eluting peaks.

21. We recommend using the offline DA Reprocessor program, as the Qualitative method in MassHunter Qualitative Analysis will take longer time and will require higher computing power.

22. MPP is a chemometric platform that can extract information from chemical systems and use the high-content mass spectral data in differential analysis to determine relationships among sample groups. MPP allows for the analysis of spectral features by importing and organizing data, identifying unique molecular features in all samples, and removing any found in sample controls and from samples lacking the gene cluster of interest.

23. This step compiles all molecular features detected in any control sample. The total number of control samples will vary depending on sample set, but it should include at least the five biological control replicates (without a functional gene cluster) and the solvent control runs. Any feature found in a control sample will be removed from samples (with gene cluster) in future analysis.

24. This selection helps to maintain a conservative analysis by only retaining the entities found in a biological sample set. This number could vary depending on the number of biological replicates. A conservative unique list for each pathway-containing sample will be built by removing any MOF found in at least one of the control replicates (no gene cluster) and keeping those that are present in all biological replicates. The strictness of the analysis can be softened by selecting for entities that are found in <"5" of the biological samples. This will tend to increase your "unique feature list" but can be used as a tool to find minor metabolites that might be under the detection limit of the sampling methodology.

25. We also recommend confirming mass-to-charge of each ion in the final network. We occasionally observe a very small percentage of false positives, minor abundance ions that are also found in controls, and these are removed in this final confirmation.

26. "Auto MS/MS" data can be used to create an untargeted mass spectral network (Fig. 3a). Method development and network construction can be performed in the same manner as is described for the data set collected for the pathway-targeted mass spectral network.

27. Depending on small molecules of interest and their molecular weights, different collision energies and isotope models might work best, as settings can affect selection of ions to fragment

by the software. Be aware that low-abundance ions might not fragment regardless of method conditions resulting in less than 100 % coverage. Also, some secondary metabolites are unstable and may degrade before MS/MS analysis. If this is the case, re-extract fresh cultures and immediately perform MS/MS analysis.

28. Stronger collision energies are generally needed to effectively fragment higher molecular weight metabolites.

29. General Mass Spectral Molecular Networking is described in great detail in GnPS (Global Natural Products Social Molecular Networking) platform (gnps.ucsd.edu). We strongly recommend searching through their online "GnPS documentation page" for detailed information. Familiarize yourself with the different hyperlinks.

30. MSConvert converts original data files into mzXML files that can be uploaded for networking analysis. To acquire MSConvert, install ProteoWizard "Windows (includes vendor reader support)"; http://proteowizard.sourceforge.net/downloads.shtml.

31. Insilicos is a life science software that allows you to view the mzXML files and confirm that they were converted correctly. Download program from www.insilicos.com.

32. Cytoscape is used to view the networks created on GnPS. Download program from www.cytoscape.org.

33. There are various free FTP programs available online as further described in the GnPS website.

34. Given settings are imputed when you start the analysis, but can be modified, Parent mass tolerance setting should be set depending on your instruments' high-resolution data. We routinely see <5 ppm error among our parent masses, and therefore, we set a lower Da error than the given setting (2.0 Da). Cosine setting dictates the connectivity strength between ion masses (similarity of fragment ions). For example, a cosine of 0.0 indicates that the ion masses are not related, while a cosine of 1.0 indicates that the ion masses are identical. In this workflow, we often build clusters based on the less conservative cosine cutoff of 0.5 (Fig. 3).

35. Further detailed information and properties of the Cytoscape software can be found on the GnPS documentation site.

36. For isotopic labeling experiments, *E. coli* auxotrophic variant strains from the Keio collection can be acquired from the Coli Genetic Stock Center (CGSC, Yale University, New Haven, CT, USA; http://cgsc.biology.yale.edu/) or other collections. Use of auxotrophic strain backgrounds is very helpful in data

analyses, as the relative label:nonlabel substrate ratio is conserved in the parent ions of, for example, an amino acid-labeled nonribosomal peptide.

37. Performing a 1:1 incorporation of nonlabeled/labeled substrate will decrease overall abundance of ions of interest. Modify concentrations as deemed necessary for HRMS analysis.

Acknowledgments

We thank Crawford lab members T. Tørring and C. Perez for feedback and reviewing a preliminary version of the manuscript while working through their own metabolomics data analysis. Our work on secondary metabolite discovery, biosynthesis, and mode of action has been supported by the National Institutes of Health (National Cancer Institute grant 1DP2CA186575 and National Institute of General Medical Sciences grant R00-GM097096), the Searle Scholars Program (grant 13-SSP-210), and the Damon Runyon Cancer Research Foundation (grant DFS:05-12).

References

1. Cimermancic P, Medema MH et al (2014) Insights into secondary metabolism from a global analysis of prokaryotic biosynthetic gene clusters. Cell 158:412–421

2. Newman DJ, Cragg GM (2012) Natural products as sources of new drugs over the 30 years from 1981 to 2010. J Nat Prod 75:311–335

3. Watrous J, Roach P, Alexandrov T et al (2012) Mass spectral molecular networking of living microbial colonies. Proc Natl Acad Sci U S A 109:E1743–E1752

4. Nguyen DD, Wu CH, Moree WJ et al (2013) MS/MS networking guided analysis of molecule and gene cluster families. Proc Natl Acad Sci U S A 110:E2611–E2620

5. Vizcaino MI, Engel P, Trautman E et al (2014) Comparative metabolomics and structural characterizations illuminate colibactin pathway-dependent small molecules. JACS 136:9244–9247

6. Vizcaino MI, Crawford JM (2015) The colibactin warhead crosslinks DNA. Nat Chem 7:411–417

7. Bode HB, Bethe B, Hofs R et al (2002) Big effects from small changes: possible ways to explore nature's chemical diversity. Chembiochem 3:619–627

8. Choi BK, Hercules DM, Gusev AI et al (2001) LC-MS/MS signal suppression effects in the analysis of pesticides in complex environmental matrices. Fresenius J Anal Chem 369:370–377

9. Furey A, Moriarty M, Bane V et al (2013) Ion suppression; a critical review on causes, evaluation, prevention and applications. Talanta 115:104–122

Part III

Bioinformatic Methods

Chapter 13

Annotating and Interpreting Linear and Cyclic Peptide Tandem Mass Spectra

Timo Horst Johannes Niedermeyer

Abstract

Nonribosomal peptides often possess pronounced bioactivity, and thus, they are often interesting hit compounds in natural product-based drug discovery programs. Their mass spectrometric characterization is difficult due to the predominant occurrence of non-proteinogenic monomers and, especially in the case of cyclic peptides, the complex fragmentation patterns observed. This makes nonribosomal peptide tandem mass spectra annotation challenging and time-consuming. To meet this challenge, software tools for this task have been developed. In this chapter, the workflow for using the software mMass for the annotation of experimentally obtained peptide tandem mass spectra is described. mMass is freely available (http://www.mmass.org), open-source, and the most advanced and user-friendly software tool for this purpose. The software enables the analyst to concisely annotate and interpret tandem mass spectra of linear and cyclic peptides. Thus, it is highly useful for accelerating the structure confirmation and elucidation of cyclic as well as linear peptides and depsipeptides.

Key words Nonribosomal peptides, Tandem mass spectrometry, In silico fragmentation, Annotation, Linear peptides, Cyclic peptides, Depsipeptides

1 Introduction

Nonribosomal peptides, and among these especially cyclic peptides, are a group of natural products that attracts the interest of a large number of researchers due to their intriguing structures and powerful and diverse bioactivities. Cyclic peptides are found in many bacterial and fungal strains [1, 2]. Cyanobacteria, especially, have long been known for the wealth of cyclic peptides and cyclic depsipeptides they produce [3, 4]. These compounds can be of ribosomal origin (e.g., the microviridins found in *Microcystis* strains) [5, 6], but predominantly they are synthesized by nonribosomal peptide synthetases (NRPSs) [7–10]. Nonribosomal peptides are fascinating for the natural product chemist, as they are not only composed of proteinogenic amino acids but mostly contain unusual and highly modified amino acids or monomers constructed by polyketide synthases (PKSs). These peptides are often derived from a mixed PKS/NRPS

Bradley S. Evans (ed.), *Nonribosomal Peptide and Polyketide Biosynthesis: Methods and Protocols*, Methods in Molecular Biology, vol. 1401, DOI 10.1007/978-1-4939-3375-4_13, © Springer Science+Business Media New York 2016

biosynthetic pathway as described in great detail, e.g., for the cyanobacterial microcystins [11–13]. In addition to the 20 proteinogenic amino acids, over 500 different monomers are found in the currently known nonribosomal peptides and are compiled in the publically accessible NORINE database [2, 14]. Still, additional monomers are frequently being described in the literature.

Among the most prominent examples of pharmaceutically exploited nonribosomal cyclic peptides are cyclosporine, an immunosuppressant from *Tolypocladium inflatum* [15, 16], and vancomycin, an antibiotic from *Amycolatopsis orientalis* [17, 18]. Several other compounds or compound mixtures are either used directly (e.g., tyrothricin, daptomycin, and actinomycin D) or serve as lead structures for current drug development (e.g., the cryptophycins [19]).

The structural characterization of newly isolated or synthesized bioactive peptides is the basis for their further exploitation. Mass spectrometry often serves as a valuable tool for the first steps in dereplication, structure confirmation, and elucidation of peptides, when limited amount of material complicates direct characterization by NMR spectroscopy. To validate proposed structures, theoretical in silico fragmentation is commonly used for comparison with experimental data. This, however, is a challenge especially in the case of cyclic peptides, where complex fragmentation patterns are observed. Few software tools are capable of in silico fragmentation and tandem mass spectrum annotation of nonribosomal and cyclic peptides [20–23]. The following features make mMass the most concise and user-friendly software for this purpose:

- Stand-alone tool running on multiple operating systems
- Spectrum manipulation and processing abilities (e.g., support for various data formats, baseline correction, peak picking, etc.)
- Ability to save spectra, sequences, and interpretation results for future use
- Monomer database and editor for easy monomer and sequence management
- Suitability for both linear and cyclic peptides as well as depsipeptides and other structures composed of individual monomers
- Inclusion of all known cyclic peptide fragmentation pathways
- Customizable neutral losses allowing for inclusion of side-chain losses
- Allowance of multiple neutral losses from one fragment
- Matching to and annotation of experimental data

With these features, the software can conveniently be used in early steps of structure elucidation/dereplication (matching of in silico fragmentation of compounds in databases with experimental tandem MS spectra) and as one of the last steps of structure confirmation (extent of assignable peak intensities) (*see* **Note 1**).

In this chapter, the workflow for using the software mMass for the annotation of experimentally obtained peptide tandem mass spectra is described, taking the nonribosomal cyclic peptide Microcystin LF as an example. For more information on the fragmentation of cyclic peptides as well as on the development and validation of this software tool please refer to [23], which should also be cited in case mMass is used for this purpose.

2 Materials

2.1 mMass

The current version of the Python source code, executables for MS Windows, Mac OS X, and Linux as well as test data can be downloaded free of charge from the project webpage http://www.mmass.org. The executables do not require installation. All download packages contain a comprehensive manual.

2.2 Data

The raw data (mzXML file format) used for the step-by-step instructions below as well as the resulting mMass data file (msd format) can be downloaded free of charge at http://dx.doi.org/10.6084/m9.figshare.988774.

3 Methods

3.1 Import and Process Data in mMass, Enter Postulated Sequence

1. Start mMass and open data file (File → Open… or Ctrl + O; *see* **Note 2**). In case of LC-MS data, select one scan you want to work with.

2. Pick peaks either automatically or by hand. For this example, peaks were picked by hand, disregarding isotope peaks (*see* **Note 3**).

3. Open sequence editor (Sequence → New…).

4. Refer to Fig. 1.

   (1) Enter sequence name.

   (2) As non-proteinogenic monomers will be used, choose "Custom" from the Sequence type dropdown menu.

   (3) Microcystin LF is a cyclic peptide; thus, the respective checkbox needs to be ticked.

5. Refer to Fig. 2.

   (1) Enter the monomers the peptide is composed of (*see* **Note 4**). The chain length for nonribosomal peptides or other compounds containing non-proteinogenic building blocks is limited to 20 monomers. If an entered monomer abbreviation is not present in the Monomer Library, the cell is shaded red.

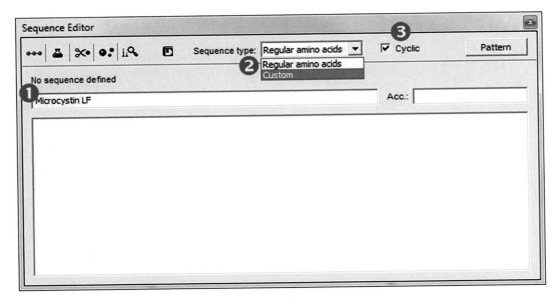

Fig. 1 Define sequence type in the Sequence Editor window

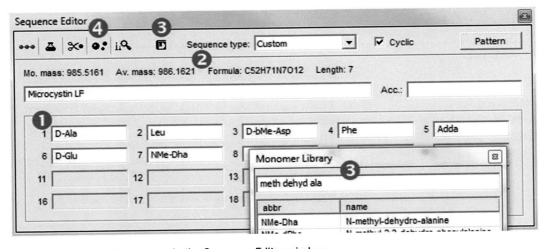

Fig. 2 Enter postulated sequence in the Sequence Editor window

(2) The mass of the peptide and the numbers of monomers it is composed of are updated automatically on the fly.

(3) A searchable monomer list can be opened to look up the required monomer abbreviations. Monomers can conveniently be dragged and dropped from this list into the Sequence Editor.

(4) When the sequence is completed, click on the Peptide Fragmentation tool icon in the Sequence Editor toolbar.

3.2 Perform In Silico Fragmentation

1. Refer to Fig. 3. All theoretical fragments resulting from the chosen fragmentation pathways are calculated and displayed in a list. In this list, the type of the ion (M-ion, A-ion, B-ion,

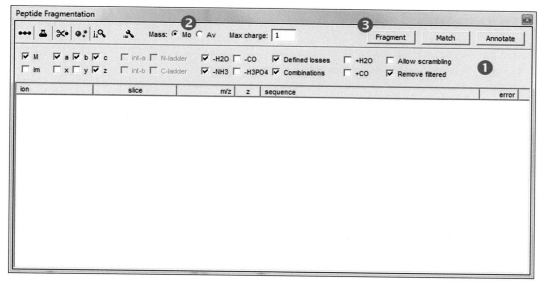

Fig. 3 Choose options for in silico fragmentation in the Peptide Fragmentation window

etc.), the slice of the cyclic peptide that is the basis for the respective fragment ion, the calculated m/z for the ion, its charge state, and the complete sequence of the fragment are displayed (*see* **Note 7**).

(1) Choose the fragments that are to be calculated (in this example M-, a-, b, c-, and z-ions as well as fragments with neutral losses) (*see* **Note 5**).

(2) If the masses that are calculated shall be monoisotopic or average masses (*see* **Note 6**).

(3) Click on the button "Fragment" in the Peptide Fragmentation toolbar.

3.3 Match In Silico Fragmentation Data to Experimentally Obtained Data

1. Click on the button "Match" in the Peptide Fragmentation toolbar.

2. Refer to Fig. 4.

   In the Match Fragments window,

   (1) indicate the peak tolerance in Da or ppm according to the expected accuracy of your experimental data (*see* **Note 8**) and

   (2) click the button "Match." This results in a first overview on the experimental peaks that can be matched with calculated theoretical fragments.

   (3) For a statistical analysis of the matching quality, click on the "Match summary" icon (*see* **Note 9**).

3. Return to the Peptide Fragmentation window. Note that all matched ions are now typeset in bold and green. Also, the

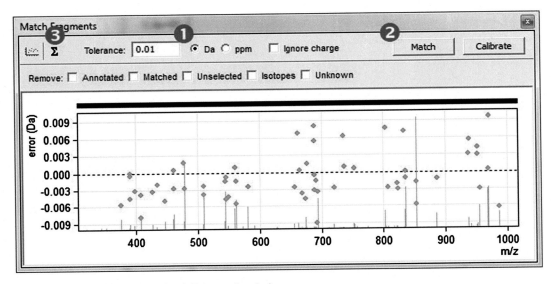

Fig. 4 Match peaks using the Match Fragments window

error between the experimental m/z and the calculated m/z is now given in the list. It is possible to only show matched peaks by right-clicking in the list and choosing the option "Show Matched Only" (*see* **Note 10**).

3.4 Annotate Experimental Tandem Mass Spectrum

1. Click on the button "Annotate" in the Peptide Fragmentation toolbar.

2. In the Documents Panel on the left-hand side in the main window, a summary list of all annotated peaks can now be found below the compound name. Note that all annotated peaks are now indicated by a green diamond in the Spectrum Viewer (*see* **Note 11**).

3. After selection of a match by clicking on it, a match can be edited or deleted by right-clicking on the match and selecting the appropriate action. Deletion of matches might be necessary if matches are theoretically possible only but are not plausible from an experimental point of view (*see* **Note 12**).

4. A report can be created using the built-in report generator (File → Analysis Report… or Shift + Ctrl + R). A spectrum image can be exported via File → Export… or Ctrl + E.

4 Notes

1. mMass can calculate theoretically possible fragments from a given sequence and match these calculated fragments to an experimentally obtained tandem mass spectrum. The software is not suitable for the prediction of tandem mass spectra, as no fragment ion intensities or the likeliness of fragment occurrences

of under experimental conditions can be calculated. Furthermore, the in silico fragmentation capabilities do not cover branched peptides or cyclic peptides with more than one cycle.

2. mMass supports several popular XML-based data formats such as mzDATA, mzXML, and mzML. A comprehensive software manual is included in all download packages. For more details on how to open and process spectra, please refer to the manual. Please note that many options for spectrum preprocessing (e.g., baseline correction, cropping, smoothing, calibration, etc.) as well as evaluation (e.g., the peak differences tool) are available in the software but are not discussed within the scope of this short description.

3. Isotope peaks will not necessarily be taken into account as such during annotation. Thus picked isotope peaks might influence the % matched peak intensity. A convenient way to unpick picked isotope peaks is using the "Deisotoping" tool (see mMass manual).

4. If monomers are missing, these can conveniently be added using the Monomer Library (Libraries → Monomers...). In case of linear peptides, it is important to begin entering the sequence with the N-terminus as monomer 1, as it is common in peptide/protein notation. Monomers such as fatty acids and sugars can also be used in a sequence.

5. For a concise review of all currently known fragmentation mechanisms of cyclic peptides as well as for a more detailed description of these options, see reference [23]. Option "Defined losses": take into account the neutral losses defined for the individual monomers in the Monomer Library; option "Combinations": allow more than one neutral loss; option "Allow scrambling": calculate scrambled fragments; and option "Remove filtered": remove less likely fragments that are mostly only theoretically possible.

   Likely fragmentation pathways for an individual compound need to be taken into account for choosing the settings; e.g., cyclic depsipeptides often show pronounced water adducts of b-series ions; thus, exclusion of the "+H2O" option in these cases would result in an incomplete annotation. Furthermore, the user needs to check whether all possible neutral losses for the individual monomers are included in the Monomer Library.

6. Usually, the calculation of monoisotopic fragment masses is most sensible here, especially if the calculations will later be matched to high-resolution MS data.

7. This list can be sorted (e.g., for ascending m/z values) by clicking on the column heads.

8. Use experimental data with as high mass accuracy as possible. The higher the mass accuracy, the lower the ambiguity of assignments to calculated fragments. If low-resolution (e.g.,

IT or QqQ) data are used, many different in silico calculated fragments might match with an experimentally observed fragment.

9. In the case of cyclic peptides, in our experience assigning >80 % of the total observed peak intensity is acceptable.

10. Many peaks will be matched to more than one calculated fragment. As during a tandem MS experiment many possible fragmentation pathways take place in parallel, observed signals are most likely caused by several individual fragment species also in reality. Of course, e.g., the formation of b-ions will be more likely than the formation of isobaric c-NH_3 ions, but mMass cannot distinguish between these fragments.

11. Peak label appearance in a spectrum can be adjusted using View → Notations → Show Notations and choosing between the options Labels (show/hide annotations), *m/z* (show/ hide *m/z* values), and marks (toggle green diamonds).

12. The software is not able to check if calculated fragments are likely to be observed under experimental conditions. The resulting annotations thus need the attentive assessment of the analyst! mMass – as all software tools – can only help the analyst to identify possibilities. It is the responsibility of the analyst to judge if possibilities indicated by mMass indeed do make sense under the experimental circumstances and with a specific compound!

Acknowledgment

The author thanks R. Pozzi, T. Schafhauser and M. Strohalm for critically reading the manuscript and suggesting improvements.

References

1. Tiburzi F, Visca P, Imperi F (2007) Do nonribosomal peptide synthetases occur in higher eukaryotes? IUBMB Life 59:730–733

2. Caboche S, Leclère V, Pupin M, Kucherov G et al (2010) Diversity of monomers in nonribosomal peptides: towards the prediction of origin and biological activity. J Bacteriol 192:5143–5150

3. Tidgewell K, Clark BR, Gerwick WH (2010) The natural products chemistry of cyanobacteria. In: Mander L, Liu H-W (eds) Comprehensive natural products II: chemistry and biology. Elsevier, Oxford, pp 141–188

4. Niedermeyer T, Brönstrup M (2012) Natural-product drug discovery from microalgae. In: Posten C, Walter C (eds) Microalgal biotechnology: integration and economy. de Gruyter, Berlin, pp 169–200

5. Ziemert N, Ishida K, Liaimer A, Hertweck C et al (2008) Ribosomal synthesis of tricyclic depsipeptides in bloom-forming cyanobacteria. Angew Chemie Int Ed 47:7756–7759

6. Velásquez JE, van der Donk WA (2011) Genome mining for ribosomally synthesized natural products. Curr Opin Chem Biol 15:11–21

7. Marahiel MA (2009) Working outside the protein-synthesis rules: insights into nonribosomal peptide synthesis. J Pept Sci 15: 799–807

8. Schwarzer D, Finking R, Marahiel MA (2003) Nonribosomal peptides: from genes to products. Nat Prod Rep 20:275–287

9. Strieker M, Tanović A, Marahiel MA (2010) Nonribosomal peptide synthetases: structures and dynamics. Curr Opin Struct Biol 20: 234–240

10. Finking R, Marahiel MA (2004) Biosynthesis of nonribosomal peptides. Annu Rev Microbiol 58:453–488

11. Tillett D, Dittmann E, Erhard M, von Döhren H et al (2000) Structural organization of microcystin biosynthesis in Microcystis aeruginosa PCC7806: an integrated peptide-polyketide synthetase system. Chem Biol 7: 753–764

12. Dittmann E, Neilan BA, Börner T (2001) Molecular biology of peptide and polyketide biosynthesis in cyanobacteria. Appl Microbiol Biotechnol 57:467–473

13. Christiansen G, Fastner J, Erhard M, Börner T et al (2003) Microcystin biosynthesis in planktothrix: genes, evolution, and manipulation. J Bacteriol 185:564–572

14. Caboche S, Pupin M, Leclère V et al (2008) NORINE: a database of nonribosomal peptides. Nucleic Acids Res 36:326–331

15. Dreyfuss M, Härri E, Hofmann H et al (1976) Cyclosporin A and C: new metabolites from Trichoderma polysporum. Microbiology 133: 125–133

16. Von Wartburg A, Traber R (1988) Cyclosporins, fungal metabolites with immunosuppressive activities. Prog Med Chem 25: 1–33

17. McCormick MH, Stark WM, Pittenger GE et al (1956) Vancomycin, a new antibiotic. I. Chemical and biologic properties. Antibiot Annu 3:606–611

18. Nagarajan R (1991) Antibacterial activities and modes of action of vancomycin and related glycopeptides. Antimicrob Agents Chemother 35:605–609

19. Rohr J (2006) Cryptophycin anticancer drugs revisited. ACS Chem Biol 1:747–750

20. Rusconi F (2009) massXpert 2: a cross-platform software environment for polymer chemistry modelling and simulation/analysis of mass spectrometric data. Bioinformatics 25:2741–2742

21. Jagannath S, Sabareesh V (2007) Peptide Fragment Ion Analyser (PFIA): a simple and versatile tool for the interpretation of tandem mass spectrometric data and de novo sequencing of peptides. Rapid Commun Mass Spectrom 21:3033–3038

22. Liu W, Ng J, Meluzzi D, Bandeira N et al (2009) The interpretation of tandem mass spectra obtained from cyclic non-ribosomal peptides. Anal Chem 81:4200–4209

23. Niedermeyer THJ, Strohalm M (2012) mMass as a software tool for the annotation of cyclic peptide tandem mass spectra. PLoS One 7:e44913

Chapter 14

Bioinformatics Tools for the Discovery of New Nonribosomal Peptides

Valérie Leclère, Tilmann Weber, Philippe Jacques, and Maude Pupin

Abstract

This chapter helps in the use of bioinformatics tools relevant to the discovery of new nonribosomal peptides (NRPs) produced by microorganisms. The strategy described can be applied to draft or fully assembled genome sequences. It relies on the identification of the synthetase genes and the deciphering of the domain architecture of the nonribosomal peptide synthetases (NRPSs). In the next step, candidate peptides synthesized by these NRPSs are predicted in silico, considering the specificity of incorporated monomers together with their isomery. To assess their novelty, the two-dimensional structure of the peptides can be compared with the structural patterns of all known NRPs. The presented workflow leads to an efficient and rapid screening of genomic data generated by high throughput technologies. The exploration of such sequenced genomes may lead to the discovery of new drugs (i.e., antibiotics against multi-resistant pathogens or anti-tumors).

Key words Drug discovery, Bioinformatics tools, Nonribosomal, Antibiotic, Peptide, Genome mining

1 Introduction

The nonribosomal peptides (NRPs) are a large, structurally diverse group of natural products [1]. They constitute a source of bioactive compounds of great biomedical importance, including antibiotics such as vancomycin or daptomycin, antitumoral compounds such as actinomycin D, or immunosuppressants such as cyclosporine. In the Norine database [2], which currently is the most comprehensive database of NRPs, only 31 % of the curated peptides are linear, the remainder having branched, cyclic, or more complex primary structures [3]. Moreover, the peptides contain collectively over 525 different monomers including non-proteogenic and modified amino acids, carbohydrates, or lipids. In consequence, there is a high demand to develop specific bioinformatics tools dedicated to these compounds because they cannot be analyzed as classical peptides.

Bradley S. Evans (ed.), *Nonribosomal Peptide and Polyketide Biosynthesis: Methods and Protocols*, Methods in Molecular Biology, vol. 1401, DOI 10.1007/978-1-4939-3375-4_14, © Springer Science+Business Media New York 2016

Nonribosomal peptide synthetases (NRPSs) that build up the NRPs are working as enzymatic assembly lines. They are megaenzymes displaying an extraordinary modular architecture containing catalytic domains, essentially adenylation domains (A) for the selection of the monomers, thiolation domains (T or PCP for Peptidyl Carrier Proteins) for the covalent tethering of activated monomers onto the NRPSs, condensation domains (C) for the peptide bond formation and thioesterase domains (Te) for the release of the peptide from the NRPS. These enzymatic domains harbor highly conserved core motifs also called signatures [4]. These consensus sequences are useful to identify proteins with unknown functions as being NRPSs, considering that similar sequences are relevant to similar functions.

In most cases, one given NRPS is expected to synthesize one peptide (or close variants). Several computational tools have been developed to analyze NRPS pathways in silico (for reviews, *see* refs. 5–7). They use the amino acid sequence of NRPSs to predict their specificity towards the amino acid building blocks they assemble into the NRP. The first predictive methods were based on the work of Stachelhaus et al. [8] and Challis et al. [9], who demonstrated for the first time that eight amino acids in the active site of the NRPS adenylation domain (A-domain) are crucial for determining the substrate specificity and can be used to predict and engineer A-domain specificities. Querying these datasets is, for example, implemented in the PKS/NRPS Web Server/Predictive Blast Server [10]. Alternative approaches use machine learning based methods like profile Hidden Markov Models (pHMMs) [11] to infer substrate specificities [12, 13], Support Vector Machines (SVMs) [14, 15] or Latent Semantic Indexing [16]. In addition to predicting the A-domain substrate specificity, the isomery of each monomer can also be determined by considering the presence of domains with epimerisation activity, to give better predictions of the structure [17–19].

In this chapter, we present the different Web-based bioinformatics tools currently used for an efficient screening of genomic data to discover new NRPs, and to compare them structurally to all known NRPs.

Two study cases will illustrate the use of the bioinformatics tools presented in this chapter. The first example is cichofactin, a lipopeptide produced by *Pseudomonas* that was identified from sequencing data by following the advised in silico workflow, followed by confirmation in wet-lab experiments [20]. The GenBank accession number of the cichofactin biosynthetic gene cluster is *KJ513093*. The second example shows how it is possible to rapidly detect the potential of NRPS synthesis from a draft genome of *Pseudomonas* split in more than 1000 contigs. The strain is *Pseudomonas syringae pv. tabaci* str. ATCC 11528, with GenBank assembly ID GCA_000159835.2.

2   Materials

2.1   Software for the Analysis of NRPSs and Their Products

During the last decade, several programs have been developed to aid on the analysis of NRPSs and NRPs. Most of these tools are accessible via websites and provide user friendly interfaces offering the detection of NRPS encoding genes within genomic sequences or the analysis of the NRPS domain organization and specificities based on their protein sequences without specific bioinformatics skills (*see* **Note 1**). A summary and links to the main tools is given in Table 1.

For the example workflows described in this manuscript we refer to the most commonly used tools.

2.2   Bioinformatics Tools to Predict NRPSs from Genome Sequence

1. antiSMASH (http://antismash.secondarymetabolites.org)

   antiSMASH [21–23], the antibiotics and secondary metabolites analysis shell, is a comprehensive genome mining platform that is capable of identifying and analyzing many different types of secondary metabolites, including products of NRPS. antiSMASH integrates several algorithms dedicated to the specific analysis of NRPS, e.g., the method of Minowa et al. [12] or NRPSpredictor [14, 15] and also provides direct links to different NRP and NRPS analysis websites like NORINE [2, 28] or NaPDoS [18].

   antiSMASH, which is licensed under the Affero GNU license, is freely accessible at http://antismash.secondarymetabolites.org. In addition, the software can be downloaded and installed on Linux, MS Windows, and Mac OS X.

   (a) Input data for antiSMASH

   antiSMASH can be used alternatively with protein sequences of NRPS in FASTA format, or preferably with genomic data. Genomic data can be provided as EMBL-formatted, or GenBank-formatted files including annotation (highly preferred method), but also multi-FASTA files containing only the DNA sequence are accepted.

   In the case of FASTA nucleotides uploaded to antiSMASH, genes are automatically predicted using the gene finding programs Glimmer for bacterial sequences [29], or GlimmerHMM for fungal sequences [30].

   Alternatively, antiSMASH can directly download sequences from NCBI RefSeq [31] or GenBank [32] databases if the user provides an accession number instead of uploading a file (*see* **Note 2** for remarks on sequence/assembly quality).

   (b) Output of antiSMASH

   antiSMASH returns a comprehensive genome mining analysis on the provided input data. If whole genome sequences are uploaded, antiSMASH will identify 44 different classes of antibiotics biosynthesis pathways. Similar

Table 1
Software dedicated to the analysis of nonribosomal peptide synthetases (Table adapted from Weber [6], with permission from Elsevier)

Program/database	URL	References
Software for the analysis of NRPS pathways		
antiSMASH	http://antismash.secondarymetabolites.org	[21–23]
ClustScan Professional	http://bioserv.pbf.hr/cms/index.php?page=clustscan	[24]
NaPDos	http://napdos.ucsd.edu/	[18]
NP.searcher	http://dna.sherman.lsi.umich.edu/	[25]
NRPS-PKS/SBSPKS	http://www.nii.ac.in/~pksdb/sbspks/master.html	[26, 27]
PKS/NRPS Web Server/Predictive Blast Server	http://nrps.igs.umaryland.edu/nrps/	[10]
Tools to predict NRPS substrate specificities		
LSI based A-domain function predictor	http://bioserv7.bioinfo.pbf.hr/LSIpredictor/AdomainPrediction.jsp	[16]
PKS/NRPS Web Server/Predictive Blast Server	http://nrps.igs.umaryland.edu/nrps/	[10]
NRPSpredictor/NRPSpredictor2	http://nrps.informatik.uni-tuebingen.de	[14, 15]
NRPSSP	http://www.nrpssp.com/	[13]

clusters or conserved operons encoding the biosynthesis of conserved compounds will be automatically identified in a database containing all currently known gene clusters and building-block biosynthesis pathways. For NRPS and PKS pathways, in addition, a comprehensive analysis of the domain organization and predictions of substrate specificity is performed and leads to putative products (Fig. 1).

Most results can directly be accessed on an interactive website (see paragraph).

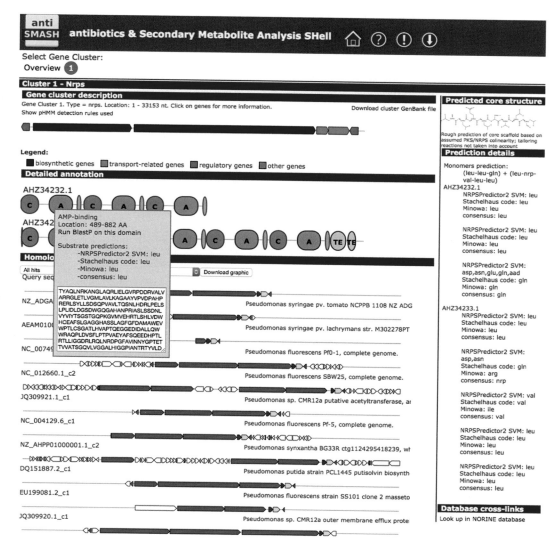

Fig. 1 Example output of antiSMASH analysis. The query was performed using the accession number of the cichofactin synthetase KJ513093, while the "non ribosomal peptides" option was selected. A pop-up window reporting the specificity of the first A-domain appeared after clicking on the considered domain. The predicted core structure of the peptide is shown on the *right column*

Alternatively, the complete antiSMASH analysis (including the HTML pages and annotated sequence files) can be downloaded using the download option "Download all results" for offline-use on the local computer (*see* **Note 3**).

2. NRPSpredictor 2 (http://nrps.informatik.uni-tuebingen.de)

NRPSpredictor [14, 15] is a Web-tool that allows the prediction of A-domain specificities. NRPSpredictor uses a Support-Vector-Machine based algorithm to classify the A-domain on four hierarchical levels (Table 2) ranging from gross physicochemical properties of the substrates down to single amino acid substrates. It uses an applicability domain model to assess the quality of the prediction (corresponding to a statistical validation). In addition, the query A-domain sequences are compared against an A-domain signature database based on the work of Stachelhaus et al. [8] and Challis et al. [9].

Table 2
Prediction levels of NRPSpredictor2

No.	Members	Description
Three clusters		
1	Arg, Asp, Glu, His, Asn, Lys, Gln, Orn, Aad	Hydrophilic
2	Gly, Ala, Val, Leu, Ile, Abu, Iva, Ser, Thr, Hpg, Dhpg, Cys, Pro, Pip	Hydrophobic-aliphatic
3	Phe, Tyr, Trp, Dhb, Phg, Bht	Hydrophobic-aromatic
Large clusters		
1	Gly, Ala, Val, Leu, Ile, Abu, Iva	Apolar, aliphatic side chains
2	Ser, Thr, Dhpg, Hpg	Aliphatic chain or phenyl group with -OH
3	Phe, Trp, Phg, Tyr, Bht	Aromatic side chain
4	Asp, Asn, Glu, Gln, Aad	Aliphatic side chain with H-bond donor
5	Cys	Cysteine
6	Orn, Lys, Arg	Long positively charged side chain
7	Pro, Pip	Cyclic aliphatic chain with polar -NH_2 group
8	Dhb, Sal	Hydroxy-benzoic acid derivatives
Small clusters		
1	Gly, Ala	Tiny size, hydrophilic, transition to aliphatic
2	Val, Leu, Ile, Abu, Iva	Aliphatic, branched hydrophobic side chain
3	Ser	Serine specific
4	Thr	Threonine specific
5	Dhpg, Hpg	Polar uncharged hydroxy phenyl
6	Phe, Trp	Apolar aromatic ring
7	Tyr, Bht	Polar aromatic ring
8	Asp, Asn	Asp-Asn hydrogen acceptor
9	Glu, Gln	Glu-Gln hydrogen bond acceptor
10	Aad	2-Amino-adipic acid
11	Orn	Orn and hydroxy-Orn specific
12	Arg	Arg specific
13	Pro	Pro-specific
14	Dhb, Sal	Hydroxy-benzoic acid derivatives

A command line version of NRPSpredictor2 is available for download at the NRPSpredictor homepage.

(a) Input data for NRPSpredictor

NRPSpredictor can be used with amino acid sequences of NRPS, which can either be pasted to a query box or uploaded in FASTA format. Alternatively, if the user does not want to upload the whole amino acid sequence due to patenting/privacy issues, one can provide manually extracted 34-aa signature sequences. The format to submit these signature sequences is "vntsfdgsvfdgfilfggeih-vygptestvyaty domain1", with a tabular character between the sequence and the domain name (*see* **Note 4**).

(b) Output of NRPSpredictor

The results of an NRPSpredictor analysis run are displayed as tables for each A-domain identified in the protein sequence(s). Each table contains the results of specificity prediction with the original NRPSpredictor1 algorithm, the improved NRPSpredictor2 algorithm and matches against the A-domain signatures of Stachelhaus et al. [8] and Challis et al. [9], which are determined by a Nearest Neighbour analysis.

NRPSpredictor2 returns prediction on four different hierarchical levels (Table 2, Fig. 2). The first level "Three clusters" makes a prediction on the general physicochemical properties of the incorporated amino acid, i.e., "hydrophilic, hydrophobic-aliphatic, hydrophobic-aromatic." The second level narrows down the prediction to large families/clusters of amino acids with similar properties. The third level tries to associate the A-domain specificity to smaller, more differentiated families. In the fourth level, a prediction is made based on matches against profiles only containing single amino acids. For all levels, a score and precision (*see* **Note 5**) are displayed. Additionally, a statistical support is notified by a green check✓ (supported) or red cross✗ (not supported). This support is the result of an applicability domain [33] analysis, which—simplified—provides information on how reliable the results are based on the similarity between the query sequence and the sequences used to train the SVM model.

3. NaPDoS (http://napdos.ucsd.edu/)

NaPDoS [18], the *N*atural *P*roducts *Do*main *S*eeker, is a Web-tool that identifies and classifies condensation (C) domains of NRPS using a phylogenetic approach. With NaPDos it is easily possible to screen (meta)genomic DNA sequences and also protein sequences for the presence of C domains and classify them according to their catalytic function (starter-C-domain, $^L C_L$, $^D C_L$, Dual-condensation-epimerization, heterocyclization) and assign them to the closest known pathway.

CifB_m1 Location: [617,753] ADomain PFAM score: 113.7			
Signatures	LWHAFDAMAWEPFLLIGGDINNYGPTEATVVATS / DAWFLGNVVK		
NRPSpredictor1	**Prediction**	**Score**	**Precision**
Large Clusters	gly=ala=val=leu=ile=abu=iva	0.810328	0.940
Small Clusters	val=leu=ile=abu=iva	0.959329	0.900
NRPSpredictor2	**Prediction**	**Score**	**Precision**
Three Clusters	hydrophobic-aliphatic	1.000342	0.974
Large Clusters	gly,ala,val,leu,ile,abu,iva	1.178141	0.947
Small Clusters	val,leu,ile,abu,iva	1.000176	0.892
Single AA	leu	1.000027	0.957
Nearest Neighbor	leu	100 %	-

CifB_m2 Location: [1663,1806] ADomain PFAM score: 195.1			
Signatures	LWSTFDASVWEWQFVCGGGHNVYGPTETTVDCSV / DAWQVGVVDK		
NRPSpredictor1	**Prediction**	**Score**	**Precision**
Large Clusters	asp=asn=glu=gln=aad	0.902309	0.969
NRPSpredictor2	**Prediction**	**Score**	**Precision**
Three Clusters	hydrophilic	0.412958	0.940
Large Clusters	asp,asn,glu,gln,aad	0.248288	0.969
Small Clusters	asp,asn	0.146582	0.969
Nearest Neighbor	gln	70 %	-

Fig. 2 Prediction of A-domain specificity using NRPSpredictor2. The analysis was performed with the N-terminal part of the cichofactin synthetase B, including two modules

In the first step of the NaPDoS analysis, the NRPS condensation domains are identified by a BlastP or BlastX, respectively for protein or coding DNA input, search against a hand-curated database containing 648 KS and C-domains or a combination of profile-HMM based domain detection and BlastP searches (*see* **Note 6**). After trimming, they are inserted into a manually curated alignment of NRPS C-domains, which is used to reconstruct a phylogenetic tree.

(a) Input data for NaPDoS

NaPDoS accepts three types of input: protein sequences, coding DNA sequences or DNA sequences of (meta) genomes/contigs. All sequences have either to be copy/pasted to the query box or uploaded as FASTA files.

In addition, there is the possibility to directly submit NaPDoS jobs from the info-box of C- within antiSMASH.

(b) Output of NaPDoS

The results of the domain identification step are displayed in a table at the first NaPDoS results page. This table contains information about the best hit(s), the identity of the query sequence with the hit, the length of the alignment, an *e*-value and an assignment to the next known pathway in the NaPDoS database. In addition, the classification of the domain (see above or Tutorial page) is shown. The table can also be downloaded as tabulator delimited text file.

Based on these results, the user can (1) select to download the identified and trimmed domain sequences in FASTA format, (2) select to download the alignment of the domain sequences with the best hit(s) of the NaPDoS database in PileUp-format, or use the alignment to calculate a maximum-likelihood tree, which can be downloaded as a SVG graphics file or as Newick formatted text (Fig. 3).

2.3 Bioinformatics tools to analyze NRPs

Norine (http://bioinfo.lifl.fr/norine/) [3, 28] is a unique resource dedicated to nonribosomal peptides that includes a database together with computational tools for data analysis. Norine currently contains more than 1100 peptides, coming from the

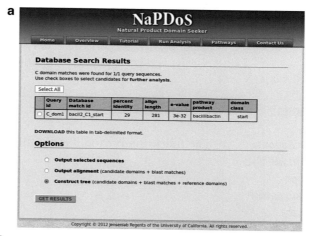

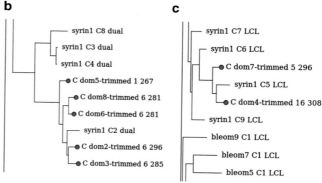

Fig. 3 C-domain subtypes predicted using NaPDoS. The complete sequence of cichofactin synthetase (including CifA and CifB) was analyzed. (**a**) The results presented as a table show that the first C-domain is predicted to be a C-starter. (**b-c**) Results presented as a phylogenetic tree for the seven remaining C-domains. They can be considered as ^LC_L (*see* part c) or dual-C/E (*see* part b) depending on the clade they are nested in

scientific literature or submitted directly by researchers. It provides detailed annotations such as structure, activity, producing organisms or bibliographical references. Norine also offers visualizer and editor for monomeric structures, as well as tools to search for monomeric structures [34, 35], which is a unique feature.

Norine can be queried either by annotations (through "general search" tab) or by structural information (through "structure search" tab) of the peptides. Searches among annotations are useful to obtain information about all peptides harboring a given activity or produced by a given organism, for example.

Structure-based search allows to find peptides containing given monomers with or without considering their 2D structure. In Norine, a specific format is used to represent the NRPs. It is based on the monomeric structure that is the monomers incorporated by the synthetases with the chemical bonds between them. The codes designating the more than 500 different monomers are mainly based on the IUPAC nomenclature (*see* **Note 7** for more details). An editing applet is provided so that this structure can be easily drawn and the correct graph format is generated automatically (Fig. 4). Whatever the query (on annotations or structures), Norine returns a peptide list and supports several displays.

3   Methods

3.1   From Genome Sequence to NRPSs: Identification and Analysis of NRP Gene Clusters with antiSMASH

The following protocol is for antiSMASH version 2.

3.1.1   Submitting Analysis Jobs to antiSMASH

1. Open the antiSMASH start page http://antismash.secondarymetabolites.org in a modern Web-browser (e.g., Mozilla Firefox or Google Chrome).

2. Upload your own genomic data using the "Choose file" button or enter NCBI GenBank/RefSeq accession number in the "NCBI ACC#" field. The extension of the file must be ". fasta" for FASTA/multi-FASTA format, ".gbk" for GenBank format, or ".embl" for EMBL format. It is highly recommended to provide an e-mail address in the e-mail field to receive a notice and a direct link to the antiSMASH result when the computation is finished. If an e-mail address is not provided, it is important to bookmark the job status page, as this is the only way to access the results.

3. If the sequence to analyze is of fungal origin, select the "DNA of eukaryotic origin" checkbox; leave it empty if the sequence is of bacterial origin. A correct selection of this option is

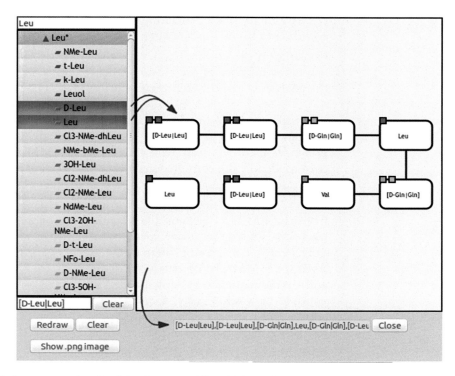

Fig. 4 Norine query using the "structure search" tool. The editor enables to draw candidate peptides. The monomers can be chosen from the menu at the left side of the editor and created by a simple click on its main window. The links between monomers are created by clicking on a monomer, then clicking on a different one that should be connected to the first. The graph representation shown under the editor is updated with every modification of the peptide

crucial, as it determines the gene prediction algorithm and also the model selection for some of the specificity predictors.

4. While the pre-selected options are suitable for most analysis requests, specific analysis options can be selected/deselected with the checkboxes.

5. Start antiSMASH job by pressing the "Submit" button at the end of the page. A new webpage showing the job status is displayed. Depending on the server load and the size of the genome, the antiSMASH analysis normally takes 2–4 h to complete.

3.1.2 Browsing the antiSMASH Results

The most important results of antiSMASH are displayed on an interactive webpage.

1. Once the computation is finished, you are redirected to an overview page with a list of all identified gene clusters of the analyzed genome. You can come back to this list by clicking on "Overview".

2. In the top panel "Select Gene Cluster" each identified gene cluster is represented as a colored circle. Each hit is also

referred in the table, including the compound type and its coordinates in the input sequence.

3. Clicking on the colored circles, or the colored "Cluster nb" (nb is the number of the cluster) labels displays the details of the analysis of the different gene clusters (Fig. 1).

3.1.3 Navigating the antiSMASH Cluster Result Page

1. In the top panel, an overview of the identified gene cluster and its coordinates are displayed. By clicking on "Show pHMM detection rules used" you can get a list of profile-HMMs that were used to identify the consulted gene cluster and assign its class.

2. Arrows represent each gene of a cluster. By clicking on these arrows, additional information is displayed, e.g., annotation from the EMBL/GenBank file; classification of the enzymes into SMCOGs (Secondary metabolite—clusters of orthologous genes), conserved protein domains/hits against the antiSMASH HMM profile database, direct links to NCBI blast and—if the sequence was downloaded from NCBI—the NCBI genome browser. Finally, there is a text box displaying the amino acid sequence of the gene product, which can be used to easily copy/paste the sequence to third party Web-servers.

3. For pathways containing PKS or NRPS genes, a detailed analysis of the domain organization is provided in the second panel. Additional information on the identified domains can be displayed by clicking on the respective cartoons. For NRPS A-domains, the results of the amino acid specificity prediction are included. For the C-domains, the precise catalytic function is provided (starter-C-domain, $^{L}C_{L}$, $^{D}C_{L}$, Dual-condensation-epimerization), together with a direct link to NaPDoS for more detailed analysis.

4. If the antiSMASH search was performed by submitting a whole genome/whole gene cluster sequence, homologous gene clusters of other organisms are identified by an integrated MultiGeneBlast [36] analysis against a custom database containing conserved operons involved in the biosynthesis of common secondary metabolite building-blocks, e.g., biosynthetic genes to produce non-proteinogenic amino acids like β-hydroxyphenylglycine. The results are displayed graphically in the bottom panel of the result page. Information about the hits can be obtained by clicking on the respective arrows representing similar genes. The color-coding is consistent for all results, i.e., similar genes always have the same color.

5. On the right side of the antiSMASH result page, a predicted core structure is displayed for PKS and NRPS gene clusters. Very important: The structure displayed is only a rough prediction of the core structure of the secondary metabolite and must not be confused with the structure of the final product of the pathway (*see* also **Note 8**).

6. Below the structure prediction panel, details of the substrate prediction are displayed for the individual genes/enzymes and detection methods (e.g., NRPSPredictor2 SVM, Stachelhaus code and Minowa) for each monomer if available; at the bottom of the panel a direct link to NORINE (*see* Subheading 2.3 or 3.4) is provided.

1. antiSMASH-results on the Web-server antismash secondary metabolites.org are deleted after 4 weeks. Therefore, it is recommended to download the antiSMASH results to a local computer. antiSMASH currently provides four types of downloads, which can be accessed by pressing the <<down arrow>> button on top of the results page:

2. *Download all results*: Downloads a ZIP compressed file containing all analyses (including the files described below) and output files. To use this archive, uncompress it at a convenient directory. The most important files for further analysis are:

   (a) index.html: This file contains the antiSMASH result page, which can be viewed with a Web-browser (*see* **Note 3**) and contains exactly the same information than the online result page described above.

   (b) <accession-number>.final.embl and <accession-number>.final.gbk: These are EMBL/GenBank formatted summary files containing the detailed antiSMASH annotation (see below)

   (c) <accession-number>.clusterXXX.gbk: Annotated files of the identified gene clusters in GenBank format ("XXX" in the file name indicates cluster number.).

3. *Download XLS overview file*: Downloads a table with an overview on all identified gene clusters and the involved genes (accession numbers) in MS Excel format.

4. *Download EMBL summary file/Download GenBank summary file*: Download the sequence file (in EMBL or GenBank format) containing all the antiSMASH annotations. This file can be used with most programs allowing import of annotated sequence data, for example Artemis [37] or standard bioinformatics libraries like Bioperl [38] or BioPython [39].

While antiSMASH provides a comprehensive analysis of secondary metabolite gene clusters on whole genome level, it sometimes may be of interest just to predict the monomers incorporated by NRPS proteins. In such cases, it is convenient to use the website of NRPSpredictor.

An overview on all options can be retrieved by clicking the "Help" button on top of the page.

3.2.1 Submitting NRPSpredictor2 Jobs

1. Open the NRPSpredictor webpage http://nrps.informatik. uni-tuebingen.de in the Web-browser.

2. Either copy and paste the FASTA-formatted amino acid sequence of the NRPS into the text box, or upload a FASTA-file via the "Choose file" button. Alternatively, paste the manually extracted Signature sequences into the search field/upload a file and select the "Signatures" checkbox as "Filetype".

3. As a standard, NRPSpredictor uses the bacterial models for prediction. If the sequence to analyze is of fungal origin, select "fungal" in the "Type of predictor" checkbox.

4. Start the NRPSpredictor calculation by pressing the "Submit" button.

3.2.2 Browsing NRPSpredictor2 Results

1. A table containing all the prediction results is displayed for every A-domain that was identified in the query sequence (*see* Subheading 2.2 for details on the output) (Fig. 2).

2. A text file containing a table of the results can be downloaded by clicking on the "Report file" link at the top of the NRPSpredictor2 results page.

3.3 Additional Analysis of NRPS Condensation Domains by Phylogenetic Classification with NaPDoS

In addition to the information provided by antiSMASH and NRPSpredictor, phylogenetic analyses of the NRPS can provide additional information on the function and the relation to known biosynthetic pathways.

A detailed description on all NaPDoS functions can be retrieved by clicking the "Tutorial" tab on the NaPDoS homepage.

3.3.1 Submitting NaPDoS Jobs

1. Open the NaPDoS webpage at http://napdos.ucsd.edu/ in the Web-browser.

2. Select the "Run Analysis" tab.

3. Select the domain type you want to analyze ("KS-domain" or "C-domain").

4. Select the query type depending on the data you want to analyze: choices are (1) "Predicted protein sequences (amino acid)" for amino acid queries; (2) "Predicted coding sequences or PCR products (DNA)" for DNA sequences of PKS or NRPS genes; (3) "Genome or metagenome contigs (DNA)" for DNA sequences of (meta)genomes.

5. Either copy and paste the query data into the "Query sequence" in protein or nucleotide FASTA format into text box, or select a file containing the data for upload.

6. Start NaPDos analysis (with standard parameters) by clicking the "SEEK" button, then "Submit Job" in the next page.

7. By clicking on "Advanced Settings" you have the possibility to influence the parameters used for domain identification, i.e., defining *e*-values, alignment lengths and number of displayed hits (*see* **Note 6**).

3.3.2 Browsing and Downloading the NaPDoS Results

1. Shortly after submitting the NaPDoS job, the results of the Database Search are displayed in a table. In "Genome or metagenome contigs (DNA)"-mode, first a table of the identified domains and their coordinates is presented. By clicking on "Get more info" a more detailed table is shown. If "Predicted protein sequences (amino acid)" or "Predicted coding sequences or PCR products (DNA)" was chosen as search mode, an equivalent detailed table containing information on the identified domains, similar pathways and a classification (*see* also Subheading 2) is displayed directly.

2. If a similar pathway was identified, information about this pathway can be obtained by clicking on the link in the table.

3. For further analyses, individual hits or all hits can be selected.

4. To download the table as tab-delimited text file use the "DOWNLOAD" link.

5. At the bottom of the page, options can be selected to download.

   (a) The identified NRPS C-domains in FASTA format.

   (b) The NaPDoS alignment containing the query sequences and the best hits against the NaPDoS database.

   (c) The phylogenetic trees as SVG graphics files or in Newick format for use in other phylogenetic software, for example Dendroscope [40].

3.4 Structure-based search against All NRPs Contained in Norine

This step can be performed either directly from the antiSMASH result pages by using the direct link to fingerprint search in Norine, or using the "structure search" tab of the Norine tool.

3.4.1 Submitting Structure Search to Norine

1. Open http://bioinfo.lifl.fr/norine in the Web-browser.

2. Select "structure search" tab on top of the page. The editor shown in Fig. 4 is open in the page.

3. Draw your peptide in monomer representation using the editor (Fig. 4).

   (a) Choose the monomers composing the query peptide. To search for a specific monomer, you can use the search box at the top of the menu at the left side of the editor. Clicking on one or several monomers will select them. It clusters them at the same position of the peptide, enabling, for example, searching for both isomers or for the monomer list obtained by analysing a NRPS. The list of all chosen monomers is shown in the box at the bottom of the menu. Once the monomer(s) are chosen, click on the main window of the editor to place it.

(b) Add bonds between the monomers. To connect a monomer to another one, click on it, then go to the monomer you want to connect to. To create a double link, when the monomers are linked by two different chemical bonds, repeat the process. Clicking on the link will erase it.

4. The designed peptide is automatically translated in a graph representation (*see* **Note 7** for a detailed description) in the text box under the editor. Once the peptide is complete, click on "Submit" button. Two searches are performed in one submission. The structure-based search [34] outputs all the peptides containing exactly the complete query structure, called pattern. One pattern can be composed of one or more fragments. The "monomer composition fingerprint search" [35] is a matching performed by calculating a distance between the query and the peptides represented as fingerprints.

3.4.2 Browsing and Downloading Norine Structure Search Results

1. As a result, a table containing the peptides matching the query structure is displayed.

2. The table contains the names of the peptides, a calculated distance-metric between the peptide and the query, the number of matched fragments and their list. The results can be sorted by clicking on the arrows located in the column names.

3. The graphical representation allows filtering of the results by one of the following criteria at a time: putative or curated peptides, 2D structure types (linear, with one or two cycles, etc.), categories (peptide, peptaibol, lipopeptide, …), activities, number of monomers. Observation of the common features of the peptides similar to a query can help predicting some of its features such as its category, its 2D structure type or its activity.

4. The table view enables results manipulation and refinement by customizing the displayed columns.

5. The peptide lists can be downloaded in text (CSV that can be opened in any spreadsheet application), XML, or JSON formats.

6. Clicking on one line of the table in the distance or fragment count column, gives access to the comparison of the query and the concerned peptide.

7. Clicking on a peptide name gives access to the description of this peptide.

3.4.3 Browsing and Downloading Norine Peptide Description

1. Detailed annotations are provided in the description page of the peptide, about including general features, its structure, the producing organisms, relevant articles concerning their structure, links to other databases if available: producing synthe-

tases in UniProt [41], 3D structure in PDB [42], chemical structure in PubChem [43].

2. Annotations can be downloaded in XML or JSON format.

In this study case, we assume, that nothing is known about the genes/functions encoded.

1. *Analyzing the cichofactin NRPS domain organization and predicting the produced peptide with antiSMASH*

   If a comprehensive analysis of an unknown biosynthetic gene is required, antiSMASH provides a huge collection of different tools for this task. In this study case, we use the pre-annotated sequence of the cichofactin biosynthetic gene cluster directly downloaded into antiSMASH.

2. Start an antiSMASH analysis at http://antismash.secondarymetabolites.org by entering your e-mail address in the "Email-address" field and the accession number "*KJ513093*" (=cichofactin biosynthetic gene cluster) in the search field and clicking on the submit button.

3. After the analysis is completed, the antiSMASH cluster overview page is displayed. As our demo sequence only contains one cluster, there is only one entry in the table. Click on the hit "Cluster 1".

4. The result page of the antiSMASH analysis is displayed (Fig. 1).

5. To get information on the identified gene click on arrows in the "Gene cluster description" panel. From the information window, you can directly forward the sequence to NCBI Blast or start the NCBI genome viewer (the latter only works for sequences downloaded from GenBank).

6. In the "Detailed annotation" panel, the domain organization of the cichofactin NRPSs is displayed. The cluster is composed of two NRPS coding genes. The first one, called AHZ34232, encodes three modules, and the second one, called AHZ34233, encodes five modules (numbered from 4 to 8).

7. To get information on the catalytic function of the C-domains, click on the respective C-domain cartoon. A pop-up window with additional information opens. From this pop-up, you can also directly submit the extracted domain sequence to NaPDos using the "Analyze with NaPDoS" link. The predictions are, in nearly all cases, the same between antiSMASH and NaPDoS, but NaPDoS additionally provides a phylogenetic analysis of the domain. The first gene starts with a "Condensation_Starter" that may incorporate a lipid moiety at the beginning of the produced NRP. The two other C-domains of this gene are annotated as "Condensation_Dual" which means that they epimerize the amino acid they add to the growing peptide

(selected by the previous module). The second gene ends by two thioesterase (TE) domains. This feature is mainly observed in NRPSs producing lipopeptides. In this protein, the modules 4, 6 and 8 contains "Condensation_Dual" C-domains. The two genes form a complete and coherent synthetase expected to produce a lipopeptide composed of eight amino acids with a the first, second, third, fifth and seventh ones in D-configuration.

8. At the right side of the result page, you find the prediction of the core structure and a text report on the various specificity predictions. The amino acid composition of the predicted peptide is reported as (leu-leu-gln) + (leu-nrp-val-leu-leu), where nrp stands for undetermined amino acid that is assigned when the three integrated predictors (Minowa, NRPSpredictor, Stachelhaus code) disagree.

3.5.2 Confirming the A-Domain Specificities of the Cichofactin NRPS Proteins with NRPSPredictor

1. Download the amino acid sequences of the cichofactin NRPS CifA and CifB from NCBI.

    (a) Open http://www.ncbi.nlm.nih.gov/ in your Web-browser.

    (b) Select "Protein" at the selection box on top of the screen.

    (c) Search for "cichofactin nrps"; you should get two results, "CifA" and "CifB".

    (d) Select both sequences by clicking on the checkbox.

    (e) Press the "Display Settings" Link, which is displayed on top of the results list; select "FASTA(text)" and confirm by clicking "Apply".

2. Open a second browser window/tab and go to http://nrps.informatik.uni-tuebingen.de

3. Copy the sequences into the text NRPSpredictor "Sequence to analyse" text-box.

4. Start NRPSpredictor analysis with "Submit Job".

5. The results of NRPSpredictor are displayed as separate tables for each identified A-Domain (*see* Fig. 2 for the two first A-domains of CifB). The outputs are more detailed than the ones obtained in antiSMASH and also include statistic information. We can notice that a low score (displayed in orange) is obtained for the second A-domain of the CifB protein. The "small cluster" hit is "asp,asn" and has a score of only 0.140582 (the lowest is 0), while the "nearest neighbour" (equivalent to "Stachelhaus code"-results in antiSMASH) is "gln" and has a score of 70 % (best is 100 %). Those scores are congruent with the fact that the three predictors give different results in antiSMASH, so with the "nrp" prediction for the fifth amino acid of the peptide.

The features described in this chapter correspond to the Norine version of November 2015.

1. A direct link to Norine is provided at the right end of the anti-SMASH results page, in the frame "Database cross-links". This link automatically generates a pattern search of the two peptide fragments produced by the two studied NRPS proteins, including the potential derivatives of the monomers (for example their D-configuration). A similarity between the fingerprints is also calculated.

2. Click on the graphical output icon to obtain an overview on the NRPs that contain the peptides most similar to the query by selecting them. In our study case, only lipopeptides are found confirming the prediction from the NRPS with the presence of a C-starter. As the peptides most similar to the query share a linear structure and a surfactant activity, those features can be assigned to the studied peptide.

3. In particular, the six members of the syringafactin family have a similarity higher than 0.7 with the combination of the two fragments. By clicking on the table output icon, interesting features of those peptides, such as the 2D structure, can be visualized together. They are all produced by the same strain, *Pseudomonas syringae* pv. tomato DC3000 [44] and are composed of nine monomers, including a lipid moiety that can vary. Their structures vary also in the seventh position where Val, Leu or Ile are observed. This variation occurs because the A-domain can select different amino acids, according to the substrates available in the culture media. Those amino acids are part of the "small cluster" in NRPSPredictor2 results.

4. The monomeric composition of cichofactin was determined experimentally by Pauwelyn et al. [20] as Fatty-acid_Leu_Leu_Gln_Leu_Gln_Val_Leu_Leu. So the unpredicted X was demonstrated to be a glutamine (Gln) pointing out a single difference with syringafactin, in which a threonine (Thr) is found at the same sixth position (the first monomer being the fatty acid). The bioinformatic study of the NRPS genes/proteins also considering the conformation of the amino acids allows predicting the following linear structure: Fatty-acid_D-Leu_D-Leu_D-Gln_Leu_D-Gln_Val_D-Leu_Leu.

In this study case, we propose a workflow to infer the presence of NRPS genes in draft genome or metagenomic sequences. Those sequences have the particularities to be split in many fragments. They are contained in multi-FASTA files (each fragment is separated from the previous one with a line starting by > character). The studied sequences are from *Pseudomonas syringae pv. tabaci* str. ATCC 11528 [45], with GenBank assembly ID GCA_000159835.2.

1. *Screening for the presence of NRPS genes with NaPDoS*

   (a) Open NaPDoS (at http://napdos.ucsd.edu/), and select "Run Analysis" tab. Select "C domains" under Domain type and "Genome or metagenome contigs (DNA)" as "Query Type". Paste the DNA sequences of all contigs into the text box or upload the corresponding file in multi-FASTA format. Start analysis with "Seek" and then "Submit job" in the second page if the parameters are correct.

   (b) The result table provides the list of contigs containing C-domains in the "parent seq" column. Clicking on "Get more information", gives more details and the domains sorted by contig order. So it is easily possible to manually extract the contigs with at least one C-domain from the multi-FASTA file. In our study case, 21 C-domains were predicted, in 20 different contigs.

   (c) With a text editor software, search for the contig names and copy paste the corresponding sequence in another file. You can also extract two (or more) contigs before and after, and the ones between two close contigs. For example, NaPDoS finds C-domains in the following contigs (the numbers refer to the order in the genome): ACHU02000685, ACHU02000688, ACHU02000691. So you can extract all contigs between ACHU02000683 and ACHU02000693. This gives 59 contigs in our study case.

2. *Submission of the most interesting contigs to antiSMASH*

   The extracted sequences can then be submitted to antiSMASH.

   (a) In the submission form, give your e-mail and the fragments in multi-Fasta file. Open the "Restrict which of the 24 supported secondary metabolite types to detect" parameter to select only "nonribosomal peptides" as you extract contigs related to C-domains. This selection reduces the calculation time.

   (b) In the study case, 19 clusters (2 of the type "Other") are annotated among the 59 extracted.

   (c) Those contigs only consist of truncated NRPS genes with two to four domains. Cluster 1 and 11 also contain other genes such as a dioxygenase or an ABC transporter. The observation of the "Homologous gene clusters" given help in identifying the possible category of the peptide.

   (d) Genes in cluster 1 align with clusters of other *pseudomonas* strains annotated as pyoverdin producers so we can presume that the studied strain also produce a pyoverdin. But we cannot predict its structure or monomeric composition as the NRPS genes are incomplete. The number of genes in cluster 11 is too small to predict a peptide category.

(e) In clusters 14, 16, 17, and 19, dual C/E condensation domains are predicted. Those domains are, until now, only observed in lipopeptides. So we can infer that the studied strain also produces a lipopeptide.

4 Notes

1. The use of these Web services is intended for the analysis of a limit number of genomes or protein sequences. If you plan do large-scale analyses, for example analyze all NRPS sequences from GenBank or analyze large metagenomic datasets, please contact the providers/authors before submitting the jobs via the website or scripts, as there may be better ways to run such big analyses independent of the Web service.

2. The quality of the input data has very important effect on the quality of the antiSMASH predictions (and generally of all predictive tools): As antiSMASH uses rules to detect the biosynthetic gene clusters require the presence of conserved domains (e.g., a NRPS is identified by detecting condensation, adenylation and PCP domains encoded by in the same gene), good predictions can only be made if the input data has a sufficient length. When using antiSMASH with draft sequence data, sequence scaffolds should always be preferred over analyzing the contig data. The use of antiSMASH with metagenomic datasets is technically possible, but due to the mostly very short length of the assembled sequences, most gene clusters will be missed as the conserved domains used for cluster and biosynthetic type identification are encoded on different contigs. Therefore it is disencouraged to use antiSMASH for this kind of analyses.

3. antiSMASH can be used with any major up-to date Web-browsers, like for example Mozilla Firefox or Google Chrome. However, for browsing downloaded antiSMASH results, it is recommended to use Mozilla Firefox, as very strict Javascript security settings in Google Chrome prevent the interactive display of the antiSMASH (Sub)Clusterblast results for locally saved HTML pages.

4. The signature data should be prepared in a text editor and uploaded as a file or transferred into the text box by copy/paste as pressing the [tab] key in most Web-browsers switches to the next element of the HTML page instead of inserting a "tabulator" character to the text in the text box.

5. The precision (=number of true positive hits/(number of true positives + false positives) on the validation test set of NRPS sequences) is a measurement of the prediction quality of the hit-SVM model.

6. Depending on the type of input, NaPDoS uses a blast or HMM based approach to identify the C-domains. Thus, the results obtained may differ, i.e., there may be domains identified with one method while the standard thresholds for the domain identification criteria prevent the detection with the other method. It therefore is highly recommended to try out different parameters. For some classes of NRPS (e.g., glycopeptides) it is worth to select less stringent domain identification criteria (i.e., shorter minimal alignment size, increased e-value) in the NaPDoS "Advanced Settings" options of the "Run Analysis" page.

7. Description of the formatted strings adopted by Norine for the monomeric structures: the monomers composing the peptide are separated by commas. The monomers are symbolized by three letter codes for amino acids and carbohydrates, with their substituent groups or configuration if needed; lipid numbers for long carbon chains such as fatty acids; commonly used short names for other compounds. For pattern search, wildcard is allowed: X stands for any monomer, Code* for the derivatives of the 'Code' monomer, names of the clusters defined in Norine (monomers are clustered based on their chemical properties, for example "fatty acid") (for a complete list consult the "monomers list"), personal set of alternatives represented between brackets by a monomer list separated by "|" (pipe character). The links between monomers are cited after the monomer sequence, isolated by "@". The monomers are ranked from 0 in the order they appear in the sequence and their respective neighborhoods are displayed.

8. Currently, it is not yet possible to predict an accurate chemical structure of the final product just based on sequence analysis. The structure displayed in antiSMASH is solely based on the prediction of the substrate specificities of PKS and NRPS; in the current version antiSMASH2, it does not take into account information about the stereochemistry, any post-PKS/post-NRPS modifications or unusual features of the PKS/NRPS enzymes. As such modifications are very common, the displayed structure must only be regarded as a hint on the final product of the biosynthetic pathway.

Acknowledgements

TW is supported by a grant of the Novo Nordisk Foundation. The work on Norine is supported by PPF Bioinformatique of Lille 1 University and Bilille.

We thank Hyun Uk Kim for testing the workflows and careful reading of the chapter.

References

1. Sieber SA, Marahiel MA (2003) Learning from nature's drug factories: nonribosomal synthesis of macrocyclic peptides. J Bacteriol 185:7036–7043

2. Caboche S, Leclère V, Pupin M et al (2010) Diversity of monomers in nonribosomal peptides: towards the prediction of origin and biological activity. J Bacteriol 192:5143–5150

3. Caboche S, Pupin M, Leclère V et al (2008) NORINE: a database of nonribosomal peptides. Nucleic Acids Res 36:D326–D331

4. Conti E, Stachelhaus T, Marahiel MA et al (1997) Structural basis for the activation of phenylalanine in the non-ribosomal biosynthesis of gramicidin S. EMBO J 16:4174–4183

5. Fedorova ND, Moktali V, Medema MH (2012) Bioinformatics approaches and software for detection of secondary metabolic gene clusters. Methods Mol Biol. 944:23–45

6. Weber T (2014) In silico tools for the analysis of antibiotic biosynthetic pathways. Int J Med Microbiol. 304:230–235

7. Boddy CN (2014) Bioinformatics tools for genome mining of polyketide and nonribosomal peptides. J Ind Microbiol Biotechnol. 41(2):443–50

8. Stachelhaus T, Mootz HD, Marahiel MA (1999) The specificity-conferring code of adenylation domains in nonribosomal peptide synthetases. Chem Biol 6:493–505

9. Challis GL, Ravel J, Townsend CA (2000) Predictive, structure-based model of amino acid recognition by nonribosomal peptide synthetase adenylation domains. Chem Biol 7:211–224

10. Bachmann BO, Ravel J (2009) Chapter 8 Methods for in silico prediction of microbial polyketide and nonribosomal peptide biosynthetic pathways from DNA sequence data. In: Hopwood DA (ed) Methods in enzymology. Academic, New York, pp 181–217

11. Eddy SR (1998) Profile hidden Markov models. Bioinformatics 14:755–763

12. Minowa Y, Araki M, Kanehisa M (2007) Comprehensive analysis of distinctive polyketide and nonribosomal peptide structural motifs encoded in microbial genomes. J Mol Biol 368:1500–1517

13. Prieto C, García-Estrada C, Lorenzana D et al (2012) NRPSsp: non-ribosomal peptide synthase substrate predictor. Bioinformatics 28:426–427

14. Röttig M, Medema MH, Blin K et al (2011) NRPSpredictor2—a web server for predicting NRPS adenylation domain specificity. Nucleic Acids Res 39:W362–W367

15. Rausch C, Weber T, Kohlbacher O et al (2005) Specificity prediction of adenylation domains in nonribosomal peptide synthetases (NRPS) using transductive support vector machines (TSVMs). Nucleic Acids Res 33:5799–5808

16. Baranašić D, Zucko J, Diminic J et al (2014) Predicting substrate specificity of adenylation domains of nonribosomal peptide synthetases and other protein properties by latent semantic indexing. J Ind Microbiol Biotechnol. 41:461–467

17. Rausch C, Hoof I, Weber T et al (2007) Phylogenetic analysis of condensation domains in NRPS sheds light on their functional evolution. BMC Evol Biol 7:78

18. Ziemert N, Podell S, Penn K et al (2012) The natural product domain seeker NaPDoS: a phylogeny based bioinformatic tool to classify secondary metabolite gene diversity. PLoS One 7:e34064

19. Caradec T, Pupin M, Vanvlassenbroeck A et al (2014) Prediction of monomer isomery in Florine: a workflow dedicated to nonribosomal peptide discovery. PLoS One 9:e85667

20. Pauwelyn E, Huang C-J, Ongena M et al (2013) New linear lipopeptides produced by *Pseudomonas cichorii* SF1-54 are involved in virulence, swarming motility, and biofilm formation. Mol Plant Microbe Interact 26:585–598

21. Blin K, Medema MH, Kazempour D et al (2013) antiSMASH 2.0—a versatile platform for genome mining of secondary metabolite producers. Nucleic Acids Res 41:W204–W212

22. Medema MH, Blin K, Cimermancic P et al (2011) antiSMASH: rapid identification, annotation and analysis of secondary metabolite biosynthesis gene clusters in bacterial and fungal genome sequences. Nucleic Acids Res 39:W339–W346

23. Weber T, Blin K, Duddela S et al (2015) antiSMASH 3.0—a comprehensive resource for the genome mining of biosynthetic gene clusters. Nucl Acids Res 43:W237–W243. doi:10.1093/nar/gkv437

24. Starcevic A, Zucko J, Simunkovic J et al (2008) ClustScan: an integrated program package for the semi-automatic annotation of modular biosynthetic gene clusters and in silico prediction of novel chemical structures. Nucl Acids Res 36:6882–6892. doi:10.1093/nar/gkn685

25. Li MH, Ung PM, Zajkowski J et al (2009) Automated genome mining for natural products. BMC Bioinformatics 10:185. doi:10.1186/1471-2105-10-185

26. Anand S, Prasad MVR, Yadav G et al (2010) SBSPKS: structure based sequence analysis of

polyketide synthases. Nucl Acids Res 38:W487–W496. doi:10.1093/nar/gkq340

27. Ansari MZ, Yadav G, Gokhale RS, Mohanty D (2004) NRPS-PKS: a knowledge-based resource for analysis of NRPS/PKS megasynthases. Nucl Acids Res 32:W405–W413. doi:10.1093/nar/gkh359

28. Flissi A, Dufresne Y, Michalik J, et al (2016) Norine, the knowledgebase dedicated to non-ribosomal peptides, is now open to crowd-sourcing. Nucl Acids Res (in press)

29. Delcher AL, Harmon D, Kasif S et al (1999) Improved microbial gene identification with GLIMMER. Nucleic Acids Res 27:4636–4641

30. Majoros WH, Pertea M, Salzberg SL (2004) TigrScan and GlimmerHMM: two open source ab initio eukaryotic gene-finders. Bioinformatics 20:2878–2879

31. Pruitt KD, Tatusova T, Brown GR et al (2012) NCBI Reference Sequences (RefSeq): current status, new features and genome annotation policy. Nucleic Acids Res 40:D130–D135

32. Benson DA, Clark K, Karsch-Mizrachi I et al (2014) GenBank. Nucleic Acids Res 42:D32–D37

33. Schölkopf B, Platt JC, Shawe-Taylor J et al (2001) Estimating the support of a high-dimensional distribution. Neural Comput 13:1443–1471

34. Caboche S, Pupin M, Leclère V et al (2009) Structural pattern matching of nonribosomal peptides. BMC Struct Biol 9:15

35. Abdo A, Caboche S, Leclère V et al (2012) A new fingerprint to predict nonribosomal peptides activity. Journal of computer-aided molecular design 26:1187–1194

36. Medema MH, Takano E, Breitling R (2013) Detecting sequence homology at the gene cluster level with MultiGeneBlast. Mol Biol Evol 30:1218–1223

37. Rutherford K, Parkhill J, Crook J et al (2000) Artemis: sequence visualization and annotation. Bioinformatics 16:944–945

38. Stajich JE, Block D, Boulez K et al (2002) The Bioperl toolkit: perl modules for the life sciences. Genome Res 12:1611–1618

39. Cock PJA, Antao T, Chang JT et al (2009) Biopython: freely available Python tools for computational molecular biology and bioinformatics. Bioinformatics 25:1422–1423

40. Huson DH, Richter DC, Rausch C et al (2007) Dendroscope: an interactive viewer for large phylogenetic trees. BMC Bioinformatics 8:460. doi:10.1186/1471-2105-8-460

41. The UniProt Consortium (2013) Update on activities at the Universal Protein Resource (UniProt) in 2013. Nucleic Acids Res 41:D43–D47

42. Berman HM, Kleywegt GJ, Nakamura H et al (2013) The future of the protein data bank. Biopolymers 99:218–222

43. Bolton EE, Wang Y, Thiessen PA et al (2008) PubChem: integrated platform of small molecules and biological activities. In: Wheeler RA, Spellmeyer DC (eds) Annual reports in computational chemistry. Elsevier, Amsterdam, pp 217–241

44. Berti AD, Greve NJ, Christensen QH et al (2007) Identification of a biosynthetic gene cluster and the six associated lipopeptides involved in swarming motility of Pseudomonas syringae pv. tomato DC3000. J Bacteriol 189:6312–6323

45. Studholme DJ, Ibanez SG, MacLean D et al (2009) A draft genome sequence and functional screen reveals the repertoire of type III secreted proteins of Pseudomonas syringae pathovar tabaci 11528. BMC Genomics 10:395

Chapter 15

The Use of ClusterMine360 for the Analysis of Polyketide and Nonribosomal Peptide Biosynthetic Pathways

Nicolas Tremblay, Patrick Hill, Kyle R. Conway, and Christopher N. Boddy

Abstract

Polyketides and nonribosomal peptides constitute two large families of microbial natural products. Over the past 20 years a broad range of microbial polyketide and nonribosomal peptide biosynthetic pathways have been characterized leading to a surfeit of genetic data on polyketide and nonribosomal peptide biosynthesis. We developed the ClusterMine360 database, which stores the antiSMASH-based annotation of gene clusters in the NCBI database, linking the structure of the natural product to the biosynthetic gene cluster. This database is searchable and enables the user to access multiple sequence files for phylogenetic analysis of polyketide and nonribosomal peptide biosynthetic genes. Herein we describe how to add compound families and gene clusters to the database and search it using key words or structures to identify specific gene clusters. We also describe how to download multiple sequence files for specific catalytic domains from polyketide and nonribosomal peptide biosynthesis.

Key words Genome, Gene cluster, Polyketide, Nonribosomal peptide, Natural product, Biosynthesis, Database

1 Introduction

Polyketides and nonribosomal peptides constitute two large families of microbial natural products. These natural products play major roles in the biology of microbes that produce them, functioning as signaling molecules and virulence factors, as well as many other roles [1, 2]. Due to their ability to selectively modulate the activity of biological targets, many of these compounds have been developed into pharmaceutical agents including anti-infectives, anticancer agents, immunosuppressive agents, and cholesterol-lowering drugs [3].

A broad range of microbial polyketide and nonribosomal peptide biosynthetic pathways have been characterized over the past 20 years [4–7]. This work has shown that the genes responsible for the biosynthesis of these natural products are clustered together on bacterial and fungal genomes and that the genes are

Bradley S. Evans (ed.), *Nonribosomal Peptide and Polyketide Biosynthesis: Methods and Protocols*, Methods in Molecular Biology, vol. 1401, DOI 10.1007/978-1-4939-3375-4_15, © Springer Science+Business Media New York 2016

typically collinear. In collinear gene clusters, the core polyketide synthase or nonribosomal peptide synthetase genes occur in the order in which the proteins function [8]. These two properties have facilitated identification and prediction of the function of many of these pathways. The biochemical and genetic characterization of polyketide and nonribosomal peptide biosynthetic pathways has enabled the development of powerful bioinformatics tools [9–14] for the rapid and detailed annotation of the proteins encoded in these biosynthetic gene clusters. Notable among these tools is the web-based tool antiSMASH [9, 10] that enables users to easily identify and annotate these pathways directly from DNA sequence data. This tool has become indispensable in the day to day activities of researchers in the polyketide and nonribosomal peptide natural product biosynthesis fields.

The explosion of biosynthetic gene cluster sequencing and bacterial genome sequencing has led to a surfeit of genetic data on polyketide and nonribosomal peptide biosynthesis in the NCBI nucleotide database. A consequence of this is that it has become increasingly difficult to keep up with which pathways have been sequenced. With the rapidly expanding amount of sequence data it is also challenging to locate specific gene clusters from the vast NCBI database. Add to this the observation that many sequenced pathways have no known polyketide or nonribosomal peptide product. The assignment of potential products to these orphan pathways is thus a major unaddressed question. To aid with these challenges, we developed the ClusterMine360 database, which stores the antiSMASH annotation of known gene clusters, linking them to the structure of the natural product produced by the gene cluster [15]. This database is searchable and enables the user to access multiple sequence files for phylogenetic analysis of polyketide and nonribosomal peptide biosynthetic genes.

Herein we describe how to add compound families and gene clusters to the ClusterMine360 database. We explain how to search the database using key words or structures to identify specific gene clusters. We also describe how to download multiple sequence files for specific catalytic domains from polyketide and nonribosomal peptide biosynthesis. Lastly, we demonstrate how to perform a phylogenic analysis of downloaded multisequence data and demonstrate the utility of phylogenetic analysis for the analysis of adenylation domains from nonribosomal peptide biosynthetic pathways.

2 Materials

ClusterMine360 can be accessed on a networked Windows or Mac computer with an Internet connection. A Web browser like Mozilla Firefox, Google Chrome, or Apple Safari is required. While

ClusterMine360 can be accessed using Internet Explorer, the antiSMASH-based cluster annotation data does not display correctly in many versions of Internet Explorer and thus the use of this browser is not recommended. The database can be found at the following address www.clustermine360.ca.

To deposit sequence data in ClusterMine360, DNA sequence data is required. The data can be supplied to the database by use of an accession number from the International Nucleotide Sequence Database Collaboration, which includes the DNA databank of Japan (DDBJ), European Nucleotide Archives (ENA), and GenBank, the National institutes of Health's genetic sequence database. Alternatively a RefSeq accession number can be used. RefSeq projects are curated sequence annotation projects that are not part of DDBJ/ENA/GenBank and can be distinguished from GenBank accessions by presence of an underscore as the third character in the accession number. Only these two types of deposited nucleotide datasets can be added to the database. In addition to the sequence data, these datasets contain key information such as the source organism and the PUBMED identification number for the journal article disclosing the sequencing results. This data is required by ClusterMine360, which uses an autocuration approach to populating many fields in the database.

3   Methods

3.1   Create an Account

While no account is required to view or download sequence or annotation data from ClusterMine360, an account is required to add content to the database.

Making an account:

1. Go to Log In (link in upper right corner of all ClusterMine360 pages).
2. Click on register.
3. Fill in the information required on the form and click on Create User.

Accounts are active immediately after their creation, enabling users to add or modify data.

3.2   Add a Compound Family

The database is partitioned into two categories of information, compound families and gene clusters. Typically a compound family is linked to a gene cluster. When a biosynthetic pathway for a known compound is discovered, a compound family is added to the database and the gene cluster is then linked to this compound family (*see* Fig. 1).

Adding a compound family to the database is done using the following steps.

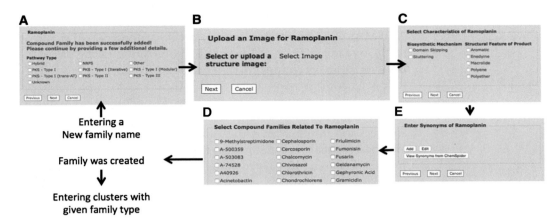

Fig. 1 Screen shots of workflow used to add a new compound family, in this example Ramoplanin. (*A*) Clustermine360 prompts the user to add pathway types that would best suit this family. (*B*) ClusterMine360 offers to add an image of a compound to represent the family. (*C*) Clustermine360 then offers a choice of different possible biosynthetic mechanisms and structural features that characterize the compound family. (*D*) Synonyms can be added to allow the users querying the database to better find a given compound. (*E*) The list of compound families presented represents all of the families in the database. Selecting related compound families leads to richer search experience when users query the database for related compounds

1. Login to ClusterMine360.

2. Navigate to the add tab.

3. Click on the Add a Compound Family link.

4. Enter the new compound family name and press the next button.

5. The Add a Compound Family Tool lists similar compound family names to ensure that the new compound family name to be added is unique. If there are similar compound family names in the database, the user is asked to click the check box preceding the most appropriate name, either the names in the database or the newly added compounds family name (*see* **Note 1**), and press next. This adds the compound family name to the database.

6. The Add a Compound Family Tool then requests the user to identify the type of pathway thought to be responsible for the biosynthesis of the compound. Click on the most appropriate check box and press next (*see* **Note 2**).

7. The Add a Compound Family Tool prompts the users to add an image depicting the chemical structure of the new Compound Family. Click on the Select Image. This opens the add image window where the user can query the ChemSpider database for a preexisting image, generate an image from a SMILES string, or upload an image of their own in png, gif, or jpg format.

8. To aid in searching the database, the Add Compound Family Tool requests the user identify if unusual biosynthetic mechanisms such as domain skipping or stuttering are thought to occur during the biosynthesis of the compound or if there are notable structural features present in the molecule. Click all checkboxes that apply and press next.

9. The Add Compound Family Tool then requests that synonyms for the Compound Family name be added. These can be entered manually or ChemSpider can be queried and the synonyms imported directly from there. There is no limit to the number of synonyms that can be added.

10. As many natural product Compound Families in the database are related, the Add Compound Family Tool provides a list of all Compound Families in the database and asks the user to select related compounds. There is no limit to the number of related compounds that can be added. Click on the appropriate checkboxes and press next to complete the addition of a new compound family (*see* **Note 3**).

In addition to the manually curated data added by the user, adding a compound family triggers autocuration of the database. Compound Families are automatically searched against the PubChem database [16, 17] to obtain Medical Subject Heading (MeSH) pharmacological identifiers that identify the bioactivity of the Compound Family and Simplified molecular-input line-entry system (SMILES) strings [18–20] describing the structure of the Compound Family. This data enables searching of the database by structure and substructure (*see* **Note 4**).

3.3 Add a Gene Cluster to the Database

Gene clusters can be added to the database directly from sequence data, either NCBI RefSeq or GenBank sequences. Gene clusters can be linked to a compound family or added to the database as orphan gene clusters with no known compound family.

1. Adding a gene cluster to a compound family
   To add a gene cluster to a compound family, the compound family must already be added to the database. In addition the RefSeq or GenBank accession number of the gene cluster or accession number of a larger sequence file such as a genome in which the clusters can be found is required. In the case of larger sequence files the nucleotide positions corresponding to the approximate beginning and end of the gene cluster are also needed.

2. Login to ClusterMine360.

3. Navigate to the Add tab.

4. Click on the Add a Cluster link.

5. Using the dropdown menu select the Compound Family name for the gene cluster to be associated with.

6. Enter the RefSeq or GenBank accession number. If the full sequence in the sequence file is to be used, click on the entire sequence checkbox. If a subset of the sequence file is to be used, indicate the range of sequence data to be used by adding the nucleotide sequence start and stop numbers. Press submit (*see* **Note 5**).

   Adding a Gene Cluster triggers autocuration of the database. Using the RefSeq or GenBank accession number, ClusterMine360 accesses the NCBI nucleotide database and retrieves information about the sequence. This information includes its description, the name and lineage of the organism it was obtained from, and references that are associated with the record. Following this, the sequence data is submitted to the bioinformatics tool antiSMASH to be analyzed and annotated. Once complete, the data is automatically imported into the ClusterMine360 database. In addition the antiSMASH data is parsed to extract information on biosynthetic pathway type. This updates the pathway type field of the Compound Family. In addition, if antiSMASH has identified any polyketide synthase or nonribosomal peptide synthetase domains, the amino acid sequence of those domains will be extracted and stored in ClusterMine360's sequence repository along with key information, such as domain substrate specificity, stereochemistry, and activity of the domain. The progress of the newly added gene cluster can be tracked by going to the Sequence Repository page and viewing the Currently Processing link. Typically processing of a gene cluster is complete within 2–4 h; however, processing time is dependent both on the length of the sequence submitted and load on the antiSMASH server.

7. Adding a gene cluster without a compound family

   There are an ever growing number of orphan gene clusters in the NCBI database. Orphan gene clusters are biosynthetic pathways for which no product is yet known. Genome sequencing and metagenome sequencing projects have produced an enormous reservoir of these clusters. Recent estimates of suggest that there are at least 1000 orphan modular polyketide biosynthetic gene clusters in the NCBI database [21] and this number is likely to increase. As these clusters have no known product, they cannot be added to a Compound Family in ClusterMine360. Thus to accommodate these data sets, the gene clusters are added as Orphan Gene Clusters.

8. Login to ClusterMine360.

9. Navigate to the Add tab.

10. Click on the Add an Orphan Cluster Sequence to Repository link.

11. Enter the RefSeq or GenBank accession number and press submit (*see* **Note 6**).

Unlike Gene Clusters added to Compound Families in the database, Orphan Gene Clusters cannot be easily browsed. Gene Clusters associated with Compound Families can be browsed by users in a number of different modes (*see* Subheadings 3.4 and 3.5). Orphan gene clusters, however, are only accessible through the sequence repository.

3.4 Viewing Data

Both Compound Family and Gene Cluster data can be easily viewed in ClusterMine360. Compound Family data can be readily accessed from the Compound Family page (Fig. 2). This page provides an alphabetic list of the compound families in the database and an image of their chemical structure. Clicking on the chemical structure or name navigates to the particular Compound Family data page. This page provides data on synonyms, related Compound Families, pathway type information, Pharmacological identifiers, extracted from PubChem, a link to the PubChem entry for that compound, an image of the chemical structure and information on the user who added the compound to the database. If there is a Cluster associated with the Compound Family, this information is provided at the bottom of the page (Fig. 2B). The Cluster data field contains the RefSeq or GenBank accession number for the sequence data, which is also a link to the RefSeq or GenBank page, the organism lineage as extracted from the RefSeq or GenBank file, the pathway type which is obtained from antiSMASH, a description of the sequence data, also extracted from the RefSeq or GenBank page, a link to the antiSMASH results page, and information on the user who contributed the Gene Cluster. Lastly a link to the Gene Cluster Details page provides further details on the cluster. An edit tab is present on each Compound Family page to edit the information in the database (Fig. 2C). Editable fields include Compound Family name, Pathway Type, Characteristics, Synonyms, PubChem substance and compound identifiers, and related Compound Families. As the PubChem database is dynamic and new information is continuously added to many of the compounds present in the PubChem database, there is also a checkbox to trigger ClusterMine360 to re-query PubChem for information on that particular PubChem identification number and extract the data into the ClusterMine360 database. This prevents the data in ClusterMine360 from becoming outdated (Fig. 2).

Gene Cluster data can be viewed by navigating to the Clusters page. This page provides a Table of the Gene Clusters in the database, excluding orphan gene clusters. The table includes the RefSeq or GenBank Accession number, which is also a link to

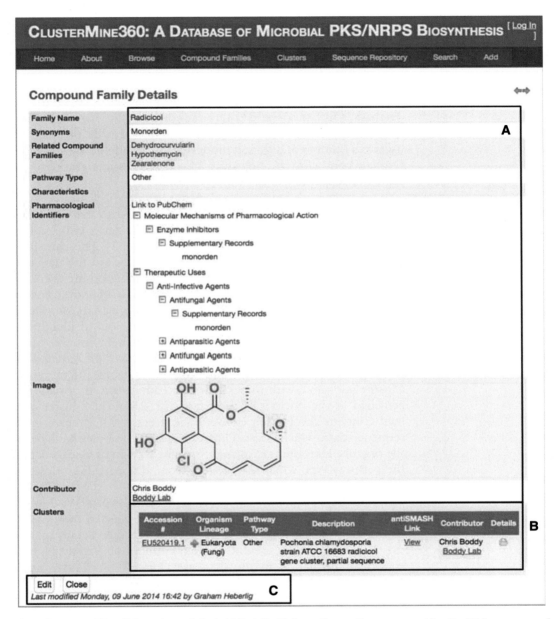

Fig. 2 Compound Detail Page for radicicol. (*A*) Detailed information on the compound family. (*B*) A summary of gene cluster data if available for the compound. (*C*) Compound Family Pages can be edited by login into the website and using the edit button

RefSeq or GenBank page, the organism lineage, and the Compound Family the cluster is associated with. The table can be sorted alphabetically by any of these three categories by clicking on the column heading. Twenty Gene Cluster entries are displayed per page. In addition to these first three columns, the table also displays the pathway type as extracted from the antiSMASH analysis, the

RefSeq or GenBank sequence description, a link to the antiSMASH results page, the name of the user who added the Gene Cluster to ClusterMine360, and a link to the Gene Cluster Details page for additional information on the cluster.

The Gene Cluster Details page provides extensive details on the Gene Cluster as well as some key information on the associated Compound Family (Fig. 3). These details includes the RefSeq or GenBank accession number and sequence range if the cluster is a subsection of a larger GenBank data set, Compound Family name and synonyms, related compounds, the pathway type, characteristics of the pathway, the producing organism and its lineage, the sequence description from the RefSeq or GenBank data file, a link to the reference describing the sequencing data, and a field for additional references. A key feature of the Gene Cluster Details page is the domains field (Fig. 3B). Clicking on the view domains toggle switch displays all the core biosynthetic domains identified in the polyketide synthase and nonribosomal peptide synthetase ORFs found in the gene cluster. For each domain a domain type identifier is provided. These include typical domains such as β-ketosynthase (KS), acyltransferase (AT), acyl carrier proteins (ACP), ketoreductases (KR), dehydratases (DH), enoyl reductase (ER) from polyketide biosynthetic pathways and condensation (C), adenylation (A), peptidyl carrier proteins (PCP), and epimerization (E) domains from nonribosomal peptide biosynthetic pathways. Domains common to both pathways such as methyltransferase (MT), aminotransferase, and thioesterase (TE) domains are also identified. The name of the corresponding gene from the RefSeq or GenBank file, the length in amino acids of the domain, and any predictable substrate selectivity, such as amino acid selectivity for adenylation domains is also provided. Each domain has a download link for downloading the protein sequence of each domain in fasta format for more detailed analysis. In addition to the individual domain information, the Gene Cluster Details page provides a link to the antiSMASH results page, status of the antiSMASH result parsing (important for files still undergoing processing), and the user who contributed the gene cluster to the database (Fig. 3C).

Similar to individual Compound Family pages, the Gene Cluster Detail page can be edited (Fig. 3D). Fields such as producing organism, phylum, organism lineage, sequence description, which are extracted from the RefSeq or GenBank pages are not editable. However, since these pages are updated in the NCBI database, it is possible to trigger a re-extraction of the data from these pages using the edit page function. Similarly data extracted from antiSMASH results cannot be edited in ClusterMine360. This ensures the consistent quality of the annotated data in the database. However, antiSMASH is continually updating its ability to analyze polyketide and nonribosomal peptide biosynthetic pathways. To be able to capture the improved data analysis, gene

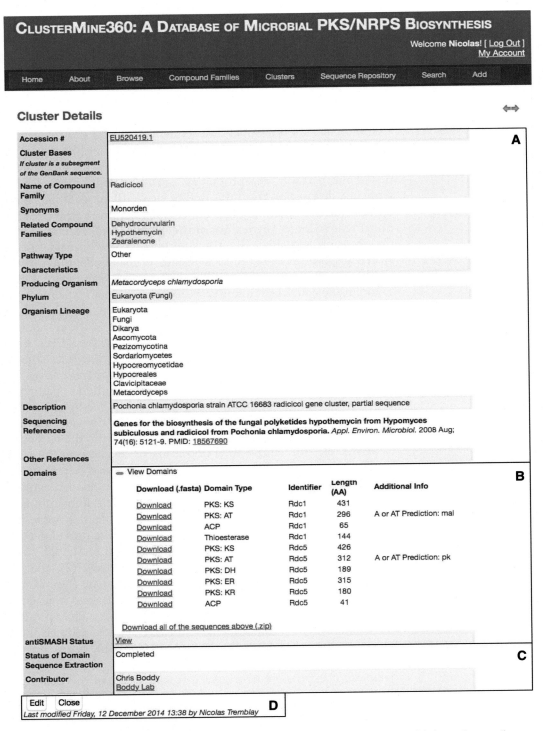

Fig. 3 Gene Cluster Detail page for the radicicol biosynthetic pathway. (*A*) Detailed information on the gene cluster is provided here including a link to the published article describing the sequencing of the gene cluster. (*B*) All the domains from the polyketide synthase or nonribosomal peptide synthetase genes in the pathway can be viewed and downloaded. (*C*) The page describes the status of domain extraction from the antiSMASH analysis. (*D*) The gene cluster data page is editable using the edit button

clusters can be resubmitted to antiSMASH (trigger antiSMASH data reset) and the domains can be re-extracted from the antiS-MASH results (trigger domain sequence extraction reset) using the edit function. These tools enable ClusterMine360 to retain the overall high quality, user independent curation while still being flexible enough to adapt to changes in the sequence data or bioinformatics analysis tools.

3.5 Searching the Database

Because of the number of compound families and gene clusters in ClusterMine360, it is necessary to be able to effectively search and browse the data. This enables researchers to ask key questions about newly discovered pathways and known pathways. For example, a researcher who wants to determine the biogenesis of a particular functional group may do a substructure search to identify if any known natural product with that structure is in the database. Similarly, a researcher who has sequenced a gene cluster and found a potential novel biosynthetic reaction may want to do a substructure search and browse the gene clusters to determine if any known cluster possess genes to catalyze that chemistry. To facilitate the diverse queries of natural products community, ClusterMine360 has been designed to be searchable by text and substructure and browsed by compound family, pathway type, and producing organism.

3.5.1 Search by Substructure

A substructure search can be performed with a simple fragment of a structure or a complete structure. The built in structure to SMILES converter allows users to draw the fragment or structure they wish to search with and obtain the SMILES descriptor of this fragment to search the database with.

1. Navigate to the search tab.
2. Select substructure search.
3. Draw the molecule using structure drawing interface.
4. Convert Structure to SMILES. If you already have a SMILES descriptor you can insert this text directly in place of **steps 3** and **4** (*see* **Note 4**).
5. Click on search.

   As an example, using cyanide in the substructure search (SMILES descriptor of N#C) retrieves two Compound Families, borrelidin and saframycin (Fig. 4). Clicking on the names or structures of these hits navigates to the Compound Family page where the Gene Cluster Details or antiSMASH result pages can be easily accessed.

3.5.2 Searching by Text

ClusterMine360 can be search by text as well (Fig. 4B). The fields that are queried in the text search include RefSeq and GenBank accession numbers (without version number), organism names,

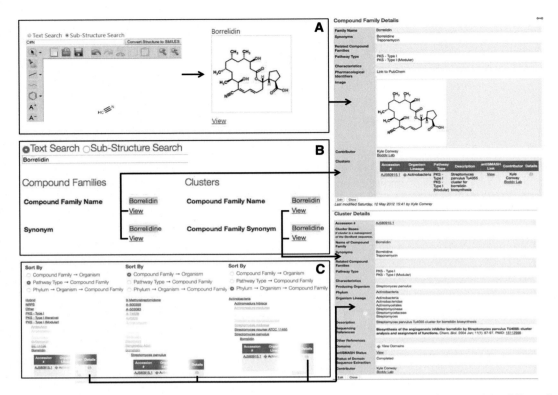

Fig. 4 Search of ClusterMine360. (*A*) A substructure search for a cyanide brings up two hits, borrelidin and saframycin. Only borrelidin is shown. Clicking on the structure or view link navigates to the Compound Family page for borrelidin. (*B*) A text search for borrelidin identifies a Compound Family hit and a Gene Cluster hit. The link for the Compound Family hit navigates to the borrelidin Compound Family details page. The Cluster hit navigates to the Cluster details page. (*C*) The browse link allows users to browse the database by Compound Family, Pathway type, and producing organisms. Links from the browse page navigate to Cluster details pages

lineage, sequence file description, Compound Family name, Compound Family synonyms.

1. Navigate to the search tab.

2. Enter a text based query into the search window.

3. Click on search.

For example searching the database with the text search of "natalensis" retrieves two hits, one for the organism name of *Streptomyces natalensis*, the producing organism for pimaricin, and one for the GenBank sequence description of the pimaricin biosynthetic gene cluster. The links for both results open the Gene Cluster details page for the pimaricin biosynthetic gene cluster.

3.5.3 Browsing Search

This type of searching can be useful when the search criteria are not well defined. The browsing search enables searching of the database by compound family, pathway type, and producing organism (Fig. 4C). Browsing by compound family provides a list of the compound family names. As some compound families are produced by more than one organism and can have more than one sequenced gene cluster in the database, clicking on the compound family provides a list of the Gene Clusters for that compound family in the database listed by producing organism genus and species. For example, clicking on the Compound Family erythromycin, reveals links to the gene clusters form *Aeromicrobium erythreum* [22] and *Saccharopolyspora erythraea* NRRL 2338 [23]. The database can also be browsed by pathway type. Selecting a pathway type, such as PKS-type II, displays all the compound families that are associated with that pathway type. Clicking on the compound family name reveals a summary of the gene cluster data, showing RefSeq or GenBank accession number, producing organism lineage, pathway type, sequence file description, antiSMASH results page link, the contributor who added the gene cluster to ClusterMine360 and a link to the Gene Cluster Detail page. Lastly the database can be browsed by producing organism. This data is arranged via top down approach for phylum to genus and species. Selecting a phylum reveals the genus and species of producing organisms. Selecting the genus and species then reveals the Compound Families with associated gene clusters present in the organism. Clicking on the Compound Family provides a summary of the gene cluster data as was seen in browsing by pathway type.

3.6 Downloading Sequences from the Sequence Repository

The sequence repository in ClusterMine360 provides a unique resource for the polyketide and nonribosomal peptide biosynthesis community. The sequence repository contains over 950 gene clusters, including approximately 700 orphan gene clusters, from over 290 organisms and over 18,000 polyketide synthase or nonribosomal peptide synthetase domains extracted from the antiSMASH analysis of these gene clusters. A summary of the gene clusters types and the domains present in ClusterMine360 can be found by navigating to the Sequence Repository page and clicking on the Overview of Repository Contents link.

The Details of Clusters/Sequences in Repository link on the Sequence Repository page brings the user to a list of all the gene clusters, including orphan gene clusters, in ClusterMine360. Because of the large number of gene clusters in the database, dropdown menus for Compound Family, Phyla, Organism, and Pathway Type are available to refine the gene clusters displayed. A summary of the gene cluster data is shown in table format for all the gene clusters matching the dropdown menu search criteria. The summaries include RefSeq or GenBank accession number, pathway type, compound family (for orphan gene clusters the compound family is unknown), producing organism phylum and genus and species, the contributor who added the gene cluster to ClusterMine360, and a link to antiSMASH results page. In addition a link revealing the domains present in the gene cluster is also present.

Navigating to the Download Sequences link on the Sequence Repository page enables the user to download multisequence files from the repository. All domains identified by antiSMASH in polyketide synthase and nonribosomal peptide synthetase open reading frames are available to be downloaded from the sequence repository. The user can select a subset of these 18,000 domains to download using checkboxes on the Download Sequence page. Selectable options include Phyla (all or any subset of phyla), Cluster Type (all or any subset of those identified by antiSMASH), and domain types (all or any subset of those identified by antiSMASH). In addition there are check boxes to select for selectivity of domains for which antiSMASH predicts selectivity. These include the adenylation, acyltransferase, and ketoreductase domains. For example all the adenylation domains with Cys amino acid selectivity can be selected. Pressing the submit button initiates an extraction of the domains with the selected features and generation of a text file including all the selected sequences in fasta format. This mutisequence fasta file can be saved and used as input for a number of different tools, such as phylogenetic analysis software, like the webserver Phylogeny.fr [24]. A key feature of the downloaded sequences from the Sequence Repository is the information rich title for each sequence. These titles contain accession number, producing organism, gene identifier, pathway type, domain type, and any predicted properties of that domain. This information rich title greatly facilitates analysis of data generated from these multisequence files, such as phylogenetic trees.

4   Example of Using Downloaded Sequence Data for Pathway Analysis

The Sequence Repository enables rapid construction of phylogenetic trees from polyketide synthase or nonribosomal peptide synthetase domains. This type of analysis can be essential for the analysis of these biosynthetic pathways. For example phylogenetic analysis of KS domains from *trans* AT pathways revealed that the KS domains cluster based on the substitution pattern at the alpha and beta carbons of the KS substrate [25]. This observation has been crucial in predicting the products formed by *trans* AT pathways. An example of how to use ClusterMine360 to generate and analyze multisequence files, a cluster analysis of the adenylation domain of nonribosomal peptide biosynthetic pathways is shown here.

Adenylation domains select and activate particular amino acids to be added to the growing nonribosomal peptide. Recently, several studies [26, 27] have used the A3/A7 primers of Ayuso-Sacido and Genilloud [28], which amplify part of the adenylation domain, to characterize metagenomic DNA. In type I polyketides, the analogous acyltransferase domain is known to cluster by substrate specificity rather than phylogeny [29]. Cluster analysis of adenylation domains whose phylogenetic origin and amino acid specificity are known can be used to evaluate if the phylogenetic origin and amino acid specificity can be determined for environmental sequences that cluster close to these sequences.

Adenylation domains of known specificity were downloaded from ClusterMine360. Sequences were aligned with Muscle [30] in MEGA [31] with default parameters. Adenylation domains alignments were curated with GBLOCKS [32] with the minimum length of a block reduced from 10 to 5 and up to 20 contiguous nonconserved positions allowed. Curated alignments were submitted to the PhyML site using the LG substitution model [33]. Tree searching used both the nearest neighbor interchange (NNI) and the subtree pruning and regrafting (SPR) algorithms. The phylogenetic tree is shown in Fig. 5.

Branch support values at nodes are approximate likelihood ratio test (aLRT) values. Adenylation domain specificity is given for the Cysteine Actinomycete and non-Actinomycete clades in linked tables.

Adenylation domains selective for either dihydroxybenzoate (dhb) or Cys cluster together regardless of phylogeny. For many other known adenylation domains there was no clear clustering pattern, with domains from different phylogenetic groups and substrate selectivity clustering together in well supported clades. There were, however, two large clusters that seemed to be predominantly Actinomycetal or non-Actinomycetal. These each contained adenylation domains selective for 18 different amino acid substrates. Within these two clusters there were subclusters that were specific to adenylation domains with a specific amino acid selectivity, for example Ser or hydroxyphenylglycine selective domains in the Actinomycetal group. While this cluster analysis is not unambiguous, when combined with information on the GC content of the sequence data, it may be useful for identifying both the selectivity and phylogeny of environmental adenylation domains.

5 Notes

1. It can be challenging to determine if a new Compound family should be added or not. Polyketides and nonribosomal peptides are typically produced as families of compounds by a particular organism. For example *Sorangium cellulosum*, the producing organism of epothilone an anticancer polyketide natural products, produces epothilone A–D in appreciable quantities (Fig. 6) [34, 35]. All these compounds should be included in a single compound family as they are produced by the same biosynthetic pathway. In some cases it is less clear if compounds should be classified as different Compound Families. For example monocillin II, pochonin D, and radici-

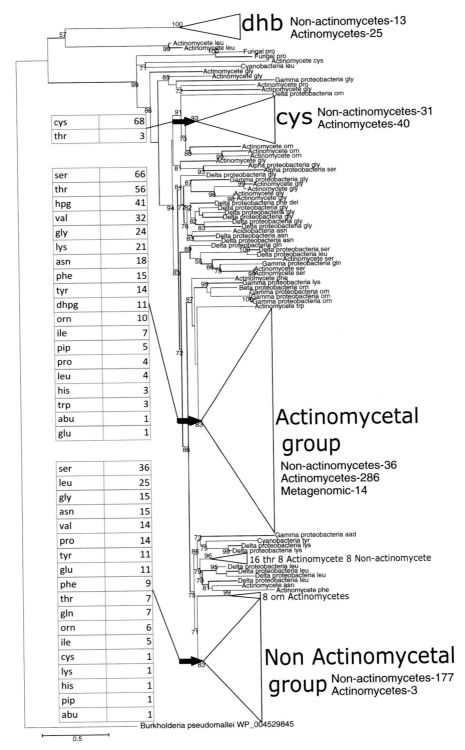

Fig. 5 Rooted maximum likelihood phylogenetic amino acid tree of nonribosomal peptide synthetase adenylation domains using PhyML [28]. The LG substitution model was used. Tree searching was carried out using the nearest neighbor interchange (NNI) and subtree pruning and regrafting (SPR) algorithms. Branch support values at nodes are approximate likelihood ratio test (aLRT) values. Adenylation domain specificity is given for the Cysteine, Actinomycete and non-Actinomycete clades in linked tables

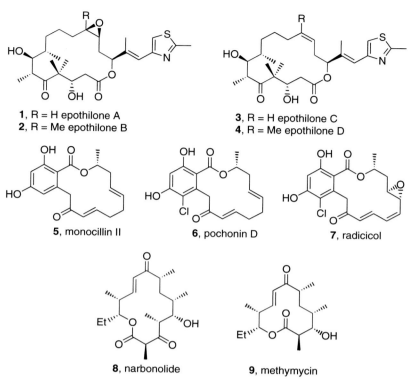

Fig. 6 Structures of polyketide products. Identifying if a new compound family should be added can be challenging. **1–4** are all part of the epothilone family. **5–7** are all part of the radicicol family. **8** and **9** are produced by pikromycin biosynthetic pathway but vary substantially in their structure

col all vary chemically (Fig. 6), however, since they are produced by the same polyketide biosynthetic pathway [36, 37], it is reasonable to include them together in a single compounds family, the radicicol Compound Family. The same is true for narbonolide and 10-deoxymethymycin (Fig. 6), which are also produced by the same pathway [38] via a module skipping mechanism [39] and are best treated as a single Compound Family. Adding individual Compound Families for each named compound will not negatively impact the database since Gene Cluster data can be linked to multiple Compound Families. Therefore if in doubt, add a new Compound Family (Fig. 6).

2. In some cases identifying the pathway type can be trivial, in others very challenging. For example a typical nonribosomal peptide, such as vancomycin or daptomycin, can be easily identified as a nonribosomal peptide by the presence of the non-proteinogenic amino acids. However, distinguishing between types of polyketide synthase pathways can often be impossible without knowledge of the biosynthetic pathway. For example it is not possible to predict before analysis of the gene cluster if, for example, psymberin [40] was a *trans* AT type I polyketide

synthase pathway or a typical modular polyketide synthase pathway. As this field is updated with data from the antiSMASH analysis of gene clusters associated with the Compound Family, it can be left blank when creating the Compound Family.

3. Many compounds in the database are clearly related. For example teicoplanin and vancomycin are both highly related glycopeptides antibiotics [41]. If unsure about the relatedness of Compound Families, leave this field blank. This field can be updated as more compounds are added to the database and can be informed by examining the related biosynthetic gene clusters feature provided in the antiSMASH analysis of the associated gene clusters.

4. If a compound is not present in the PubChem database when the compound is added to the database, no SMILES string will be imported into the Compound Family data. This will prevent the compound from being identified in structure and substructure searches. SMILES (simple molecular-input line-entry system) strings typically do not explicitly include hydrogen atoms. Thus when entering a substructure for searching the database, there is no need to provide atoms with incomplete substitution to indicate sites of attachment to the rest of the structure. For example drawing cyanide with or without explicitly including the hydrogen bound to the carbon will produce the same SMILES string search query (C#N).

5. With the increase in bacterial and fungal whole genome sequencing projects, many gene clusters are now found in whole genomes GenBank entries. For example the erythromycin biosynthetic pathway can be found in the genome of *Saccharopolyspora erythraea* NRRL2338 (accession number AM420293). As this genome contains a number of polyketide and nonribosomal peptide biosynthetic pathways, submitting the genome without identifying the location of the erythromycin biosynthetic pathway will be problematic. To identify the location of the pathway from a well annotated gene cluster the GenBank file can be search for a key identifier of the pathway. For example searching AM420293 with the text erythromycin identifies 37 hits. The first of these in the features section of the entry are for a putative erythromycin esterase starting at nucleotide 778214 and the last in the same area of the chromosome is for an erythromycin biosynthesis: transaminase, which ends at nucleotide 832825. Using these two numbers as the start and stop of the cluster, enables just the erythromycin gene cluster from *Saccharopolyspora erythraea* NRRL2338 to be added to the database. In case where the genome has not been sufficiently annotated to directly identify the start and stop of gene clusters from the GenBank file, the genome can

be submitted first to antiSMASH for analysis. By examining the gene clusters predicted by antiSMASH, identifying the gene cluster of interest, the start and stop nucleotide number can then be accessed.

6. As most genomes over the size of 5 Mb contain multiple gene clusters, when whole genomes are deposited as Orphan Clusters, each cluster identified by antiSMASH is added to the database. The individual clusters in the database are identified numerically by the order in which they are found in the GenBank file.

Acknowledgements

This work has been supported by NSERC and an Early Researcher Award from the Province of Ontario.

References

1. Romero D, Traxler MF, López D et al (2011) Antibiotics as signal molecules. Chem Rev 111:5492–5505

2. Davies J (2013) Specialized microbial metabolites: functions and origins. J Antibiot 66: 361–364

3. Newman DJ, Cragg GM (2012) Natural products as sources of new drugs over the 30 years from 1981 to 2010. J Nat Prod 75:311–335

4. Gulder TAM, Freeman MF, Piel J (2011) The catalytic diversity of multimodular polyketide synthases: natural product biosynthesis beyond textbook assembly rules. Top Curr Chem. doi:10.1007/128_2010_113

5. Hertweck C (2009) The biosynthetic logic of polyketide diversity. Angew Chem Int Ed Engl 48:4688–4716

6. Hur GH, Vickery CR, Burkart MD (2012) Explorations of catalytic domains in non-ribosomal peptide synthetase enzymology. Nat Prod Rep 29:1074–1098

7. Donadio S, Monciardini P, Sosio M (2007) Polyketide synthases and nonribosomal peptide synthetases: the emerging view from bacterial genomics. Nat Prod Rep 24:1073–1109

8. Callahan B, Thattai M, Shraiman BI (2009) Emergent gene order in a model of modular polyketide synthases. Proc Natl Acad Sci U S A 106:19410–19415

9. Medema MH, Blin K, Cimermancic P et al (2011) antiSMASH: rapid identification, annotation and analysis of secondary metabolite biosynthesis gene clusters in bacterial and fungal genome sequences. Nucleic acids Res 39:W339–W346

10. Blin K, Medema MH, Kazempour D et al (2013) antiSMASH 2.0—a versatile platform for genome mining of secondary metabolite producers. Nucleic Acids Res 41:W204–W212

11. Li MHT, Ung PMU, Zajkowski J et al (2009) Automated genome mining for natural products. BMC Bioinformatics 10:185

12. Ziemert N, Podell S, Penn K et al (2012) The natural product domain seeker NaPDoS: a phylogeny based bioinformatic tool to classify secondary metabolite gene diversity. PLoS One 7:e34064

13. Kim J, Yi G-S (2012) PKMiner: a database for exploring type II polyketide synthases. BMC Microbiol 12:169

14. Khaldi N, Seifuddin FT, Turner G et al (2010) SMURF: genomic mapping of fungal secondary metabolite clusters. Fungal Genet Biol. doi:10.1016/j.fgb.2010.06.003

15. Conway KR, Boddy CN (2013) ClusterMine360: a database of microbial PKS/NRPS biosynthesis. Nucleic Acids Res 41: D402–D407

16. Bolton EE, Wang Y, Thiessen PA et al (2008) Chapter 12—PubChem: integrated platform of small molecules and biological activities. Annu Rep Comput Chem 4:217–241

17. Wang Y, Xiao J, Suzek TO et al (2009) PubChem: a public information system for analyzing bioactivities of small molecules. Nucleic Acids Res 37:W623–W633

18. Weininger D (1988) SMILES, a chemical language and information system. 1. Introduction to methodology and encoding rules. J Chem Info Model 28:31–36

19. Weininger D, Weininger A, Weininger JL (1989) SMILES. 2. Algorithm for generation of unique SMILES notation. J Chem Info Model 29:97–101

20. Weininger D (1990) SMILES. 3. DEPICT. Graphical depiction of chemical structures. J Chem Info Model 30:237–243

21. O'Brien RV, Davis RW, Khosla C et al (2014) Computational identification and analysis of orphan assembly-line polyketide synthases. J Antibiot 67:89–97

22. Brikun IA, Reeves AR, Cernota WH et al (2004) The erythromycin biosynthetic gene cluster of Aeromicrobium erythreum. J Ind Microbiol Biotechnol 31:335–344

23. Oliynyk M, Samborskyy M, Lester JB et al (2007) Complete genome sequence of the erythromycin-producing bacterium Saccharopolyspora erythraea NRRL23338. Nat Biotechnol 25:447–453

24. Dereeper A, Guignon V, Blanc G et al (2008) Phylogeny.fr: robust phylogenetic analysis for the non-specialist. Nucleic Acids Res 36:W465–W469

25. Nguyen T, Ishida K, Jenke-Kodama H et al (2008) Exploiting the mosaic structure of trans-acyltransferase polyketide synthases for natural product discovery and pathway dissection. Nat Biotechnol 26:225–233

26. Charlop-Powers Z, Owen JG, Reddy BVB et al (2014) Chemical-biogeographic survey of secondary metabolism in soil. Proc Natl Acad Sci U S A 111:3757–3762

27. Luo K, Du G-P, Zhao Z-X et al (2010) Phylogenetic analysis of type I polyketide synthase and non-ribosomal peptide synthase genes from Mila Mountain in Tibet plateau. J Hunan Agric Univ 36:506–511

28. Ayuso-Sacido A, Genilloud O (2005) New PCR primers for the screening of NRPS and PKS-I systems in actinomycetes: detection and distribution of these biosynthetic gene sequences in major taxonomic groups. Microb Ecol 49:10–24

29. Jenke-Kodama H, Sandmann A, Müller R et al (2005) Evolutionary implications of bacterial polyketide synthases. Mol Biol Evol 22:2027–2039

30. Edgar RC (2004) MUSCLE: multiple sequence alignment with high accuracy and high throughput. Nucleic Acids Res 32:1792–1797

31. Tamura K, Peterson D, Peterson N et al (2011) MEGA5: molecular evolutionary genetics analysis using maximum likelihood, evolutionary distance, and maximum parsimony methods. Mol Biol Evol 28:2731–2739

32. Castresana J (2000) Selection of conserved blocks from multiple alignments for their use in phylogenetic analysis. Mol Biol Evol 17:540–552

33. Guindon S, Dufayard J-F, Lefort V et al (2010) New algorithms and methods to estimate maximum-likelihood phylogenies: assessing the performance of PhyML 3.0. Syst Biol 59:307–321

34. Gerth K, Bedorf N, Höfle G et al (1996) Epothilons A and B: antifungal and cytotoxic compounds from Sorangium cellulosum (Myxobacteria). Production, physico-chemical and biological properties. J Antibiot 49:560–563

35. Hardt IH, Steinmetz H, Gerth K et al (2001) New natural epothilones from Sorangium cellulosum, strains So ce90/B2 and So ce90/D13: isolation, structure elucidation, and SAR studies. J Nat Prod 64:847–856

36. Wang S, Xu Y, Maine EA et al (2008) Functional characterization of the biosynthesis of radicicol, an Hsp90 inhibitor resorcylic acid lactone from Chaetomium chiversii. Chem Biol 15:1328–1338

37. Zhou H, Qiao K, Gao Z et al (2010) Insights into radicicol biosynthesis via heterologous synthesis of intermediates and analogs. J Biol Chem 285:41412–41421

38. Xue Y, Zhao L, Liu HW et al (1998) A gene cluster for macrolide antibiotic biosynthesis in Streptomyces venezuelae: architecture of metabolic diversity. Proc Natl Acad Sci U S A 95:12111–12116

39. Xue Y, Sherman DH (2000) Alternative modular polyketide synthase expression controls macrolactone structure. Nature 403:571–575

40. Fisch KM, Gurgui C, Heycke N et al (2009) Polyketide assembly lines of uncultivated sponge symbionts from structure-based gene targeting. Nat Chem Biol 5:494–501

41. Nicolaou K, Boddy C, Bräse S et al (1999) Chemistry, biology, and medicine of the glycopeptide antibiotics. Angew Chem Int Ed Engl 38:2096–2152

Chapter 16

Alignment-Free Methods for the Detection and Specificity Prediction of Adenylation Domains

Guillermin Agüero-Chapin, Gisselle Pérez-Machado, Aminael Sánchez-Rodríguez, Miguel Machado Santos, and Agostinho Antunes

Abstract

Identifying adenylation domains (A-domains) and their substrate specificity can aid the detection of nonribosomal peptide synthetases (NRPS) at genome/proteome level and allow inferring the structure of oligopeptides with relevant biological activities. However, that is challenging task due to the high sequence diversity of A-domains (~10–40 % of amino acid identity) and their selectivity for 50 different natural/unnatural amino acids. Altogether these characteristics make their detection and the prediction of their substrate specificity a real challenge when using traditional sequence alignment methods, e.g., BLAST searches. In this chapter we describe two workflows based on alignment-free methods intended for the identification and substrate specificity prediction of A-domains.

To identify A-domains we introduce a graphical-numerical method, implemented in *TI2BioP version 2.0* (*t*opological *i*ndices *to bio*polymers), which in a first step uses protein four-color maps to represent A-domains. In a second step, simple topological indices (TIs), called spectral moments, are derived from the graphical representations of known A-domains (positive dataset) and of unrelated but well-characterized sequences (negative set). Spectral moments are then used as input predictors for statistical classification techniques to build alignment-free models. Finally, the resulting alignment-free models can be used to explore entire proteomes for unannotated A-domains. In addition, this graphical-numerical methodology works as a sequence-search method that can be ensemble with homology-based tools to deeply explore the A-domain signature and cope with the diversity of this class (Aguero-Chapin et al., PLoS One 8(7):e65926, 2013).

The second workflow for the prediction of A-domain's substrate specificity is based on alignment-free models constructed by transductive support vector machines (TSVMs) that incorporate information of uncharacterized A-domains. The construction of the models was implemented in the NRPSpredictor and in a first step uses the physicochemical fingerprint of the 34 residues lining the active site of the phenylalanine-adenylation domain of gramicidin synthetase A [PDB ID 1 amu] to derive a feature vector. Homologous positions were extracted for A-domains with known and unknown substrate specificities and turned into feature vectors. At the same time, A-domains with known specificities towards similar substrates were clustered by physicochemical properties of amino acids (AA). In a second step, support vector machines (SVMs) were optimized from feature vectors of characterized A-domains in each of the resulting clusters. Later, SVMs were used in the variant of TSVMs that integrate a fraction of uncharacterized A-domains during training to predict unknown specificities. Finally, uncharacterized A-domains were scored by each of the constructed alignment-free models (TSVM) representing each substrate specificity

Bradley S. Evans (ed.), *Nonribosomal Peptide and Polyketide Biosynthesis: Methods and Protocols*, Methods in Molecular Biology, vol. 1401, DOI 10.1007/978-1-4939-3375-4_16, © Springer Science+Business Media New York 2016

resulting from the clustering. The model producing the largest score for the uncharacterized A-domain assigns the substrate specificity to it (Rausch et al., Nucleic Acids Res 33:5799–5808, 2005).

Key words Alignment-free models, Adenylation domains, NRPS, Topological Indices, Transductive support vector machines

1   Introduction

Adenylation domains (A-domains) are mandatory components of nonribosomal peptide synthetases (NRPS) and are responsible for the activation and selection of the amino acids that are sequentially incorporated during nonribosomal peptide biosynthesis. A-domains constitute therefore a signature that could be used for genome-wide screenings as well as for the elucidation of oligopeptide structure with relevant biological activities [1]. However, the high sequence divergence among A-domains, i.e., from 10 to 40 % of sequence identity, even between A-domains belonging to the same NRPS cluster, prevents an easy BLASTp detection when a single template is used [2]. A-domains also activate 50 different natural/unnatural amino acids with a frequent cross-specificity over several of them. This fact is motivating for the genetic engineering of novel peptides libraries but also makes the prediction of A-domain's substrate specificity a real challenge. In this chapter, we describe two workflows based on alignment-free methodologies for the detection and prediction of the substrate specificity of A-domains.

1.1   An Alignment-Free Method for A-Domain Detection (Workflow 1)

Graph theory has facilitated the development of chemical graph theory (CGT), which allows the combinatorial and topological exploration of the chemical molecular structure through the calculation of molecular descriptors [3]. In CGT, the molecular topology is represented by graphs where atoms are the vertices and bonds the edges. Such graphs that constitute an approximation of a molecular structure can be characterized numerically through the so-called molecular descriptors. The notion of molecular descriptors is not new, and they have been traditionally used in medicinal chemistry to assist the process of drug design and to model drug-receptor relationships [4, 5].

During the last decades, the CGT has been extended to bioinformatics enabling the characterization of biomolecules, e.g., DNA/RNA and proteins, as part of comparative studies that do not have to rely on the use of sequence alignments. To make this possible, biomolecules have been represented as graphs in which nucleotides or amino acids are the nodes and the bonds between them are the edges of the graph [3].

Several bidimensional (2D) representations, also called graphs or maps, have been reported for DNA, RNA, and proteins such as the spectrum-like, star-like, Cartesian-type, and four-color maps

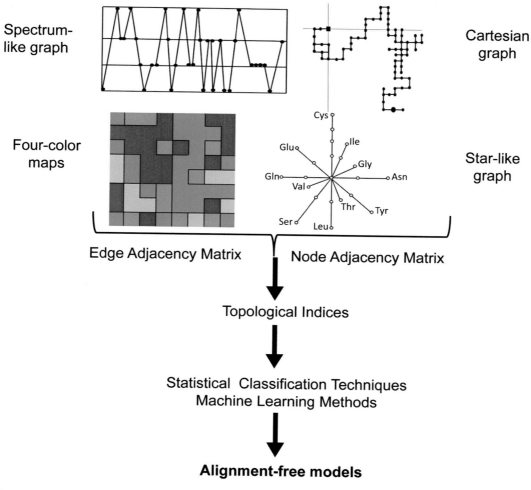

Fig. 1 Workflow for the calculation of the topological indices from several 2D graphs for DNA, RNA, and proteins

[3, 6–9] (Fig. 1). Despite the fact that these 2D maps do not represent the "real structure" of natural biopolymers, they have been very effective in highlighting the similarities/dissimilarities that exist among sequences by either direct visualization or numerical characterization [3].

The 2D graphs are built from the linear sequence that is arranged into certain 2D spaces (Cartesian, star-like, spectrum-like, and squared spiral spaces). The resulting graphs can be numerically described using the so-called topological indices (TIs). TI calculation requires the definition of the adjacency matrix which translates the connectivity/adjacency relations between nodes (n) or edges (e) in the graph to a matrix arrangement [10]. Once the adjacency matrix is built, there are several algorithms that could be used for TI estimation [11] (Fig. 1). These TIs can capture quantitatively structural information beyond the linearity of the sequence, i.e., long-distance interactions occurring in the pseudo 2D folding. Thus, TIs derived from a gene/protein family graphical profile (positive set) are used to develop alignment-free models with the contribution of a negative set, allowing the search of a certain functional signature in large databases [12–15].

Recently, protein four-color maps and its numerical characterization through TIs were implemented in the graphical-numerical approach called *TI2BioP* (*t*opological *i*ndices *to biop*olymers), which was used as an alignment-free tool to detect the A-domain signature in the proteome of the cyanobacteria *Microcystis aeruginosa* [12]. This method calculates simple TIs called "spectral moments," which were defined by Estrada [16] for small organic molecules, and they are being currently extended to 2D graphs of biopolymers through *TI2BioP* software which is described in this chapter, specially its application to the detection of the A-domain signature.

1.2 An Alignment-Free Method to Predict the Substrate Specificity of A-Domains (Workflow 2)

Predicting the specificity of A-domains could be seen as deciphering the nonribosomal code of NRPS for oligopeptide biosynthesis. Similar to the genetic code, which dictates the translation of DNA triplets (codons) to proteins, there is a nonribosomal system to select the proper amino acid that will be incorporated during the oligopeptide biosynthesis [17]. This nonribosomal code has been elucidated by analyzing the A-domain's sequence signatures and relating them to their substrate-binding pocket specificities via multiple alignments [18].

The elucidation of the crystal structure of two members of the superfamily of adenylate-forming enzymes, the firefly luciferase and the phenylalanine-adenylation domain of gramicidin synthetase A (GrsA), provided insight into the structural basis of substrate recognition [19, 20]. Specially, the structure of the N-terminal adenylation subunit in a complex with adenosine monophosphate (AMP) and l-phenylalanine (PheA) determined by Conti and co-workers revealed the residues involved in substrate selection [20]. This fact allowed narrowing down the specificity-determining region of this protein family to a stretch of about 100 amino acid residues. This specificity-determining region is located between two highly conserved core motifs (A4-A5) as it was made clear by conventional alignment strategies. Stachelhaus et al. in 1999 examined such ~100 amino acids and found nine out of the ten residues lining the PheA binding pocket. The remaining residue is the highly conserved Lys517 placed at the core motif A10. The consecutive order of these ten residues represents the signature sequence of an A-domain that determines different substrate specificities. From clusters of A-domains that activate the same amino acid (cross-specificity), the consensus sequences for the recognition of various substrates were elucidated and called "codons" which can be also interpreted as the "code" of NRPS. This code, in similarity to the genetic code, appears to be degenerated (i.e., four different signature sequences were defined for leucine-activating domains, three for valine, and two for cysteine). Thus, the rules that connect signature sequence and amino acid specificity were explored for deducing substrate specificity of several uncharacterized or unknown A-domains [18].

Briefly, another very similar prediction technique based on alignments was developed by Townsend and co-workers [21]. They identified only eight amino acid residues within the binding pocket of the GrsA A-domain as relevant to deduce a specificity-conferring code to assign the substrate specificity for uncharacterized A-domains [21]. So far, the specificity prediction of uncharacterized A-domains has been manually inferred based on the "code" of domains with known specificity combined with sequence comparisons [18, 21]. In addition, the reliability of A-domain's specificity predictions based on the "specificity-conferring code" concept is affected by the degeneracy of the NRPS code, which introduces noise in alignments. These methods also depend heavily on the available characterized data to achieve an accurate prediction.

More recently, Rausch et al. allowed the automation of the prediction of A-domain's specificity using support vector machines (SVMs) as alignment-free models [22]. SVMs are the newest supervised machine learning technique playing and increasingly important role in computational biology [23]. They are very suitable alignment-free tools for classification purposes since they are based on the concept of decision planes that define decision boundaries to separate data having different class memberships [24]. Rausch et al. considered the physicochemical fingerprint of the 34 residues lining the active site of the GrA A-domain (8 Å around the bound substrate) and the corresponding positions of other A-domains with known (characterized) and unknown (uncharacterized) specificities into a real-feature vector for a "transductive support vector machine" (TSVMs). A-domains with known specificities to similar substrates, also known as cross-specificities, were clustered by the physicochemical properties of the amino acids (AA). In the first step of TSVM building, optimal parameters were set for each resulting cluster made up of characterized A-domains. Secondly, TSVMs incorporate a fraction of uncharacterized A-domains during training to label uncharacterized A-domains giving a considerable improvement in generalization over SVMs [22] (Fig. 2).

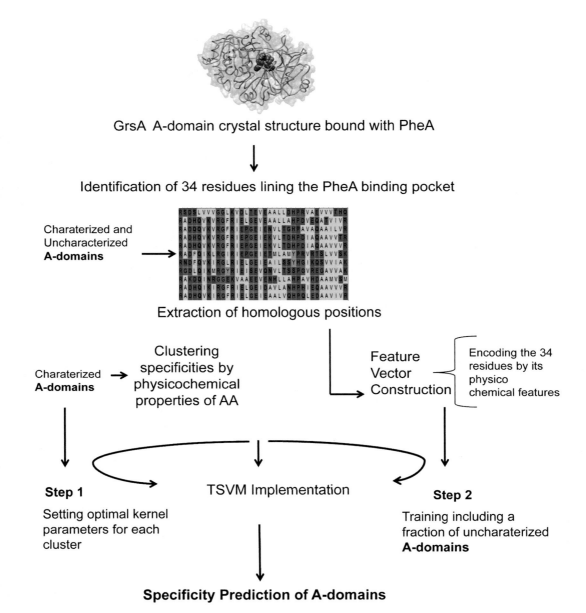

Fig. 2 Workflow for the specificity prediction of A-domains using the method reported by Rausch et al. The method identifies the physicochemical fingerprint (feature vector) of the 34 residues lining the A-domain active site of gramicidin synthetase A (GrsA). Such feature vector is extracted in other characterized and uncharacterized A-domains by identifying homologous positions. Feature vectors are used to implement TSVMs as alignment-free models for A-domain's specificity prediction. Before TSVM implementation, characterized A-domains with similar specificities are clustered by physicochemical properties of their substrates (amino acids), and the resulting clusters that represent different substrate specificity were used in the first step of TSVM implementation (optimization) providing as much alignment-free models as the number of clusters. In a second step, a fraction of the uncharacterized data is included during TSVM training to achieve a better generalization during substrate specificity prediction

2   Materials

Workflow 1: Identification of A-domains

2.1   Database Used in Graphical- Numerical Method for A-Domain's Identification

To obtain reliable and predictive alignment-free models, the training and predicting steps during model building require characterized data. Consequently, characterized A-domains (positive set) supported by literature reports and a benchmark dataset made up of CATH domains (negative set) were used to train and test the classification models. Data is detailed as follows:

Positive set:

Training set: 109 A-domain sequences from NRPS were collected from the major NRP-PKS database (http://www.nii.res.in/nrps-pks.html).

Test set: An independent subset of 29 A-domain sequences was gathered from the subset of the NRPS-PKS hybrids (http://www.nii.res.in/nrps-pks.html).

Negative set: The starting group was made up of 8 871 protein sequences downloaded from the *CATH* (*c*lass, *a*rchitecture, *t*opology, and *h*omology) domain database of protein structural families (version 3.2.0) (http://www.cathdb.info). We select the FASTA sequence database for all CATH domains sharing lesser than <35 % of sequence identity. The starting data was reduced to 8 854 CATH domains: 17 cases were removed because they showed the A-domain signature when an *hmmsearch* was performed against the AMP-binding profile HMM (PF00501).

Test set: The members of the test set (2 213 sequences) were selected taking out at random the 20 % from the 8 854 CATH domains.

Training set: The remaining CATH domains (6641) were used to train the models.

Workflow 2: Prediction of A-domain's substrate specificity

2.2   Database Used for the Specificity Prediction of A-Domains Using TSVMs

Selection of A-domains with known and unknown specificity

A total of 397 A-domains with known specificity were collected. The first 160 A-domain sequences were extracted from Stachelhaus et al. [18]. Secondly, 227 A-domains were found by scanning the UniProt/TrEMBL/Swiss-Prot protein database [25, 26] using profile HMMs for complete NRPS modules containing at least one AMP binding (A-domain), one condensation, and one peptidyl carrier domain. The specificity of the detected A-domains was obtained directly from the literature or by following other sources of information (PubMed links, gene name, organism, authors, etc.). The rest of characterized A-domains were taken from J. Ravel's NRPS BLAST server [21] to complete a total of 397. A-domain specificity annotation was only based on experimental evidence, by an ATP-PP*i*-exchange reaction [27].

In addition to the 397 specificity annotated A-domains, 646 uncharacterized A-domains were gathered from UniProt [26] following the same search procedure. All A-domains were processed to get feature vectors as follows:

Extraction of homologous positions of A-domains

The sequence of GrsA A-domain was reduced to 34 amino acids which are placed at distance of up to 8 Å from the bound PheA in GrsA. At this distance, all residue positions that might have an interaction with the substrate in the lining active site are comprised. To extract these 34 residues from the GrsA-Phe crystal structure [PDB ID 1 amu, [20]], the "Biochemical Algorithms Library" [BALL, [28]] was used within a simple Python script. These 34 residue stretches were initially aligned with 397 A-domains with known specificities in order to extract homologous positions (signature of 34 amino acids). It was verified that all extracted residues lie in conserved gap-free segments, to allow for a reliable inference of their structural and functional relevance. Afterwards, the same alignment strategy was carried out for the 646 uncharacterized A-domains to extract the signature of 34 amino acids.

Clustering A-domains with cross-specificities

Due to the fact that A-domains frequently show high-side substrate specificities that lead to alternate peptide products, A-domain's specificities were clustered according to the physico-chemical similarities of their substrates (amino acids). The reasoning behind clustering A-domain substrate specificities based on the properties of the substrate comes from the work of Townsend and co-workers who deduced that specificities for physicochemically similar substrates often only differ in single residues [21]. The results of the clustering of 397 A-domain sequences with known specificities are shown in Fig. 3. Grouping cross-substrate specificities of A-domains by the physicochemical properties of their substrates (amino acids) is termed composite specificities and can produce few large clusters made up of more different specificities into a wider physicochemical substrate classification or more small clusters that put together less specificities under a restricted physicochemical feature division. Larger clusters are formed by the connection of the small ones (Fig. 3).

After the extraction of the signature of 34 amino acids in all characterized A-domains, redundancy on them was checked. Finally, 282 A-domains with known specificities remained after large cluster formation. Similarly, 273 characterized A-domains were retained which formed small clusters.

A-domain's pre-processing for TSVMs

From each A-domain (characterized and uncharacterized), the signature of 34 amino acids located ≤8 Å from the bound substrate was extracted by multiple alignments with the GrsA A-domain

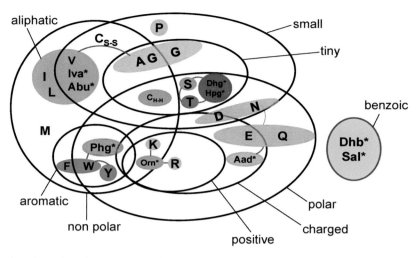

Fig. 3 A-domain substrates with cross-specificities are clustered by similar physicochemical properties of the amino acids. That is called composite specificities of A-domains (different composite specificities correspond to different colored sets). Thirteen small colored clusters with composite specificities are formed under a more restricted physicochemical amino acid classification. Eight larger clusters were made up of more different specificities, joining smaller clusters by *red lines*. An *asterisk* indicates rare non-proteinogenic amino acids. This figure was taken from the reference number [22]

homologous positions. Each amino acid was encoded by 12 different values representing its physicochemical properties extracted from AAindex [29]. These values were standardized in such a way that the interval of ±1SD (calculated from the value distribution of each AAindex file) is projected onto the interval of ±1. Standardized value = index value-mean index value/standard deviation; thus, a vector of 408 features for each A-domain sequence was obtained.

3 Methods

Workflow 1: Identification of A-domains

3.1 Protein Four-Color Maps

Protein four-color maps are inspired on Randic's DNA/RNA [30] and protein 2D graphical representations [31], but instead of using the concept of virtual genetic code, the spiral of square cells is constructed straightforward from the amino acid sequences [12] (Fig. 4). The four colors are associated with each one of the amino acid groups: polar (green), nonpolar (red), acid (yellow), and basic (blue), similar to the amino acid division used in the protein Cartesian representation previously reported by Agüero-Chapin et al. [32, 33].

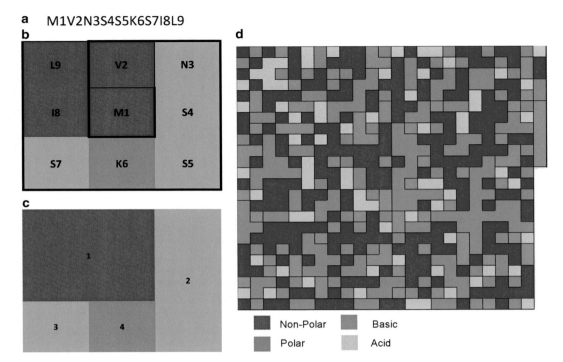

a   M1V2N3S4S5K6S7I8L9

Fig. 4 Steps for the four-color map construction for the first nine amino acids of pdb 1 amu. (**a**) The first nine amino acids of pdb 1 amu. (**b, c**) Building the four-color map for this protein fragment. (**d**) The final four-color map for pdb 1 amu

3.2 Numerical Characterization of Protein Four-Color Maps through TI's Spectral Moments

Estrada et al. introduced a new TI called spectral moments to characterize numerically the molecular topology through adjacency relationship between molecular bonds in small organic molecules [16]. Spectral moments were defined as the sum of main diagonal entries of the different powers of the bond or edge adjacency matrix (B) (Eq. 1) [34], where Tr is called the trace and indicates the sum of all the values in the main diagonal of the matrices (B)k, which are the natural powers of B. The different powers of B give the spectral moments of several orders ($\mu_0 - \mu_{15}$):

$$\mu_k = \mathrm{Tr}\left[\left(\mathrm{B} \right)^k \right] \tag{1}$$

B is a square matrix of $n \times n$ row and column where its non-diagonal entries are ones or zeroes if the corresponding bonds (b) or edges (e) share or not one atom. The elements of the adjacency matrix bij or eij are equal to 1 if i and j are adjacent otherwise take the value of 0 [35]. In analogy, when two nodes (nij) of the graph represented by atoms, nucleotides, amino acids, etc., share a common edge, they are considered adjacent and Estrada's algorithm can be applied.

To calculate spectral moments from the protein four-color maps, each region of the map was considered as a node made up

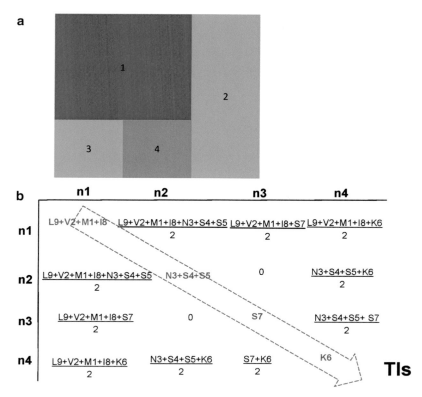

Fig. 5 The definition of the node adjacency matrix derived from the four-color map of the protein fragment $M_1V_2N_3S_4S_5K_6S_7I_8L_9$. Adjacency relationship between nodes sharing one edge in the map (i.e., *n*1 and *n*2) is weighted as the average of the properties (charge) of the amino acids made up both nodes. No adjacency relationship is set with zero value, and the main diagonal entries of the matrix are the sum of the properties of the amino acids placed in each cluster

for the amino acid clustering; two adjacent regions of the map sharing at least one edge (not a vertex) are connected. B is defined by considering the adjacency relationships between nodes. The number of nodes or clusters in the graph is equal to the number of rows and columns in B. Since a cluster is made up for several amino acids sharing similar physicochemical properties, the cluster is weighted with the sum of the individual properties (e.g., electrostatic charge (q)) of all amino acids placed in the cluster. The main diagonal of B was weighted with the average of the electrostatic charge (Q) between two adjacent clusters. The q values were taken from Amber 95 force field [36]. Figure 5 represents how the node adjacency matrix (B) is defined from the four-color map built up from the first nine amino acids of pdb 1 amu ($M_1V_2N_3S_4S_5K_6S_7I_8L_9$) to calculate the TIs. q values are represented in the matrix as the amino acid symbols ($M = 1.91$, $V = 2.24$, $N = 2.07$, $S = 2.09$, $K = 2.254$, $I = 2.02$, $L = 1.91$).

The calculation of the spectral moments until the order $k=3$ from the four colors maps is illustrated below. Expansion of expression (1) for $k=0$ gives the $^{fc}\mu_0$, for $k=1$ the $^{fc}\mu_1$, and for $k=2$ the $^{fc}\mu_2$. The node adjacency matrix derived from this 2D map is described for each case:

$$^{fc}\mu_0 = \text{Tr}\left[(B)^0\right] = \text{Tr}\left(\begin{bmatrix} 8.09 & 7.17 & 5.09 & 5.17 \\ 7.17 & 6.25 & 0 & 4.25 \\ 5.09 & 0 & 2.09 & 2.17 \\ 5.17 & 4.25 & 2.17 & 2.25 \end{bmatrix}^0\right) = 4.0 \tag{1a}$$

$$^{fc}\mu_1 = \text{Tr}\left[(B)^1\right] = \text{Tr}\left(\begin{bmatrix} 8.09 & 7.17 & 5.09 & 5.17 \\ 7.17 & 6.25 & 0 & 4.25 \\ 5.09 & 0 & 2.09 & 2.17 \\ 5.17 & 4.25 & 2.17 & 2.25 \end{bmatrix}^1\right) = 8.09 + 6.25 + 2.09 + 2.25 \tag{1b}$$

$$^{fc}\mu_2 = \text{Tr}\left[(B)^2\right] = \text{Tr}\left(\begin{bmatrix} 8.09 & 7.17 & 5.09 & 5.17 \\ 7.17 & 6.25 & 0 & 4.25 \\ 5.09 & 0 & 2.09 & 2.17 \\ 5.17 & 4.25 & 2.17 & 2.25 \end{bmatrix} \times \begin{bmatrix} 8.09 & 7.17 & 5.09 & 5.17 \\ 7.17 & 6.25 & 0 & 4.25 \\ 5.09 & 0 & 2.09 & 2.17 \\ 5.17 & 4.25 & 2.17 & 2.25 \end{bmatrix}\right) \tag{1c}$$

$$= \begin{bmatrix} 169.5 & 124.8 & 63.0 & 95.0 \\ 124.8 & 108.5 & 45.7 & 73.2 \\ 63.0 & 45.7 & 34.9 & 35.7 \\ 95.0 & 73.2 & 35.7 & 54.5 \end{bmatrix} = 169.5 + 108.5 + 34.9 + 54.5$$

TI2BioP version 2.0, available at http://ti2biop.sourceforge. net/, automatically represents all domain sequences (positive and negative sets) as four-color maps (Fig. 6) and allows the calculation of spectral moment series ($^{fc}\mu k$) to be used for statistical classification techniques in building alignment-free models.

3.3 Alignment-Free Models Development with Four-Color Map TIs for A-Domain's Detection

Spectral moment's series ($^{fc}\mu_0 - ^{fc}\mu_{15}$) derived from four-color maps representing protein sequences of training and test series were used as ordered predictors to build different classification models comprising from linear discriminant functions (LDF) until artificial neural networks (ANNs) and decision tree models (DTM). A categorical variable that takes a value of 1 for all instances in the positive set (A-domains) and of –1 for all instances in the negative set (CATH domains) was defined as dependent variable. DTM was the most simple, predictive, and robust developed model for A-domain detection. It was built using the classification tree (CT) module of the STATISTICA 8.0 for Windows [37].

In the development of the DTM, the C and RT (classification and regression tree)-style univariate split selection method was used since it examines all possible splits for each predictor variable

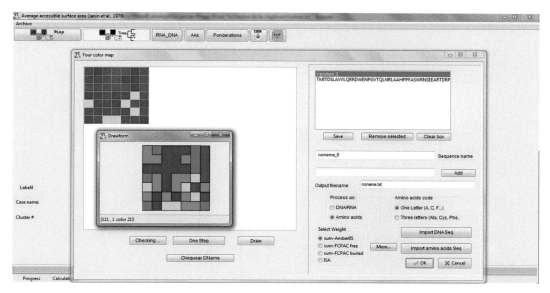

Fig. 6 Window view of *Tl2BioP* for the representation of protein four-color maps

at each node to find the split producing the largest improvement in goodness of fit. The prior probabilities were estimated for both groups with equal misclassification cost. The *Gini* index was used as a measure of goodness of fit, and the "FACT-style direct stopping" was set to 0.1 as stopping rule to select the right-sized classification tree [12]. The architecture of the developed DTM for A-domain detection is depicted in Fig. 7 where the starting data (positive and negative set) is divided into smaller subsamples (decision nodes) according to whether a selected TI ($^{fc}\mu k$) is above of a chosen cutoff value or not to finally assign a class membership in terminal nodes.

The performance of alignment-free model (DTM) was evaluated by several statistical measures commonly used for classification: accuracy, sensitivity, specificity, and F score (it reaches its best value at 1 and worst score at 0). The robustness of the classification model was verified by a cross-validation (CV) procedure. Ten random subsamples were selected from the learning sample. The CT of the specified size is computed ten times, each time leaving out one of the subsamples as a test set for CV. The CV costs computed for each of the ten test samples are then averaged to give the tenfold estimate of the CV costs [38]. In addition, a test set made up for 2242 domains was selected to evaluate the prediction power of each model [12].

Workflow 2: Prediction of A-domain's substrate specificity

3.4 Overview of Transductive Support Vector Machines (TSVMs)

SVMs produce a separating hyperplane that maximizes the margin or distance between the closest data points having different class memberships. The data points that lie on the hyperplane's margins are known as support vector (SV) points, and the decision function

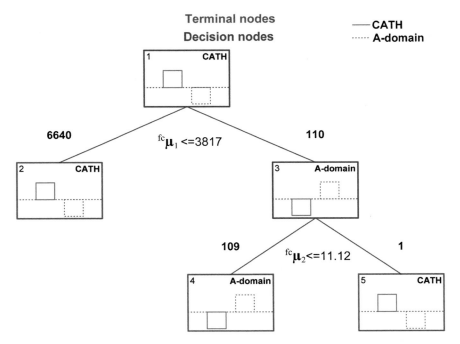

Fig. 7 Architecture for the DTM. Decision nodes are represented in *blue* and terminal nodes are in *red*. A-domains are labeled using an intermittent line. Otherwise CATH domains are signed by a continuous line. Labels at the right corner of the nodes indicate tentative membership to A- or CATH domain class. Numbers at the left corner represent the node's number

(Eq. 2) is represented as a linear combination of only these points, if data are linearly separable (Fig. 8):

$$f(x) = \sum_i y_i \alpha_i k(x, x_i) \qquad (2)$$

Therefore, the model complexity of a SVM is unaffected by the number of features encountered in the training data (the number of support vectors selected by the SVM learning algorithm is usually small). For this reason, SVMs are well suited to deal with learning tasks where the number of features is large with respect to the number of training instances.

However, most of the classification tasks are not as simple as a linear separating function; often more complex structures like curves are needed in order to make an optimal separation. Classification tasks based on nonlinear separating lines to distinguish between objects of different class memberships are possible, but SVMs instead of constructing the complex curve, the original data points (input space) are rearranged using a set of mathematical functions, known as kernels (k), into a feature space. The process of rearranging data points is known as mapping and allows that the mapped data is linearly separable (Fig. 9).

Unlike the classical SVMs, also termed "inductive" SVMs, which use labeled or annotated data to train the model, "transduc-

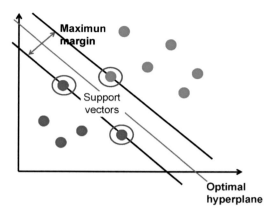

Fig. 8 Graphic representation of a linear SVM. Hyperplane that maximizes the margin between the closest data points having different class memberships

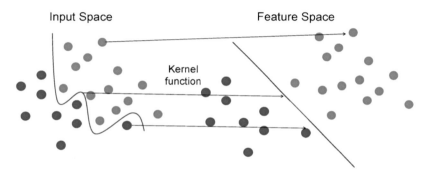

Fig. 9 The basic idea behind SVMs. The original data placed in input space is rearranged into a feature space through a kernel function to allow a linear separation of the data

tive" SVMs (TSVMs) integrate unlabeled data points, e.g., A-domains with unknown specificities in the training set. TSVMs consider that unlabeled data points can be labeled according to their positions in the hyperspace, e.g., nearby points and points on the same structure are likely to share the same label [39].

3.5 TSVM Implementation

TSVMs were built as alignment-free models using the program package SVM^light [40] which has algorithms for training large TSVMs. The algorithm proceeds by solving a sequence of optimization problems lower-bounding the solution using a form of local search. Different kernel functions such as linear, polynomial, radial basis, and sigmoid were tried. The optimal kernel function and parameters were determined in a grid search by computing leave-one-out (loo) test on training data (characterized A-domains). In the loo test, the predictive model is trained on a dataset that has been reduced by one data point. The generated model is then used to give a prediction for the removed data point. Thus, loo is the most fine-grained form of cross-validation providing unbiased estimates for the classification measures such as sensitivity, specificity, error rate, and Mathew's correlation coefficient (MCC).

Two common parameters (C and j) for all kernels were optimized. The parameter C is the penalty that is assigned to erroneous training points that cannot be classified correctly, and j determines how training errors on positives data points outweigh errors on negatives. The radial basic kernel function (RBF) has an additional parameter (γ), where $\gamma = 1/2\sigma^2$ requiring of σ^2 that is approximately the mean of the squared Euclidian distances of all pairs of data points.

The linear kernel $k(x,y) = xy$ and the RBF $k(x,y) = exp\left(-\ddagger \| x - y \|^2\right)$ functions were selected after loo test, being preferred the linear kernel when was equally good or better than RBFs due to its simplicity. After kernel function selection and parameters optimization using A-domains with known specificities, 646 uncharacterized A-domain sequences were incorporated. Thus, a fraction of this unlabeled data is included in training the model, and the rest will be used as a test set [22].

3.6 Composite Specificities and Substrate Specificity Prediction

To predict substrate specificity of A-domains, their specificities were grouped into few large clusters and into more small clusters called composite specificities that consist in clustering A-domains with cross-substrate specificities by the amino acid physicochemical features. Eight large clusters were made up of A-domains with more diverse composite specificities. Thirteen small clusters were formed using the same data but using more restricted criteria of clustering (Fig. 3). Larger clusters have several advantages over small clusters due to larger positive datasets for SVM training that can be available, covering a larger spectrum of sequence variations.

For each cluster "large and small," an alignment-free model based on TSVMs was trained for each composite specificity. Thus, a feature file containing real-valued feature vectors made up of 408 physicochemical amino acid properties for A-domains was used. A-domains to certain known specificity labeled as "+," all other A-domains different but known specificity labeled as "−," and the uncharacterized A-domains labeled as "0" were prepared as input data for TSVMs [22].

3.7 Multiclass Prediction of the Specificity of a Query Sequence

Once single SVM models were built for each composite specificity, the predictions of all models are combined to one single prediction for the "large" and "small" clusters. The scores returned from the models (distance of the classified point to the hyperplane) are combined by a max rule: the SVM that outputs the largest score is used to assign the specificity to the unknown sequence [24]. Because the quality of the single models differs, the scores were multiplied by the squared MCC value of the model. MCC is a model quality measure, close to 1 for very good models, and decreases with the reliability of the models.

An implementation of the described method, called the *NRPSpredictor*, is freely available via the server reachable at http://www-ab.informatik.unituebingen.de/software [22, 41].

4   Notes

<u>Workflow 1: Identification of A-domains</u>

Results and Discussion

The DTM built up with four-color map TIs reached a sensitivity of 100 %; it was significantly higher than that obtained by DTM supported by other alignment-free concepts like amino acid composition (53.70 %) and pseudo amino acid composition (67.89 %) [42]. Although A-domains share 10–40 % of sequence identity, it was possible to retrieve all of them (138) using four-color maps. However, the BLASTp search using a single template provided false positives (significant matches) among CATH domains at both high (E-value = 10) and relatively stringent cutoffs (E-values <0.05), which is considered statistically significant and useful for filtering easily identifiable homologue pairs [43]. HMM-based searches easily retrieved all A-domain members at expectation values (E-value $\leq$10) without reporting any false positive confirming its higher sensitive respect to simple BLASTp. Thus, sequence-search methods based either on graphical or alignment profiles demonstrated more sensitivity in the A-domain signature detection. They were successfully combined to detect new putative A-domains (remote homologous) in the proteome of *Microcystis aeruginosa* [12].

Some Considerations about Graphical-Numerical Methods

The use of artificial representations for DNA, RNA, and proteins as well as its numerical characterization has been useful for comparative sequence analyses. Although these graphs are relatively easy to derive from sequences, since they do not represent the real structure of the biopolymers, it still is hard to implement robust algorithms to process them massively. An attempt to do so is *TI2BioP* which implements TI estimation by powering matrixes, from thousands of graphs or maps that demand a high computational cost. Protein four-color maps based on amino acids clustering into four groups (Subheading 3.1) bear some redundancy or loss of sequence information that is relieved when their numerical characterization is performed [12] (Subheading 3.1.2).

<u>Workflow 2: Prediction of A-domains substrate specificity</u>

Results and Discussion

The quality of the TSVM multiclass prediction for "large and small" clusters models was comparatively evaluated by loo tests on the set of sequences with known specificity (training data).

Models derived from large cluster covering 282 A-domains out of 300 with known specificities showed a total error rate of 13 %, while models derived from small cluster covering 273 A-domains out of 300 with known specificities reached a total error rate of 23 %. In addition, the performance of BLASTp search was evaluated using the 34 amino acid signature of the 300 characterized A-domains as a sequence database. The closest BLAST matches were used to infer the specificity which resulted in an error rate of 22.3 %. Despite the fact that this simple BLAST-based strategy provided good results, it cannot be used to build generalized models and only finds the closest sequence(s). Alignment-free predictions were also compared to the original manual technique developed by both research groups of Stachelhaus and Townsend on 1230 A-domains currently detectable in UniProt. The automatic method (TSVMs) can predict the specificities for 18 % more sequences than the old one [22].

Some Considerations about TSVMs
TSVMs have been widely used as a means of improving the generalization accuracy of SVM by using unlabeled data. However, in some cases the performance of TSVMs can be substantially worse than its supervised counterpart SVM, due to the transductive training procedure may yield suboptimal models. This unstable TSVM performance can be solved by tuning parameters towards labeled data. On the other hand the SVM[light] proposed by Joachims [40] has the limitation of being just practical for few thousands of instances [44].

Acknowledgments

The authors acknowledge the Portuguese Fundação para a Ciência e a Tecnologia (FCT) for the financial support to GACH (SFRH/BPD/92978/2013). This work was partially supported by the European Regional Development Fund (ERDF) through the COMPETE—Operational Competitiveness Programme and national funds through FCT under the projects PEst-C/MAR/LA0015/2013 and the projects PTDC/AAC-CLI/116122/2009 (FCOMP-01-0124-FEDER-014029) and PTDC/MAR/115199/2009 (FCOMP-01-0124-FEDER-015451). The funders had no role in the study design, data collection and analysis, decision to publish, or preparation of the manuscript.

References

1. Jenke-Kodama H, Dittmann E (2009) Bioinformatic perspectives on NRPS/PKS megasynthases: advances and challenges. Nat Prod Rep. doi:10.1039/b810283j

2. Ansari MZ, Yadav G, Gokhale RS et al (2004) NRPS-PKS: a knowledge-based resource for analysis of NRPS/PKS megasynthases. Nucleic Acids Res. doi:10.1093/nar/gkh359

3. Randic M, Zupan J, Balaban AT et al (2011) Graphical representation of proteins. Chem Rev. doi:10.1021/cr800198j

4. Gonzalez-Diaz H, Vilar S, Santana L et al (2007) Medicinal chemistry and bioinformatics – current trends in drugs discovery with networks topological indices. Curr Top Med Chem 7:1015–1029

5. Estrada E, Uriarte E (2001) Recent advances on the role of topological indices in drug discovery research. Curr Med Chem 8: 1573–1588

6. Randic M (2004) Graphical representation of DNA as a 2-D map. Chem Phys Lett 386: 468–471

7. Randic M, Zupan J, Vikic-Topic D (2007) On representation of proteins by star-like graphs. J Mol Graph Model 26:290–305

8. Randic M, Zupan J (2004) Highly compact 2D graphical representation of DNA sequences. SAR QSAR Environ Res 15:191–205

9. Nandy A (1994) Recent investigations into global characteristics of long DNA sequences. Indian J Biochem Biophys 31:149–155

10. Randic M, Zupan J (2001) On interpretation of well-known topological indices. J Chem Inf Comput Sci 41:550–560

11. Gonzalez-Diaz H, Gonzalez-Diaz Y, Santana L et al (2008) Proteomics, networks and connectivity indices. Proteomics. Doi: 10.1002/pmic.200700638

12. Aguero-Chapin G, Molina-Ruiz R, Maldonado E et al (2013) Exploring the adenylation domain repertoire of nonribosomal peptide synthetases using an ensemble of sequence-search methods. PLoS One 8(7):e65926. doi:10.1371/journal.pone.0065926

13. Aguero-Chapin G, Varona-Santos J, de la Riva GA et al (2009) Alignment-free prediction of polygalacturonases with pseudofolding topological indices: experimental isolation from Coffea arabica and prediction of a new sequence. J Proteome Res 8:2122–2128

14. Aguero-Chapin G, Sánchez-Rodríguez A, Hidalgo-Yanes PI et al (2011) An alignment-free approach for eukaryotic ITS2 annotation and phylogenetic inference. PLoS One 6:e26638

15. Cruz-Monteagudo M, Gonzalez-Diaz H, Borges F et al (2008) 3D-MEDNEs: an alternative "in silico" technique for chemical research in toxicology. 2. Quantitative proteome-toxicity relationships (QPTR) based on mass spectrum spiral entropy. Chem Res Toxicol. Doi: 10.1021/tx700296t

16. Estrada E (1996) Spectral moments of the edge adjacency matrix in molecular graphs. 1. Definition and applications to the prediction of physical properties of alkanes. J Chem Inf Comput Sci 36:844–849

17. von Döhren H, Dieckmann R, Pavela-Vrancic M (1999) The nonribosomal code. Chem Biol 6:273–279

18. Stachelhaus T, Mootz HD, Marahiel MA (1999) The specificity-conferring code of adenylation domains in nonribosomal peptide synthetases. Chem Biol. doi:10.1016/s1074-5521(99)80082-9

19. Conti E, Franks NP, Brick P (1996) Crystal structure of firefly luciferase throws light on a superfamily of adenylate-forming enzymes. Structure 4:287–298

20. Conti E, Stachelhaus T, Marahiel MA, Brick P (1997) Structural basis for the activation of phenylalanine in the non-ribosomal biosynthesis of gramicidin S. EMBO J 16(14): 4174–4183

21. Challis GL, Ravel J, Townsend CA (2000) Predictive, structure-based model of amino acid recognition by nonribosomal peptide synthetase adenylation domains. Chem Biol 7:211–224

22. Rausch C, Weber T, Kohlbacher O et al (2005) Specificity prediction of adenylation domains in nonribosomal peptide synthetases (NRPS) using transductive support vector machines (TSVMs). Nucleic Acids Res 33:5799–5808

23. Bastanlar Y, Ozuysal M (2014) Introduction to machine learning. Methods Mol Biol. doi:10.1007/978-1-62703-748-8_7

24. Schölkopf B, Tsuda K, Vert J (eds) (2004) Kernel methods in computational biology. MIT Press, Cambridge, MA

25. Boeckmann B, Bairoch A, Apweiler R, Blatter MC, Estreicher A, Gasteiger E, Martin MJ, Michoud K, O'Donovan C, Phan I (2003) The SWISS-PROT protein knowledgebase and its supplement TrEMBL in 2003. Nucleic Acids Res 31:365–370

26. Apweiler R, Bairoch A, Wu CH, Barker WC, Boeckmann B, Ferro S, Gasteiger E, Huang H, Lopez R, Magrane M (2004) UniProt: the Universal Protein knowledgebase. Nucleic Acids Res 32:D115–D119

27. Miller G, Lipman M (1973) Release of infectious Epstein-Barr virus by transformed marmoset leukocytes. Proc Natl Acad Sci 70: 190–194

28. Kohlbacher O, Lenhof H (2000) BALL-rapid software prototyping in computational molecular biology. Biochemicals Algorithms Library. Bioinformatics 16:815–824

29. Kawashima S, Kanehisa M (2000) Aaindex: amino acid index database. Nucleic Acids Res 28:374

30. Randic M, Lers N, Plavšić D et al (2005) Four-color map representation of DNA or RNA sequences and their numerical characterization. Chem Phys Lett 407:205–208

31. Randic M, Mehulic K, Vukicevic D et al (2009) Graphical representation of proteins as four-color maps and their numerical characterization. J Mol Graph Model. doi:10.1016/j.jmgm.2008.10.004

32. Aguero-Chapin G, Gonzalez-Diaz H, Molina R et al (2006) Novel 2D maps and coupling numbers for protein sequences. The first QSAR

study of polygalacturonases; isolation and prediction of a novel sequence from Psidium guajava L. FEBS Lett 580:723–730

33. Aguero-Chapin G, de la Riva GA, Molina-Ruiz R et al (2011) Non-linear models based on simple topological indices to identify RNase III protein members. J Theor Biol 273:167–178

34. Estrada E (1997) Spectral moments of the edge-adjacency matrix of molecular graphs. 2. Molecules containing heteroatoms and QSAR applications. J Chem Inf Comput Sci 37: 320–328

35. Estrada E (1995) Edge adjacency relationships and a novel topological index related to molecular volume. J Chem Inf Comput Sci 35:31–33

36. Cornell WD, Cieplak P, IBayly C et al (1995) A second generation force field for the simulation of proteins, nucleic acids, and organic molecules. J Am Chem Soc 117:5179–5197

37. Statsoft (2008) STATISTICA 8.0 (data analysis software system for windows). Version 8.0 edn

38. Rivals I, Personnaz L (1999) On cross validation for model selection. Neural Comput 11: 863–870

39. Zhou D, Bousquet O, Lal T et al (2004) Learning with local and global consistency. Adv Neural Inform Process Syst 16: 321–328

40. Joachims T (1999) Making large-scale SVM learning practical. In: Schölkopf B, Burges C, Smola A (eds) Advances in Kernel methods. MIT-Press, Cambrige, MA, pp 169–184

41. Rottig M, Medema MH, Blin K et al (2011) NRPSpredictor2 – a web server for predicting NRPS adenylation domain specificity. Nucleic Acids Res. doi:10.1093/nar/gkr323

42. Shen HB, Chou KC (2008) PseAAC: a flexible web server for generating various kinds of protein pseudo amino acid composition. Anal Biochem 373:386–388

43. Boekhorst J, Snel B (2007) Identification of homologs in insignificant blast hits by exploiting extrinsic gene properties. BMC Bioinformatics 8:356

44. Collobert R, Sinz F, Weston J, Bottou L (2006) Large scale transductive SVMs. J Mach Learn Res 7:1687–1712

Chapter 17

Characterization of Nonribosomal Peptide Synthetases with NRPSsp

Carlos Prieto

Abstract

Bioinformatic sequence analysis allows the functional characterization of newly sequenced proteins. Nonribosomal peptide synthetases (NRPSs) are multi-modular enzymes involved in the biosynthesis of natural products. The current omics era has enabled the exponential growth of the sequenced NRPS, and it is important to characterize the final product of these synthetases. Here, how to achieve the prediction of substrates which bind to adenylation domains in NRPS with NRPSsp (www.nrpssp.com) bioinformatic tool is described.

Key words Nonribosomal, Peptide, Synthetases, Bioinformatics, Classifier, NRPSsp

1   Introduction

Nonribosomal peptide synthetases (NRPSs) are multi-modular enzymes which biosynthesize many important peptide compounds. These natural products are relevant for the industrial microbiology area and new bioinformatics protocols are being developed for their characterization. The minimal modules which form a NRPS contain the following domains: A-activation (performs the amino acid selection and adenylation), T-thiolation or acyl carrier domain (for the thioesterification), and C-condensation domain (peptide bond formation) [1]. It has been shown that the A-domain recruits a particular type of substrate and the other domains allow the elongation of the nascent peptide. Thus, the primary composition of the final polypeptide is determined by the sequential order of the A-domains along the NRPS.

Until a few years ago, the characterization and identification of NRPS metabolites were a complex process which implies the isolation, purification, and identification of bioactive molecules in order to decipher their chemical structure. Nowadays, the advances of next-generation sequencing technologies and biological databases have enabled the development of new computational biology tech-

Bradley S. Evans (ed.), *Nonribosomal Peptide and Polyketide Biosynthesis: Methods and Protocols*, Methods in Molecular Biology, vol. 1401, DOI 10.1007/978-1-4939-3375-4_17, © Springer Science+Business Media New York 2016

niques for their characterization. In particular, structural biology, sequence analysis, and machine learning have been applied to the characterization of NRPS substrates. The first main contribution was performed using the 3D structure of the adenylation domain of gramicidin synthetase A (GrsA) [2] linked to the phenylalanine amino acid. It enabled the identification of key residues in the active site pocket of the A-domain. These key residues were proposed as a specificity-conferring code for the prediction of NRPSs-binding substrates [3] and enabled the development of computational methods which predict the binding constitutive substrates and the biochemical structure of the final product.

These methods perform the analysis in two steps: (1) The constituent domains of NRPS modules are identified and the sequence of adenylation domains is extracted. Hidden Markov models (HMM) have been applied for this purpose. They allow the construction of profiles from a multiple alignment of sequences and the novo identification of these domains in an input sequence. Databases such as Pfam or InterPro have precomputed the characteristic profiles of NRPSs. These profiles are available as hidden Markov models which can be used with HMMER for the domain identification of NRPSs (PF00501 correspond with the adenylation domain). (2) Key residues or whole sequence of adenylation domains is analyzed with machine learning or classification methods, and a prediction of binding substrates is provided. For example, support vector machines have been used in the Web server NRPSpredictor2 [4]. Recent studies have proposed the use of HMM for the prediction as an alternative method to SVM [5, 6].

In this book chapter, a methodology which uses HMM for the characterization of NRPSs substrates is described. It explains how the adenylation domains are extracted and analyzed from a FASTA sequence by means of a Web server and stand-alone commands.

2   Materials

2.1   Web Site Execution

1. Prepare your sequence(s) in a FASTA format file. FASTA format is a text-based format for representing peptide sequences, in which amino acids are represented using single-letter codes. The format also allows the specification of sequence names and comments which precede the sequences. The FASTA format may be used to represent either single sequences or many sequences in a single file. A set of single sequences constitute a multisequence file. The first line of each sequence in a FASTA file starts with a ">" (greater-than) symbol. Following the initial line (used for a unique description of the sequence), the protein sequence is provided in standard one-letter code. NRPSsp allows multisequence files and the analysis is performed for all the sequences in the file.

2.2 Stand-Alone Execution

1. Download the HMM of the AMP-binding domain (ID = PF00501, http://pfam.sanger.ac.uk/family/PF00501/hmm).

2. Download the HMMs of NRPSsp which enable the classification (http://www.nrpssp.com/downloads/nrpshmms).

3. Install HMMER 3 software (http://hmmer.org/) in your local machine. Follow the instructions on its Web site.

4. Prepare your input sequence(s) in a FASTA format file.

3 Methods

3.1 Web Site Execution

1. Paste your sequence(s) or upload your FASTA file with the Web form of NRPSsp and click on the "Run Analysis" button (Fig. 1, vignette 1).

2. Once the analysis is performed, the results page is displayed (Fig. 1 vignette 2). It presents one table for each protein sequence and these tables have one row for each adenylation domain that have been identified in the sequences. The following information is provided for each result analysis:

   – Protein identifier: The sequence description which has been provided in the input FASTA file.

   – Adenylation domain start position: The position of the sequence where the analyzed adenylation domain starts.

   – Adenylation domain end position: The position of the sequence where the analyzed adenylation domain ends.

   – Substrate: The predicted substrate which binds with the adenylation domain.

   – Substrate name: The name of the predicted substrate.

   – Score: This is the HMMER bit score. The bit score tells how the input adenylation domain matches the HMM of the predicted substrate. It reflects whether the sequence is a better match of the profile model (positive score) or of the null model of nonhomologous sequences (negative score).

   – Prediction-conditioned fallout: This value is estimated as False Positives/True Positives + False Positives. The calculation is based on the error obtained in the LOO test for a particular score. It represents the estimated error that occurs for the resulting score. The font color reflects the reliability of the result (red = low, yellow = medium, and green = high).

   – Alignments: This links to the resulting HMMER alignments which have been performed to obtain this result.

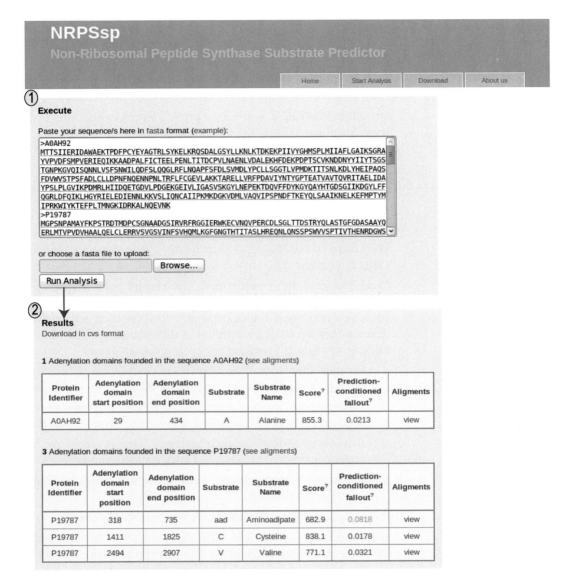

Fig. 1 NRPSsp analysis example: (1) Sequence of proteins was pasted in the text area and the "Run Analysis" button was clicked on to launch the predictor. (2) Results calculated by the application were shown. Each table corresponds to a sequence analyzed. CVS-formatted results and HMM alignments can be downloaded from this page

3.2 Stand-Alone Execution

1. Run HMMER in order to locate adenylation domains with the following command:
   # Prepare the HMM database.
   hmmpress AMP-binding.hmm
   # Scan adenylation domains in the input sequences.
   hmmscan --domtblout AMP_Domains_Coords.txt AMP-binding.hmm inputProteins.fasta > outputDomainsAlign.txt

2. Extract the adenylation sequence.
Get a FASTA file with adenylation sequences.
grep -v "^#" AMP_Domains_Coords.txt | awk '{print $4"/"$20"-"$21, $20, $21, $4}' | esl-sfetch -Cf inputProteins.fasta ->inputADomains.fasta

3. Run HAMMER in order to classify the adenylation domain.
Perform the substrate prediction.
hmmsearch -o hmmAligments.txt --tblout substrateprediction.csv nrpshmms inputADomains.fasta

4. Explore your results.
Results can be explored with spreadsheet software. The file "substrateprediction.csv" contains the results of the prediction done. The result with a bigger score should be selected as the predicted binding substrate. The file "hmmAligments.txt" contains the HMMER alignments.

4 Notes

Some previous considerations should be taken into account after the analysis of sequences. Firstly, the different types of binding substrates which can be predicted are limited to 30, and it is expected that the low score results correspond with substrates that are not considered in the predictor. Moreover, the exploration of sequences and species which are in the training database is recommendable. If input sequences come from species which have high phylogenetic similarity with training datasets, then more reliable results are expected. Finally, it is also recommendable to use other prediction software, such as NRPSpredictor2, in order to compare and evaluate the results obtained.

Acknowledgments

This work was supported by Agencia de Inversiones y Servicios de Castilla y León (record CCTT/10/LE/0001) and by Juan de la Cierva programme (JCI-2009-05444) of the Ministry of Science and Innovation (Spain).

References

1. Schwarzer D, Marahiel MA (2001) Multimodular biocatalysts for natural product assembly. Naturwissenschaften 88:93–101

2. Conti E, Stachelhaus T, Marahiel MA et al (1997) Structural basis for the activation of phenylalanine in the non-ribosomal biosynthesis of gramicidin S. EMBO J 16: 4174–4783

3. Stachelhaus T, Mootz HD, Marahiel MA (1999) The specificity-conferring code of adenylation domains in nonribosomal peptide synthetases. Chem Biol 6:493–505

4. Röttig M, Medema MH, Blin K et al (2011) NRPSpredictor2 – a web server for predicting NRPS adenylation domain specificity. Nucleic Acids Res 39:W362–W367

5. Khayatt BI, Overmars L, Siezen RJ et al (2013) Classification of the adenylation and acyltransferase activity of NRPS and PKS systems using ensembles of substrate specific hidden Markov models. PloS One 8:e62136

6. Prieto C, Garcia-Estrada C, Lorenzana D et al (2012) NRPSsp: non-ribosomal peptide synthase substrate predictor. Bioinformatics 28:426–427

INDEX

Printed in the United States
By Bookmasters